NEUROBIOLOGY AND CELL PHYSIOLOGY OF CHEMORECEPTION

ADVANCES IN EXPERIMENTAL MEDICINE AND BIOLOGY

Recent Volumes in this Series

A Continuation Order Plan is available for this series. A continuation order will bring delivery of each new volume immediately upon publication. Volumes are billed only upon actual shipment. For further information please contact the publisher.

NEUROBIOLOGY AND CELL PHYSIOLOGY OF CHEMORECEPTION

Edited by

P. G. Data

Università degli Studi "G. D'Annunzio"
Chieti, Italy

H. Acker

Max-Planck-Institut für Molekulare Physiologie
Dortmund, Germany

and

S. Lahiri

University of Pennsylvania
Philadelphia, Pennsylvania, U. S. A.

Springer Science+Business Media, LLC

Library of Congress Cataloging-in-Publication Data

Neurobiology and cell physiology of chemoreception / edited by P.G.
 Data, H. Acker, and S. Lahiri.
      p.   cm. -- (Advances in experimental medicine and biology ; v.
 337)
   "Proceedings of an International Symposium on Arterial
 Chemoreception, held June 24-28, 1991, in Chieti, Italy"--Copr. p.
   Includes bibliographical references and index.
   ISBN 978-0-306-44575-0
   1. Carotid body--Congresses. 2. Chemoreceptors--Congresses.
 I. Data, Pier Georgio. II. Acker, H. (Helmut), 1939-   .
 III. Lahiri, Sukhamay. IV. International Symposium on Arterial
 Chemoreception (1991 : Chieti, Italy) V. Series.
 QP368.8.N48 1993
 612.4'92--dc20   93-29402
                                                                    CIP

Proceedings of an International Symposium on Arterial Chemoreception, held June 24–28, 1991, in Chieti, Italy

ISBN 978-0-306-44575-0   ISBN 978-1-4615-2966-8 (eBook)
DOI 10.1007/978-1-4615-2966-8

©1993 Springer Science+Business Media New York
 Originally published by Plenum Press, New York 1993

PREFACE

This volume records the papers presented in Chieti, Italy, at the 1991 meeting of the International Society for Arterial Chemoreception (ISAC). This was the eleventh of a series of assemblies held since 1959. This field of research, which examines the critical function and mechanisms of O_2 and pH/PCO_2 sensitive cells in the body, is unique in that it encompasses diverse biological and medical areas. The reader of this book will note chapters concerning modern techniques like in situ hybridization; analysis of cell membrane channels and intracellular ion movements; immunohistochemistry of peptides, hormones, and the corresponding receptors of chemoreceptor cells; and systemic analysis of reflex pathways involving chemoreceptor cells and their meaning in health and disease. This broad spectrum will appeal to readers interested in the chemoreceptor field, as well as young scientists seeking a scientific field where not only structural analysis but also a sense for functional connections is required.

In recognition of the importance of the contribution of a new scientific generation to this field, ISAC awarded the prestigious F. de Castro-C. Heymans-E. Neil prize to A. Görlach, a young scientist. Also at this meeting, the Ferdinando Data Foundation Award, for helping scientists from non-developed countries to pursue their interest in chemoreception, was initiated.

ISAC held plenary lectures in honor of C. Eyzaguirre and R. Forster, both of whom have contributed greatly to research in the chemoreceptor field.

We are grateful for the generous financial support offered by the University of Chieti, C.N.R. (the National Council of Research), the Abruzzo Regional Government, and the Ministry of University and Scientific and Technological Research (Italy).

The skillful assistance of Plenum Press, the Publisher of the papers presented, is appreciated. Grateful thanks go to the staff of the Physiological Institute of Chieti University (Italy) for the most taxing of duties, preparation of the manuscripts and assistance in editing this book.

We hope that this volume will meet with good reception from the growing number of scientists who are interested in the fascinating field of chemoreception and chemoreceptor reflex mechanisms.

The Editors

CONTENTS

SECTION I
Morphology, Cyto- and Immunochemistry

SECTION II
Molecular Biology, Biophysics and Biochemistry

ix

SECTION III
Neurotransmitter

SECTION IV
Reflex Mechanism

SECTION V
Developmental and Adaptative Physiology and Pharmacology

SECTION I

Morphology, Cyto- and Immunochemistry

VASCULAR ANALYSIS OF THE CAROTID BODY IN THE SPONTANEOUSLY HYPERTENSIVE RAT

*J.A. Clarke, *M.de Burgh Daly and H.W. Ead*

Department of Anatomy, Queen Mary and Westfield College,
London E1 4NS and *Department of Physiology , Royal Free
Hospital School of Medicine, London NW3 2PF, U.K.

INTRODUCTION

Recent studies using a quantitative image analysis technique (Clarke, Daly & Ead 1990) have shown that the vascular compartment of the carotid body may be analysed with special reference to the vessels whose diameters lie between 5μm and 12μm. It is well known that in the spontaneously hypertensive rat (SHR) the carotid bodies are enlarged (Habeck 1991). Accordingly we have applied this technique to the analysis of the carotid body in these SHR animals to determine the contribution of the vascular compartment to the increase in volume. In this study the volumes of the carotid body and its vascular compartment were studied during postnatal maturation of the control and SHR animals. The vascular compartment was divided into that containing small vessels (diameter 5μm - 12μm) and that containing vessels greater than 12μm in diameter.

METHODS

Spontaneously hypertensive rat (SHR) of the WKY/OLA strain and controls (WKY/OLA) of either sex, and aged 3 weeks to adult were used.

They were anaesthetised with pentobarbitone sodium (Sagatal, 4 mg $100g^{-1}$ body weight) intraperitoneally. The carotid bodies were prepared for histology by the method described previously (Clarke & Daly, 1983,1985), perfusion fixation being carried out with isotonic phosphate buffered 3% glutaraldehyde at a temperature of 37° C. Perfusion pressures were 90 mmHg for three-week old animals, and 100 mmHg for all other animals. Routine histological methods were used for embedding the material, and serial sections cut at 5 μm thickness were stained with the MSB method. The criteria for defining the border of the carotid body were similar to those described previously (Clarke et al., 1990).

Volumes of the carotid body and its vascular compartment were determined using the results of an analysis of every section. The results obtained were compared with those from the same sections using Simpson's rule (Clarke et al., 1990). The carotid body in the rat is a discrete structure and is suitable therefore to this method of analysis. Paired carotid bodies from two animals were used for each of the hypertensive groups except for the 5 week old specimens where only 3 carotid bodies were analysed. Paired control carotid bodies from 4 animals from the 5-6 week group and 12 week group were used. Paired carotid bodies from five animals were used for the adult control group.

RESULTS AND DISCUSSION

Figure 1 shows that there is no difference in the body weights of the control and SHR groups of animals. When the volume of the carotid body is examined, that of the SHR animals is increased compared with controls of the same age (Fig. 2 and Table 1). In the SHR group there was a gradual increase in volume of the organ from 3 weeks onwards which appears to be "arrested" after 8 weeks (Fig. 2), so that the size of the carotid body in the neonatal (5-6 weeks) and adult animals was the same (Table 1). A similar "arrest" of growth is to be seen in the results of Habeck (1985). This observation is at variance with some of the other organs where growth continues without interruption (Habeck, 1985).

This is in contrast to the control group of animals in which the volume of the adult carotid body is 75% larger than that of the neonates (Table 1). Comparing now the sizes of the neonatal (5-6 weeks) and adult carotid bodies in the two groups of animals, the mean carotid body volume in the neonatal SHR animals is 97% larger than the controls, but only 18% larger in the adults (Table 1).

The values for the sizes of the total vascular compartment are shown in Table 1. In the 5-6 week neonates the total vascular volume of the carotid body in the control group was $1.6mm^3 \times 10^{-3}$ representing 14.2% of the total carotid body volume. Comparing the SHR group with the control animals the vascular volume was 25% larger and the carotid body volume was 97% larger. The increase in size of the carotid body in 5-6 week neonatal group of SHR animals must be due to an augmented extravascular compartment.

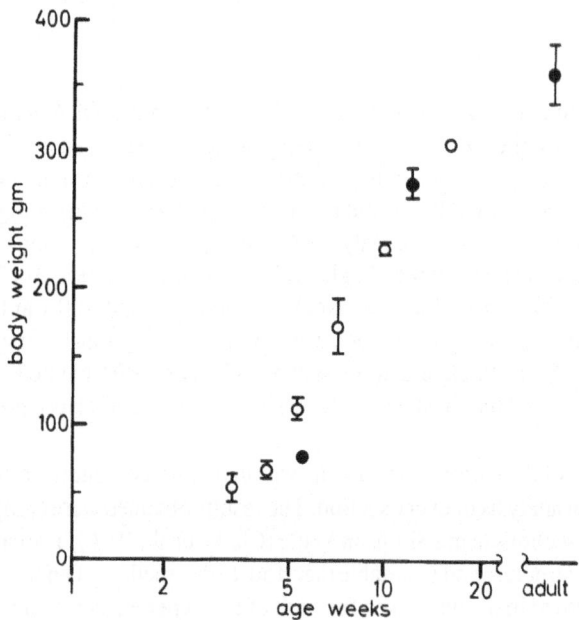

Figure 1. Changes in body weight with age. Group mean ± s.d. Controls (●), SHR (o).

Table 1. The total volume, total vascular volume and small vessel volume of the carotid body in 5-6 week old neonatal and adult control and spontaneously hypertensive rats (SHR). Cb, carotid body. Group mean values

	Total Cb volume	Total vascular volume		Small vessel volume	
	$mm^3 x 10^{-3}$	$mm^3 x 10^{-3}$	%Cb volume	$mm^3 x 10^{-3}$	%Cb volume
Neonates					
Control	11.3	1.6	14.2	0.74	6.4
SHR	22.3	2.0	8.9	1.0	4.5
Adults					
Control	19.8	3.8	19.2	1.1	5.6
SHR	23.4	3.2	13.7	1.9	8.2

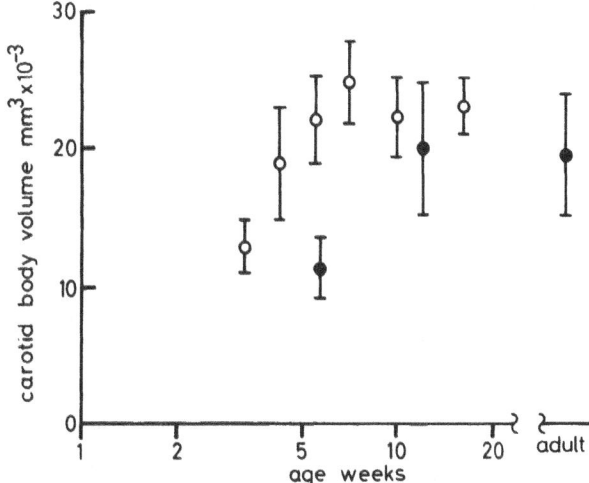

Figure 2. Growth in carotid body volume. Group means ± s.d. Controls (•), SHR (o).

In the two adult groups (Table 1) the total carotid body volume in the SHR group was larger than in the controls, whereas the volume of the vascular compartment was slightly smaller. Therefore, the larger carotid body volumes in the SHR group must again be due to an increase in extravascular volume.

The proportion of the small vessel volume to the total carotid body volume in the SHR neonates is smaller than in the controls, whereas in the adult animals it is considerably larger (Table 1). However, it will be noted that in the SHR groups, the absolute value for the small vessel volume in the adults is almost twice that of the

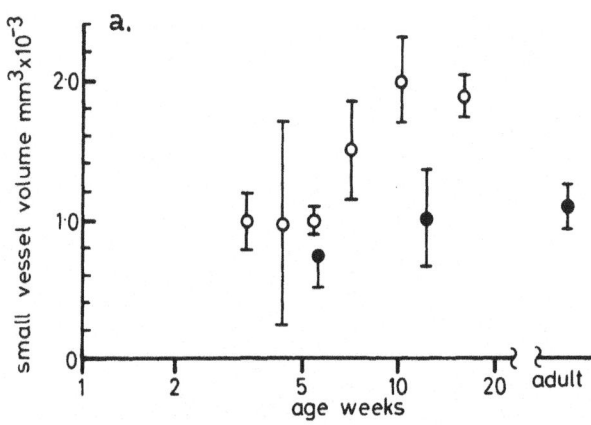

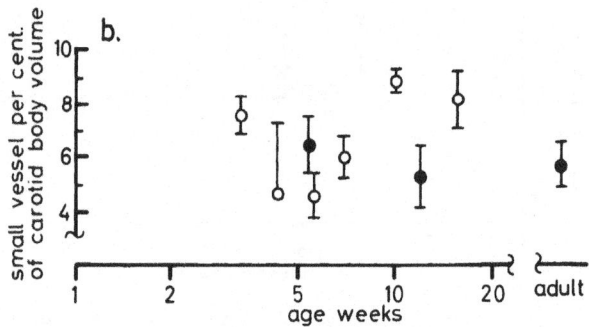

Figure 3. a. Small vessel group mean volumes b.Small vessel percentage of carotid body volume. Group mean ± s.d. Controls (●), SHR (o).

neonates, even although the total carotid body volumes were about the same. Further examination of the values indicates that the small vessel volume increases in the adult SHR animals at the expense of the larger (> 12μm) vessel volume. Since this apparent increase in the size of the small vessel bed is unlikely to be due to the influence of hypertension per se, it is interesting to speculate upon the influence of local hypoxia (generated from the pathological effects of hypertension) on small vessel growth in these SHR animals. These findings are in agreement with the observation that the activity

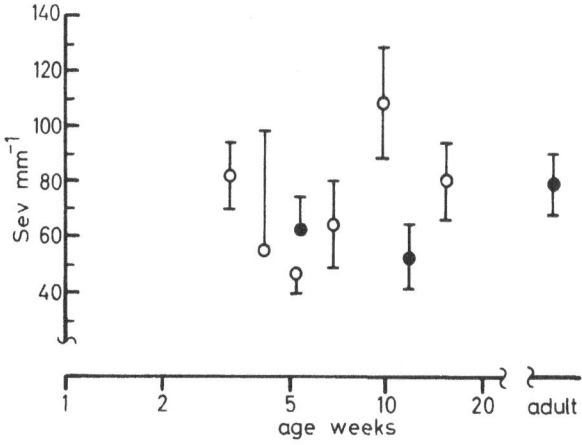

Figure 4. Ratio of small vessel endothelial surface area to the extravascular volume (Sev). Group means ± s.d. Controls (●), SHR (o).

in chemoreceptor fibres from the carotid body is increased SHR animals (Fukuda, Sato & Trzebski, 1987). Figure 3 demonstrates the growth in small vessel volume in the neonatal SHR groups.

If the ratio of the small vessel endothelial cell surface area to the carotid body extravascular volume (Sev) is calculated, it is found that the value is little changed during growth of the carotid body. This suggests that the increase in the volume of the small vessels and the increase in the extravascular volume has occurred pari passu (Fig.4).

We conclude from these results that, since no mitoses were observed in Type I and Type II cells, the major contribution to the increase in carotid body volume in adult spontaneously hypertensive rats is an augmentation of the extravascular extracellular space.

ACKNOWLEDGEMENTS

This work was supported by the Medical Research Council, the British Heart Foundation and the Special Trustees of St. Bartholomew's and St. Mark's Hospitals. Optivision Ltd supported this presentation.

REFERENCES

Clarke J.A. & Daly M. de B.: Distribution of carotid body type I cells and other periadventitial type I cells in the carotid body bifurcation regions of the cat. Anat. Embryol. 166: 169-189 (1983).

Clarke J.A. & Daly M. de B.: The volume of the carotid body and periadventitial type I and II cells in the carotid bifurcation region of the fetal cat and kitten. Anat. Embryol. 173: 117-127 (1985).

Clarke J.A., Daly M. de B. & Ead H.W: Comparison of the size of the vascular compartment of the carotid body of the fetal, neonatal and adult cat. Acta Anat 138: 166-174 (1990).

Fukada, Y, Sato, A & Trzebski, A.: Carotid chemoreceptor discharge responses to hypoxia and hypercapnia in normotensive and spontaneously hypertensive rats. J. Autonom. Nerve Syst. 19: 1-11 (1987).

Habeck, J-O., Huckstorf,C. & Honig, A.: Influence of age on the carotid bodies of spontaneously hypertensive (SHR) and normotensive rats. Exp. Path. 27: 79-89 (1985).

Habeck, J-O.: Peripheral arterial chemoreceptors and hypertension. J. Autonom. Nerve Syst. 34: 1-8 (1991).

2 ROLE OF THE CAROTID SINUS NERVE AND OF DOPAMINE IN THE BIOCHEMICAL RESPONSE OF SYMPATHETIC TISSUES TO LONG-TERM HYPOXIA

Y. Dalmaz, J.M.Pequignot, J.M. Cottet-Emard, A.Vouillarmet and L. Peyrin

URA CNRS 1195,   Faculté de Medecine   Grange-Blanche, 69373 LYON Cedex 08, France

INTRODUCTION

The sympathetic nervous system is known to react to hypoxia, but the response is not unitary and is dependent on hypoxic features (severe or moderate hypoxia, acute or chronic exposure), on the sympathetic nerve or tissue considered and on the state of animals (conscious or anesthetized) (Lee et al. 1987; Fitzgerald et al. 1990). Moreover, the mechanisms controlling the response of the sympathetic nervous system remain poorly understood and seem also highly dependent on tissues and on hypoxic conditions. Two components may be possible candidates for this regulation: one related to neural pathways coming from carotid bodies, and the other related to local dopaminergic events. In fact, previous data report that stimulation of carotid chemoreceptors by hypoxia elicits, or not, a sympathetic response indicating the involvement, or not, of a carotid chemoreceptor reflex (Matsumoto et al. 1987a, 1987b; Fukuda et al. 1989; Biesold et al. 1990; Fitzgerald and Deghani, 1990). As to concern dopamine (DA), it is able to induce an inhibitory electrophysiological and biochemical response of the postganglionic noradrenergic neurones, both in vitro and in vivo (Libet and Owman, 1974; Hanbauer, 1976; Ip et al., 1983; Brokaw et al., 1987). The role of DA located in the SIF cells of sympathetic ganglia may be evoked in this control.

The present experiments were performed on adult conscious rats exposed to long-term hypoxia (10%02; 14 days), this situation being of importance at a practical (high altitude stay) and fundamental (physiological adaptative mechanisms) standpoint. The aim of the study was to determine: 1/ the effects of long-term hypoxia on the turnover of catecholamines in adrenals, in submaxillary glands and in heart, which are representative tissues of the sympathetic system and target organs of neurones originating from the superior cervical and stellate ganglia, 2/ the role of carotid chemoreceptors and of peripheral DA in the control of the response of sympathetic tissues to hypoxia. For this purpose, in one group, the carotid sinus nerve was bilaterally transected, and in an another group, spiroperidol, a selective DA-receptor blocker was injected on the last day of hypoxia.

Neurobiology and Cell Physiology of Chemoreception
Edited by P.G. Data *et al.*, Plenum Press, New York, 1993

METHODS

Male Sprague-Dawley rats (200-220g) were used.

Exposure to normobaric hypoxia

Rats were placed for 14 days in a chamber in which the gas composition was maintained at 10% O2. Expired metabolic water was trapped in a chilled tank. Control rats were exposed to room air (21% O2).

Surgical procedure

Forty rats were subjected to bilateral transection of the carotid sinus nerve (CSN) between its glossopharyngeal branch and the carotid body, one week before experiments. Twenty rats were exposed to hypoxia and twenty rats were kept at room air.

Spiroperidol treatment

Twenty rats were pretreated with spiroperidol (Sigma) (1mg/kg, i.p.), a dopamine-receptor antagonist, on the last day of hypoxic exposure and 3 hours before sacrifice. Control rats were injected with vehicle according to the same schedule.

DA and norepinephrine (NE) content and turnover measurement

On the last day of experiments and 2h30 before sacrifice, half of the rats were injected with alpha-methyl-para-tyrosine methylester (250mg/kg; i.p.) for turnover measurement, the other half being injected with 0.9% saline. Rats were killed by cervical dislocation. The adrenals, the submaxillary glands and the heart were rapidly dissected out. Deproteinated supernatants were analyzed for their catecholamine content using the high liquid performance chromatography coupled with electrochemical detection procedure (Favre et al. 1986). The turnover of catecholamines was estimated by measuring the decrease in their content after alpha-methyl-para-tyrosine.

RESULTS

Effects of hypoxia on the content and turnover of DA and NE in sympathetic tissues

Adrenals: after 14 days of normobaric hypoxia, the content and the turnover of DA were significantly increased 1.3 fold.

Heart: 14 days of hypoxia elicited a decrease in the NE content in heart, together with an increase in NE turnover (2.1-fold) (Fig. 1).

Submaxillary glands: no change in the NE content, but a significant decrease in the NE turnover (0.5-fold) was observed after 14 days of hypoxia (Fig. 2).

Effects of bilateral transection of the CSN on the hypoxia-induced response of sympathetic tissues

Adrenals: bilateral chemodenervation did not abolish the increase in the DA turnover of rats exposed to hypoxia.

Heart: in hypoxic rats, the NE turnover in heart was still increased in spite of chemodenervation (Fig. 1).

Submaxillary glands: the hypoxia-induced decrease in the NE turnover disappeared following chemodenervation. The effects of hypoxia were abolished by transection of the CSN (Fig. 2).

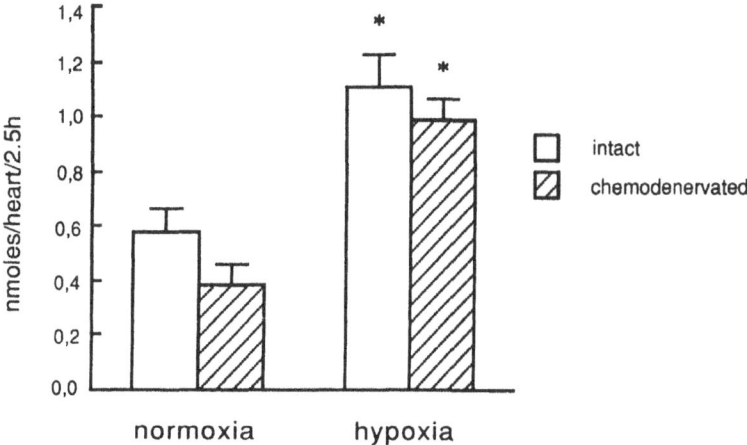

*Fig.1. Effects of hypoxia (10% O2; 14 days) on the turnover of NE (nmoles/heart/2.5h) in heart of intact rats or of bilaterally chemodenervated rats. * Significantly different from normoxic data of intact rats at p < 0.05 (Student's t-test for unpaired data)*

Effects of spiroperidol on the hypoxia-induced response of sympathetic tissues

Heart: spiroperidol treatment enhanced the hypoxia-induced increase in the NE turnover. At the end of hypoxia, the turnover of NE set at a higher level in spiroperidol-treated rats than in intact rats (Fig. 3).

Submaxillary glands: spiroperidol treatment abolished the hypoxia-induced decrease in the NE turnover and, in contrast, elicited an increase in the NE turnover. At the end of hypoxia, the turnover of NE was greatly enhanced (3.4-fold) in rats treated with spiroperidol when compared to intact rats (Fig. 3).

DISCUSSION

The present results demonstrate that long-term hypoxia (10% 02; 14 days) elicits differential changes in DA and NE turnover in adrenals, heart and submaxillary glands; the turnover of DA increases in adrenals, the turnover of NE increases in heart and decreases in submaxillary glands, evidencing that hypoxia enhances the activity in

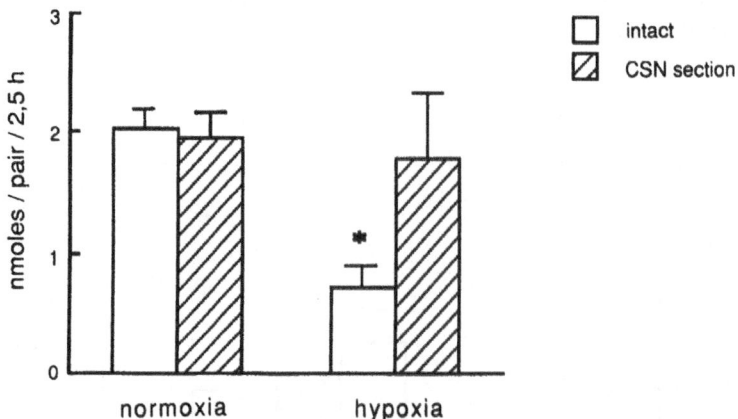

Fig. 2. Effects of hypoxia (10% O2; 14 days) on the turnover of NE in submaxillary glands of intact rats or of bilaterally chemodenervated rats. * Significantly different from normoxic data of intact rats at p<0.05 (Student's t-test for unpaired data)

adrenals and in heart, whereas it reduces the activity in submaxillary glands. Such a difference in the turnover of catecholamines has been reported by Miwa et al. (1987) using exposure of conscious rats to acute and severe hypoxia (8%02; 4 hours); in the same way, a qualitative difference in the regional vascular resistance to acute hypoxia, according to the tissues, has been observed in anesthetized cats by Fitzgerald and Deghani (1990). It is concluded that the response of the sympathetic nervous system to long-term hypoxia is tissue-dependent.

Another conclusion drawn from these experiments is that the mechanisms regulating the response of sympathetic tissues to hypoxia differ according to the tissues: the carotid sinus nerve does not seem to be involved in the hypoxia-induced increase of DA turnover in adrenals and of NE turnover in heart, but seems to exert some control in the decrease of NE turnover in submaxillary glands. During long-term hypoxia, the carotid chemoreceptors have no significant influence on the biochemical activity of adrenals and heart; in contrast, the submaxillary glands show a dependency on the presence of carotid body input, which sends some inhibitory influence under hypoxia. The control exerted by the carotid chemoreceptor stimulation differs according to the organs, which agrees with the view of Fitzgerald and Deghani (1990). However, our conclusions drawn for adrenals and heart are not in accordance with those of Fukuda et al. (1989) and Biesold et al. (1990), showing that the excitatory response of the cardiac and of the adrenal sympathetic nerves to acute hypoxia is elicited via a carotid body chemoreflex; in contrast, severe hypoxia is able to activate a part of the sympathetic nerves, even in the absence of carotid chemosensory inputs in rats (Fukuda et al., 1989), in cats (Matsumoto et al., 1987a; 1987b) or in rabbits (Bower, 1975), thus suggesting the presence of central or of preganglionic influences. One possible explanation for these discrepancies may be that the mechanisms involved in the sympathetic response differ not only according to the degree of hypoxia, but also according to the duration of hypoxia; it may be speculative to extrapolate the mechanisms involved in the response to acute and severe hypoxia to the mechanisms involved in the response to long-term moderate hypoxia. Our data suggest that time-dependent mechanisms may develop slowly independently of carotid chemoreceptors and that other elements than carotid bodies may acquire some chemosensitive properties.

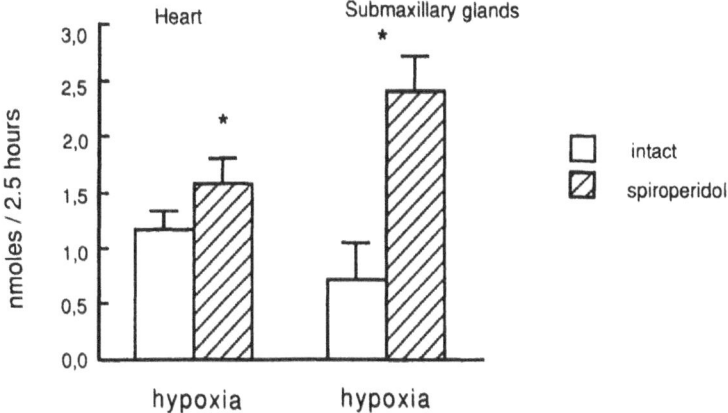

Fig. 3. Effects of hypoxia (10% O2; 14 days) on the turnover of NE in heart and in submaxillary glands of intact rats and of rats injected with spiroperidol (1mg/kg; 3h before sacrifice). * Spiroperidol data significantly different from intact values at p <0.05 (Student's t-test for unpaired data).

The effects of a DA receptor antagonist were investigated in heart and submaxillary glands of rats exposed to hypoxia, considered particularly as target organs of neurones whose cell bodies were located in the superior cervical and stellate ganglia. Because hypoxia induces no change in the NE turnover at the cell body levels (Dalmaz et al. 1988), but elicits changes at the terminal levels, i.e. in heart and in submaxillary glands (this study) it seems more adapted to study the effects of a DA blocker in the target organs. After spiroperidol treatment, the turnover of NE in heart and submaxillary glands of hypoxic rats is significantly enhanced, suggesting that dopamine may exert some inhibitory control on the sympathetic activity during hypoxia. The dopaminergic ganglionic SIF cells being activated during hypoxia (Borghini et al. 1991; Dalmaz et al. 1992), it is tempting to suggest that the ganglionic SIF cells are able to neuromodulate, in vivo, the activity of postganglionic neurones soon triggered by hypoxia. Such a conclusion has been drawn by Brokaw and Hansen (1987), showing that, under acute and severe hypoxia (20min; 5% 02), the synthesis of NE is increased in the superior cervical ganglion, only in rats pretreated with spiroperidol.

ACKNOWLEDGEMENTS

Supported by grants from INSERM (884007), from DRET (88069) and from Région Rhône-Alpes (Neurosciences).

REFERENCES

Biesold, D., Kurosawa, M., Sato, A., and Trzebski, A., 1990, Responses of sympatho-adrenal medullary system to hypoxia and hypercapnia in anesthetized artificially ventilated rats, in "Chemoreceptors and chemoreceptor reflexes", Acker, H., Trzebski, A., and O'Reagan, R.G., eds., Plenum Press, New York, 3:193.

Borghini, N., Dalmaz, Y., and Peyrin, L., 1991, The responsiveness of SIF cells in the rat superior cervical ganglion to prolonged hypoxia, J. Auton. Nerv. Syst., 33:150.

Bower, E. A., 1975, The influence of hypercapnia and hypoxia on the activity recorded from intestinal sympathetic nerves in the rabbit with cut sinus nerves, J. Physiol. (Lond.), 249:68P.

Brokaw, J. J., and Hansen, J. T., 1987, Evidence that dopamine regulates norepinephrine synthesis in the rat superior cervical ganglion during hypoxic stress, J. Auton. Nerv. System., 18:185.

Dalmaz, Y., Borghini, N., Pequignot, J.M., and Peyrin, L., 1992, Presence of chemosensitive SIF cells in sympathetic ganglia: a biochemical, immunocytochemical and pharmacological study, in " Neurobiology and cell physiology of chemoreception", Data, P. G., Acker, H., and Lahiri, S., eds., Plenum Press, New York, (this book).

Dalmaz, Y., Pequignot, J.M., Tavitian, E., Cottet-Emard, J.M., and Peyrin, L., 1988, Long-term hypoxia increases the turnover of dopamine but not norepinephrine in rat sympathetic ganglia, J. Auton. Nerv. System., 24:57.

Favre, R., De Haut, M., Dalmaz, Y., Pequignot, J. M., and Peyrin, L., 1986, Peripheral distribution of free dopamine and its metabolites in the rat, J. Neural Trans., 66:13.

Fitzgerald, R., S., and Deghani, G. A., 1990, Chemoreceptor control of organ vascular resistance during acute systemic hypoxia, in "Chemoreceptors and chemoreceptor reflexes", Acker, H., Trzebski, A., and O'Reagan, R.G., eds., Plenum Press, New York, 4:217.

Fukuda, Y., Sato, A., Suzuki, A., and Trzebski, A., 1989, Autonomic nerve and cardiovascular responses to changing blood oxygen and carbon dioxide levels in the rat, J. Auton. Nerv. Syst., 28:61

Hanbauer, I., 1976, Induction of tyrosine hydroxylase in the superior cervical ganglia of rats: Opposite influence of muscarinic and nicotinic receptor agonists, Neuropharmacology, 15:85.

Ip, N. Y., Perlman, R. L., and Zigmond, R. E., 1983, Acute transsynaptic regulation of tyrosine 3-monooxygenase activity in the rat superior cervical ganglion: Evidence for both cholinergic and noncholinergic mechanisms, <u>Proc. Natl. Acad. Sci.</u>, 80:2081.

Lee, K., Miwa, S., Fujiwara, M., Magaribuchi, T., and Fujiwara, M., 1987, Differential effects of hypoxia on the turnover of norepinephrine and epinephrine in the heart, adrenal gland, submaxillary gland and stomach, <u>J. Pharmacol. Exper. Ther.</u>, 240:954.

Libet, B., and Owman, C., 1974, Concomitant changes in formaldehyde-induced fluorescence of dopamine interneurons and in slow-inhibitory postsynaptic potentials of the rabbit superior cervical ganglion induced by stimulation of the preganglionic nerve or by muscarinic agent, <u>J. Pysiol. (Lond.)</u>. 327:635.

Matsumoto, S., Mokashi, A., and Lahiri, S., 1987a, Cervical preganglionic sympathetic nerve activity and chemoreflexes in the cat, <u>J. Appl. Physiol.</u>, 62: 1713.

Matsumoto, S., Mokashi, A., and Lahiri, S., 1987b, Ganglioglomerular nerves respond to moderate hypoxia independent of peripheral chemoreceptors in the cat, <u>J. Auton. Nerv.</u>, 19:219.

3

THE EFFECTS OF ALMITRINE ON (^{3}H)5HT AND (^{125}I) ENDOTHELIN BINDING TO CENTRAL AND PERIPHERAL RECEPTORS: AN IN VITRO AUTORADIOGRAPHIC STUDY IN THE CAT

M.R. Dashwood[1], D.S. McQueen[3], R.M. Sykes[1], J.P. Muddle[2], M. de B. Daly[1], Y. Evrard[4] and K.M. Spyer[1]

Departments of Physiology[1] and Neurological Science[2], Royal Free Hospital School of Medicine, London NW3 2PF, Department of Pharmacology[3], University of Edinburgh Medical School, Edinburgh EH8 9JZ, United Kingdom, and IRIS[4], Paris, France

INTRODUCTION

Recently we have shown that the density of [^{3}H]5HT binding sites in the carotid body (CB) and within the nucleus of the tractus solitarius (NTS) of the cat is increased following unilateral chronic sectioning of the carotid sinus nerve (CSN) (Dashwood et al 1990). [^{125}I]endothelin (ET) binding sites have also been identified in these regions, but these "receptors" are unaffected by CSN section (Spyer et al 1991). There is physiological evidence that both 5HT (Kirby and McQueen 1984) and ET (Spyer et al 1991) affect chemoreflex activity in the cat. Here we report the effect of chronic chemoreceptor stimulation, by almitrine bismesylate, on the density of central and peripheral 5HT and ET receptors in the cat.

METHODS

Receptor distributions were determined autoradiographically (for description of original methodology see Young and Kuhar, 1979) in tissue (brainstem, carotid bifurcation region-including carotid body and nodose and superior cervical ganglia) from cats receiving almitrine bismesylate ("Vectarion", Servier Laboratories, France) (n=2) and control animals (n=2). In experimental animals the right carotid sinus nerve was surgically sectioned under ketamine and xylazine anaesthesia and the drug dosage (10 mg kg^{-1} p.o. daily) continued for 8 days following denervation. Animals were then anaesthetized (sodium pentobarbitone, 60mg kg^{-1} i.p.), perfused intracardially and lightly fixed with 0.1% formaldehyde in ice-cold 100 mM phosphate buffer. Tissue was removed, stored at -70°C, then serial sections were cut in a cryostat and thaw-mounted onto gelatinised microscope slides. Binding sites were identified by incubating slide-mounted sections in buffer containing 4 nM [^{3}H]5HT or 25 pM [^{125}I]ET-1 (specific activities 11 Ci mmol^{-1} and 2200 Ci mmol^{-1} respectively, from Amersham International). The degree of binding to non-specific sites was established by incubating in the presence of excess concentrations of unlabelled ligand (5HT or ET-l) (for details see Dashwood et al 1990; Spyer et al 1991). Sections were exposed to Hyperfilm ^{3}H (Amersham) or K2 nuclear emulsion (Ilford) for low- and high-resolution autoradiography (see Moody et al 1990 for details). After the appropriate exposure time (9 weeks for [^{3}H]5HT, 4 days for [^{125}I]ET-l) autoradiographs were processed, photographed and the underlying tissue stained for histological analysis. Selected autoradiographs were subjected to densitometric analysis on an IBAS 2000 image analysis system (Kontron) and receptor densities were estimated from curves generated from microscales (Amersham) which were co-exposed with incubated tissue sections.

Table 1. Specific l[^{125}I]ET-1 binding to normal and Almitrine treated cats: image analysis

Treatment.	[125I]ET-l binding (amol mg-1 protein)	
	left NTS	right NTS
Control.	4534+/-44	4516+/-44
Almitrine-treated 1.	3264+/-128	3306+/-110
Almitrine-treated 2.	3317+/-227	3188+/-283

Values=mean+/-sem of measurements from at least 6 sections.

RESULTS

Both [^{3}H]5HT and [^{125}I]ET-l binding sites (putative receptor sites) were associated with central and peripheral regions involved in the arterial chemoreceptor reflex. Both receptor types exhibited very similar distributions. Binding sites were present in the carotid bifurcation region, including the carotid body (Fig 1), superior cervical and nodose ganglia (data not shown) and brainstem, particularly NTS, raphe nucleus and olivary nuclei (Figs 2 and 3). Interestingly, within the carotid body, high-resolution autoradiographs indicated that 5HT and ET-1 receptors were present on both the microvasculature and glomus cells (Spyer et al 1991). We have previously shown that [^{3}H]5HT binding is raised in the ipsilateral NTS and carotid body following sectioning of a carotid sinus nerve (Dashwood et al 1990). Binding of [^{125}I]ET-l to these regions was unaffected by denervation. Image analysis indicated that there was a small reduction in [^{125}I]ET-l binding to the NTS in animals following chronic chemoreceptor stimulation by almitrine, but this change was not statistically significant (table 1). [^{3}H]5HT binding to the brainstem (Fig 3), and carotid body (not shown) of almitrine-treated animals, however, was reduced to a level that was barely detectable.

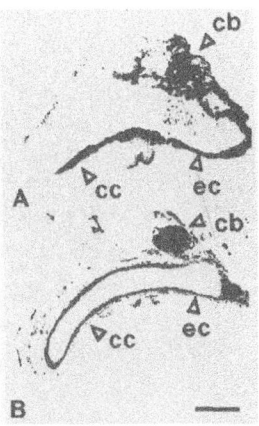

Fig. 1. [^{125}]ET-l and [^{3}H]-5HT binding to normal (non denervated) cat carotid bifurcation region.
    A. Autoradiograph from a longitudinal section of cat carotid bifurcation incubated in 25pM
    [^{125}]ET-l. B. Autordiograph from a longitudinal section of cat carotid bifurcation region
    incubated in 4nM [^{3}H]-5HT. Scale bar=1mm. cc=common carotid, cb=carotid body, ec=external
    carotid artery.

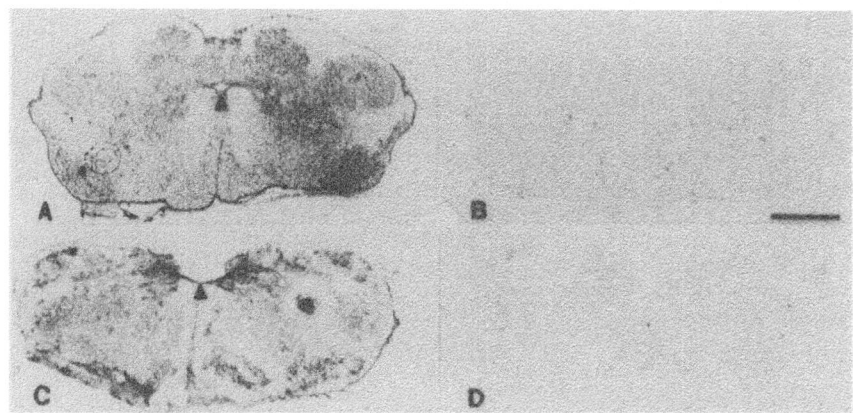

Fig.2.   *[^{125}I]ET-l binding to normal (non-denervated) cat brainstem. A. Autoradiograph from a caudal*
section of cat brainstem incubated in 25pM [^{125}I]ET-I (total binding). B) Autoradiograph from
an alternate section incubated as above, in the presence of 500nM unlabelled ET-I (non-specific
binding). C) Autoradiograph of total [^{125}I]ET-I binding to rostral brainstem. D) Non-specific
binding. Scale bar=2.5mm Binding to the caudal NTS can be seen, in A, dorsal to the central canal
(arrowed) and to the intermediate NTS, in C, lateral to obex (arrowed).

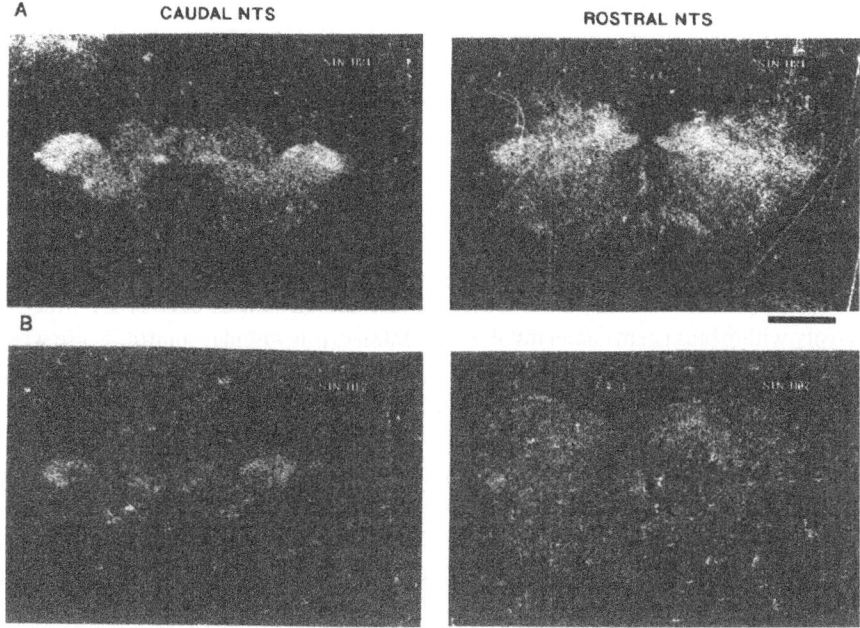

Fig.3. Image analysis of [3H]-5HT binding to brainstem of control and almitrine-treated cats.
    A. left. Image of autoradiograph from section of caudal brainstem from normal cat incubated in
    4nM [³H]-5HT. Right. Binding to rostral brainstem of above cat. B) Images from caudal and
    rostral brainstem of an almitrine-treated cat. Figures are black and white photographs taken of
    pseudocolour images generated by densitometric analysis where white indicates regions of high
    binding and black regions of low/absent binding.

DISCUSSION

Using in vitro receptor autoradiography we have demonstrated that binding sites for [^{3}H]5HT and [^{125}I]ET-1 are associated with central and peripheral regions involved in the chemoreceptor reflex (Dashwood et al 1990; Spyer et al 1991). Both receptor types exhibit very similar distributions. Within the brainstem there was high binding to the nucleus tractus solitarius, the site of termination of chemoreceptor afferents, as well as to the raphe nucleus, rostral ventrolateral medulla and olivary nuclei. [^{3}H]5HT and [^{125}I]ET-1 binding was dense at the carotid bifurcation region with binding to the smooth muscle of the common and external carotid artery, carotid sinus region and particularly high binding within the carotid body. Both receptor types were associated with the microvasculature of the carotid body as well as with glomus cells. Binding was also present on sections of superior cervical and nodose ganglia. [^{125}I]ET-1 binding often appeared patchy on the nodose ganglion with binding being to cell bodies or satellite cells. In a previous study we have shown that there is a marked increase in [^{3}H]5HT binding, within the ipsilateral NTS and carotid body, following carotid sinus nerve section (Dashwood et al 1990). This "denervation-induced receptor upregulation" does not occur with [^{125}I]ET-1 receptor binding sites.

The main purpose of the present study was to determine whether chronic chemoreceptor stimulation, evoked by almitrine, affected either 5HT or ET-1 receptor density within brainstem nuclei involved in central cardiovascular control. There was no obvious change in [^{125}I]ET-1 binding and, although image analysis provided some indication of a small decrease in [^{125}I]ET-1 binding to the NTS of almitrine-treated animals, this change was not statistically significant. There was, however, a marked reduction in [^{3}H]5HT binding to the brainstem and a suggestion that binding to other tissues (eg carotid body) was also reduced. What is particularly interesting is the fact that there was a generalised reduction in binding to the brainstem, the "down-regulation" was not restricted to specific nuclei. [^{3}H]5HT binding in almitrine-treated tissue was barely detectable and was, in fact, at the limit of resolution of our image analysis system. This made it impossible to establish whether there had been a concomitant denervation-induced increase in [^{3}H]5HT binding following sectioning of the carotid sinus nerve in almitrine-treated animals. The reduction in [^{3}H]5HT binding, caused by almitrine, may result from the chronic chemoreceptor stimulation associated with this compound. However, one other possibility should be considered. Almitrine may cause a local release of endogenous 5HT or, alternatively, have a direct effect on [^{3}H]5HT binding sites. If almitrine were to cause locally-released 5HT to bind to receptors, or itself bind to 5HT receptors, it would prevent, or reduce, the access of [^{3}H]5HT to specific binding sites during incubation of slide-mounted sections in radiolabelled ligand, and so reduce the apparent receptor density on autoradiographic images. From these results there seems no doubt that, either directly or indirectly, 5HT or 5HT receptor sites are implicated in the chemoexcitant action of almitrine. The possibility that a specific 5HT receptor subtype may be involved is presently under investigation.

ACKNOWLEDGEMENTS

We wish to thank the British Heart Foundation and Stanley Thomas Johnson Foundation for support. Support from Servier Pharmaceuticals, (IRIS) Paris, is also gratefully acknowledged.

REFERENCES

Dashwood, M.R., McQueen, D.S., Daly, M de B., Spyer, K.M., and Evrard, Y., 1990, Autoradiographic studies on the effects of chronic unilateral sectioning of a carotid sinus nerve on 5-HT and SP binding sites in the carotid body and NTS. in Chemoreceptors and Chemoreceptor Reflexes, Eds. H. Acker et al, Plenum Press New York. pp305-309.

Kirby, G.C., and Mc Queen, D.S., 1984, Effects of the antagonists MDL 72222 and Ketanserin on responses of cat carotid body chemoreceptors to 5-hydroxytryptamine. Br J Pharmacol. 83, 259-269.

Moody, C.J., Dashwood, M.R., Sykes, R.M., Chester, M., Jones, S.M., Yacoub, M.H., and Harding, S.E., 1990, Functional and autoradiographic evidence for endothelin-1 receptors on human and rat cardiac myocytes. Circ. Res. 67,764-769.

Spyer, K.M., McQueen, D.S., Dashwood, M.R., Sykes, R.M., Daly, M. de B., and Muddle, J.R., 1991, Localization of [^{125}I]endothelin binding sites in the region of the carotid bifurcation and brainstem of the cat: possible baro- and chemoreceptor involvement. J Cardiovasc Pharmacol. 17(Suppl 7):S385-389.

Young, W.S., and Kuhar, M.J., 1979, A new method for receptor autoradiography:[^{3}H]opioid receptors in rat brain. Brain Res.179, 255-279.

4   IMMUNOCYTOCHEMICAL AND NEURO-CHEMICAL ASPECTS OF SYMPATHETIC GANGLION CHEMOSENSITIVITY

B. Dinger, Z.-Z. Wang, J. Chen, W.-J. Wang, G. Hanson, L.J. Stensaas and S.J. Fidone

Department of Physiology, University of Utah School of Medicine, Salt Lake City, Utah, USA

INTRODUCTION

Previous studies by Hanson et al.[8] demonstrated that acute hypoxic stress in vivo reduced the content of the neuropeptides, substance P and met-enkephalin, in decentralized superior cervical ganglia (SCG), and elevated glucose utilization in SCG exposed to hypoxia in vitro. Because similar changes were observed in the chemosensory tissue of the carotid body and not in nonchemosensory control tissue, these authors suggested that SCG contain intrinsic chemosensory mechanisms [8]. The subsequent studies of Cheng et al.[2] and Dalmaz et al.[3] added further support to this hypothesis. Dalmaz, Pequignot and their colleagues[3,15] showed that the turnover of dopamine (DA) was significantly increased in both the superior cervical ganglia (SCG) and carotid bodies from rats chronically (2 to 28 days) exposed to hypoxic gas mixtures. Their experiments also demonstrated that the turnover of norepinephrine (NE), the dominant biogenic amine contained in postganglionic SCG neurons, was not altered by chronic hypoxia, while its turnover in the carotid body was elevated between days 7 and 28 of the chronic exposure[3,15]. A critical observation in these studies was that DA turnover in SCG remained elevated in hypoxic animals treated with guanethidine, an agent known to destroy primarily NE containing neuronal cell bodies [3]. Because nearly one-half of the DA in SCG is presumed to be contained in small intensely fluorescent (SIF) cells[12], which remain intact following guanethidine treatment, it was suggested that these peculiar cells were acting as chemosensors in sympathetic ganglia. However, despite the fact that increased DA turnover persisted in the SCG following chronic carotid sinus nerve transection and preganglionectomy[3], Dalmaz et al.[3] were unable to rule out the possibility that in their in situ preparation the changes in catecholamine (CA) metabolism were mediated by circulating hormones, such as corticosteroids, which are known to be released from the adrenal cortex and to enhance CA synthesis in response to hypoxic stress[9,20].

In previous studies in our laboratory, we have examined the response of the rabbit SCG to acute (10 min) hypoxic exposure in vitro[2], thereby avoiding the possible complications resulting from extrinsic neural and hormonal control. Our studies support the notion that low-O_2 stimulation activates the selective synthesis of DA without affecting NE levels and, in addition, they indicate that DA is released in the SCG in response to hypoxia[2]. Because pharmacological studies by other authors have demonstrated that DA mediates a slow inhibitory postsynaptic potential (s-IPSP)[13] and stimulates adenylate cyclase in rabbit noradrenergic sympathetic neurons[10], our data support a modulatory role for the DA which is released in response to changes in local tissue PO_2.

In the present study, we have explored the possibility that DA released in response to hypoxia enhances cyclic AMP (cAMP) levels in rabbit SCG incubated in vitro. We have, in addition, used immunocytochemical techniques to identify unique cell populations in the SCG which respond to hypoxia. Our results suggest that SCG SIF cells are selectivity activated by hypoxic stimuli, and that the neurochemical responses of these cells are comparable to that seen with the chemosensory type I cells of the carotid body. Furthermore, the DA released from SIF cells in response to a hypoxic challenge appears to mediate certain functionally relevant changes in the postsynaptic neurons, including the elevation of cAMP levels.

MATERIALS AND METHODS

cAMP assay. Under pentobarbital anesthesia, the superior cervical ganglia were cleaned of surrounding connective tissue in a lucite chamber containing 100% O_2^- equilibrated Tyrode's solution at 0-4°C (in mM: NaCl, 112; KCl, 4.7; CaCl$_2$, 2.2; MgCl$_2$, 1.1; Na-glutamate, 42; HEPES buffer, 5; glucose, 5.6; pH = 7.4). The tissues were weighed on a Cahn electrobalance equipped with a humidified weighing chamber and then transferred to glass scintillation vials in a waterbath-shaker for a 30 min preincubation in 1 ml of 100% O_2 Tyrode's solution at 37°C. The preincubation medium was replaced with 1 ml of Tyrode's solution equilibrated with control (100% O_2) or stimulus gas mixtures (5% O_2 or 10% O_2, balance N_2). After a 10-min incubation (37°C), the ganglia were immersed in 600 ml of cold (4°C) 6% trichloroacetic acid. The tissues were homogenized in a glass-glass homogenizer and the homogenates centrifuged at 13, 000 xg for 10 min (4°C). The supernatant was extracted 3x in 3 ml of water-saturated ethyl ether, the remaining aqueous phase was dried under vacuum (Savant), and the sealed samples stored at 4°C for up to 1 month prior to RIA. The assays were performed using commercially available cAMP RIA kits (Dupont, NEN), and the data are expressed as pmol cAMP/mg tissue.

Cyclic GMP Immunocytochemistry. Adult rats were anesthetized with sodium pentobarbital (40 mg/kg, i.p.). Carotid bodies and superior cervical ganglia were removed and placed in ice-cold (4°C) 100% O_2 equilibrated Locke's solution (NaCl, 136 mM; KCl, 5.6 mM; NaH$_2$PO$_4$, 1.2 mM; NaHCO$_3$, 20 mM; MgCl, 1.2 mM; glucose, 5.5 mM) containing theophylline (10 mM), pH 7.45. They were then preincubated in a waterbath shaker for 30 min with the same media (37°C), after which single carotid bodies and ganglia were transferred to vials containing media equilibrated with 100% O_2 or 5% O_2 and incubated for 10 min. Samples were then immediately transferred into ice-cold fixative (4% paraformaldehyde and 0.2% picric acid in 0.1 M phosphate buffer, pH 7.4) for 90 min, followed by a wash for 15 min in cold buffered saline (PBS). The tissue was cryoprotected in 20% sucrose for 1 hr at 4°C, quickly frozen to -20°C in a cryostat, and sections 6-8 µm thick were thaw-mounted onto chrome-alum gelatin-coated slides. Sections hydrated in PBS (15 min) were treated with 3% hydrogen peroxide (10 min) and 1% normal goat serum (20 min), and then incubated with the following reagents, each of which was followed by a wash (15 min) in cold PBS: a), a polyclonal cGMP antiserum (1:3000, ref. 5) in PBS containing 0.3% Triton X-100 overnight at 4°C; b), a biotinylated goat anti-rabbit secondary antibody (Vector, 1:400) at room temperature for 30 min; c), avidin-biotin-peroxidase complex (Vector) at room temperature for 60 min. Finally, the sections were incubated in a mixture of 3,3'-diaminobenzidine hydrochloride (Sigma) and hydrogen peroxide. Control sections, which were treated in the same manner except for preabsorption of the primary antibody with 10^{-3} M cGMP or replacement by normal rabbit serum, showed no specific immunoreactivity.

Tyrosine Hydroxylase Immunocytochemistry. Carotid bodies and superior cervical ganglia (SCG) were removed from adult (200-300 g) rats under pentobarbital anesthesia (40 mg/kg body weight), fixed in ice-cold PLP fixative (4% paraformaldehyde, 1.4% lysine and 0.21% sodium metaperiodate) overnight, washed in 15% sucrose (in 0.05 M Tris-buffered saline, TBS) for 2 hr, and cryoprotected overnight in 30% sucrose in TBS. Sections (8 μm thick) were cut at -20°C and mounted on chrome-alum subbed glass slides. Immunostaining utilized the avidin-biotin-peroxidase complex method. Sections were rehydrated and equilibrated in TBS for 15 min. Endogenous non-specific avidin-biotin activity was eliminated by incubating the sections in blocking agents (Vector ABC stain blocking kit). Sections were then exposed to primary antibody for tyrosine hydroxylase (TH; 1:600; Chemicon) overnight at 4°C; after a brief washing in TBS, sections were incubated with biotinylated goat anti-rabbit secondary antibody (1:400) at room temperature for 1 hr, followed by rinsing in TBS; diluted avidin-biotin-peroxidase complex (Vector ABC stain elite kit) was applied to the sections at room temperature for 1 hr. Immunopositive elements were revealed using a mixture of diaminobenzidine tetrahydrochloride (50 mg/100 ml) and hydrogen peroxide (0.03%). Hematoxylin was used to counterstain the sections. TH positive SIF cells appeared as small (7-12 μm) irregularly shaped elements which were typically aggregated into groups of 3 to 10 cells.

RESULTS AND DISCUSSION

The Effect of Hypoxia on Cyclic AMP

The basal level of cAMP in SCG incubated in 100% O_2-solution was 1.27 $\pm$ 0.13 pmole/mg tissue (X $\pm$ SEM), a value comparable to previous reports of cAMP in SCG superfused in vitro [14]. Incubation of SCG for 10 min in superfusion solutions equilibrated with 10% O_2 or 5% O_2 elevated cAMP levels to 2.00 $\pm$ 0.19 and 1.74 $\pm$ 0.16 pmole/mg tissue (p<0.05 compared to SCG incubated in 100% O_2-solution), respectively. Earlier experiments in our laboratory [17,18], as well as those of Perez-Garcia et al. [16] and Delpiano and Acker [4], have likewise shown that brief incubation of carotid bodies in hypoxic media significantly elevates the content of cAMP in the chemosensory tissue. Furthermore, in more recent studies using immunocytochemical techniques, we have localized these changes in cAMP content to carotid body type I cells [19].

Earlier studies by other authors have shown that cAMP levels in SCG increase in response to preganglionic stimulation [14] or exposure to DA [10], findings which implicate the involvement of DA and SIF cells. Because our previous studies suggested that DA is released in the SCG in response to hypoxia[2], we examined the possibility that the observed increase in cAMP is mediated by DA. These experiments utilized the dopaminergic antagonist haloperidol (10-5 M) and the selective D-1 receptor blocker SKF R83566 (10^{-5} M). In the presence of these drugs, basal cAMP levels were not altered, but more importantly, the ability of hypoxia to evoke an increase in the second messenger was completely abolished. These findings support the notion that hypoxia releases DA in the SCG, and suggest further that DA acts at specific D-1 dopaminergic receptors, which are known to be located postsynaptically on the principal ganglionic neurons (PGNs)[see 10].

Immunocytochemical Studies of SIF Cell Responses to Hypoxia

In these experiments, immunocytochemical markers were employed to evaluate the effects of hypoxia on SIF cell activity. Staining intensity was compared in SIF cells, SCG PGNs and carotid body type I cells, to determine the selectivity and specificity of the observed changes in hypoxic versus normoxic tissue. In one set of experiments, tyrosine hydroxylase (TH) immunoreactivity was assessed in two groups of rats immediately following a 3 hr exposure in a flow chamber to either air or 10% O_2. To examine long-term effects, another two groups of rats were similarly exposed, but the tissues were removed and prepared for immunocytochemistry 48 hr after the 3 hr exposure to hypoxic or normoxic gas mixtures.

In normoxic control animals, TH staining was moderately intense in carotid body type I cells and SCG SIF cells. Staining in the large PGNs was less intense and more diffuse than in the other cell types. Immediately after hypoxic exposure, the staining intensity in type I cells and SIF cells was increased somewhat, while no change was noted in PGNs. At 48 hrs after hypoxic exposure, staining in these large neurons was increased only slightly, but was greatly elevated in the type I cells and SIF cells. These findings confirm previous reports from our laboratory [6], and also by others [7], which show that TH activity in the carotid body and SCG is increased 48 hrs following an acute hypoxic episode. Furthermore, our results demonstrate an intriguing parallel in the response of chemosensory carotid body type I cells and the morphologically similar SCG SIF cells.

In a separate series of experiments, SCG and carotid bodies were removed from pentobarbital anesthetized rats and superfused in vitro. The tissue was subsequently prepared for immunocytochemical staining of the cyclic nucleotide second messenger, cyclic GMP (cGMP). Examination of tissue incubated in 100% O_2 revealed that the incidence cGMP positive cells was greater than 90% amongst carotid body type I cells and small granular (SIF) cells in SCG; PGNs in the SCG were not stained under these conditions. Staining intensity was uniformly dark in the cytoplasm of positive elements. In contrast, following a 10 min exposure to superfusion solution equilibrated with 5% O_2, both the incidence of positive carotid body type I cells and the intensity of staining was greatly reduced; similar results were found in the SCG where the number of cGMP positive small granular cells per ganglion was reduced. These results confirm our earlier studies which showed that hypoxia decreases the content of cGMP in carotid body type I cells [17]. The fact that we observe a similar change in cGMP content in the SCG and carotid body in response to a hypoxic challenge supports the notion that SCG-SIF cells, like carotid body type I cells, function as local chemosensory elements.

Numerous studies have drawn attention to the morphological and biochemical similarities of SCG-SIF cells and carotid body type cells [1,11,21]. Such observations have, quite naturally, led to speculation that SIF cells act as chemosensors in a local neuronal circuit. Although the present data do not unequivocally establish that SIF cells subserve a chemosensory role in the SCG, they do extend the list of intriguing parallels shared by SIF cells and the chemosensory type I cells of the carotid body. In addition, the observation that hypoxia elevates cAMP via a dopaminergic mechanism lends further support to the involvement of SIF cells in a chemosensory response and suggests that PGNs may be modulated by substances released during periods of lowered O_2 tension. Because SIF cells are situated as interneurons between cholinergic preganglionic terminals and noradrenergic PGNs [21], the results of our study and those of Dalmaz et al. [3] suggest the need for a detailed electrophysiological assessment of PGN responsiveness and SIF cell actions under low O_2 conditions.

REFERENCES

1. *Case, C.P. and Matthews, M.R.* A quantitative study of structural features, synapses and nearest-neighbor relationships of small granule-containing cells in the rat superior cervical sympathetic ganglion at various adult stages. Neurosci. 15: 237-282, 1985.

2. *Cheng, G.-F., Dinger, B., Hanson, G. and Fidone, S.J.* Effects of hypoxia on catecholamine storage and release in rabbit superior cervical ganglion. In: C. Eyzaguirre, S.J. Fidone, R.S. Fitzgerald, S. Lahiri and D.M. McDonald (ed.) Arterial Chemoreception. Springer-Verlag, New York, 1990, pp. 398-403.

3. *Dalmaz, Y., Borghini, N., Pequignot, J.M. and Peyrin, L.* Involvement of dopaminergic SIF cells of rat superior cervical ganglion in response to chemoreceptor stimuli. In: C. Eyzaguirre, S.J. Fidone, R.S. Fitzgerald, S. Lahiri and D.M. McDonald (eds.) Arterial Chemoreception. Springer-Verlag, New York, 1990, pp. 404-418.

4. *Delpiano, M.A. and Acker, H.* Hypoxia increases the cyclic AMP content of the cat carotid body in vitro. J. Neurochem. 57: 291297, 1991.

5. *de Vente, J., Steinbusch, H.W.M. and Schipper, J.* A new approach to immunocytochemistry of 3',5'-cyclic guanosine monophosphate: preparation, specificity and initial application of a new antiserum against formaldehyde-fixed 3',3'-cyclic guanosine monophosphate. Neurosci. 22: 361-373, 1987.

6. *Gonzalez, C., Kwok, Y., Gibb, J. and Fidone, S.* Effects of hypoxia on tyrosine hydroxylase activity in rat carotid body. J. Neurochem. 33: 713-719, 1979.

7. *Hanbauer, I., Lovenberg, W. and Costa, E.* Induction of tyrosine 3-mono-oxygenase in carotid body of rats exposed to hypoxic conditions. Neuropharm. 16: 277-282, 1977.

8. *Hanson, G., Gonzalez, C., Obeso, A., Dinger, B. and Fidone, S.* Local regulation of sympathetic ganglionic activity during acute hypoxia. In: S. Lahiri, R.E. Forster, II, R.O. Davies and A.I. Pack (eds.) Chemoreceptors and Reflexes in Breathing: Cellular and Molecular Aspects. Oxford University Press, New York, 1989, pp. 37.

9. *Hellstrom, S. and Koslow, S.H.* Effects of glucocorticoid treatment on catecholamine content and ultrastructure of adult rat carotid body. Brain Res. 102: 245-254, 1976.

10. *Kobayashi, H. and Tosaka T.* Slow synaptic actions in mammalian sympathetic ganglia, with special reference to the possible roles played by cyclic nucleotides. In: L.-G. Elfvin (ed.) Autonomic Ganglia. John Wiley and Sons, Inc., New York, 1983, pp. 281-307.

11. *Kondo, H.* Innervation of SIF cells in the superior cervical and nodose ganglia: An ultrastructural study with serial sections. Biol. Cellulaire 30: 253-264, 1977.

12. *Koslow, S.H.* Mass fragmentographic analysis of SIF cell of normal and experimental rat sympathetic ganglia. In: O. Eranko (ed.) SIF Cells. Structure and Function of Small Intensely Fluoresoent Sympathetic Cells. Fogarty Int. Center Proc., 1976, pp. 82-88.

13. *Libet, B. and Owman, Ch.* Concomitant changes in formaldehydeinduced fluorescence of dopamine interneurons and slow inhibitory postsynaptic potentials of the rabbit superior cervical ganglion, induced by stimulation of the preganglionic nerve or by a muscarinic agent. J. Physiol. (Lond.) 237: 635-662, 1974.

14. *McAfee, D.A., Schorderet, M. and Greengard, P.* Adenosine 3',5' monophosphate in nervous tissue: increase associated with synaptic transmission. Science 171: 1156-1158, 1971.

15. *Pequignot, J.M., Cottet-Emard, J.M., Dalmaz, Y. and Peyrin, L.* Dopamine and norepinephrine dynamics in rat carotid body during long-term hypoxia. J. Autonom. Nerv. Syst. 21: 9-14, 1987.

16. *Perez-Garcia, M.T., Almaraz, L. and Gonzalez, C.* Effects of different types of stimulation on cyclic AMP content in the rabbit carotid body: functional significance. J. Neurochem. 55: 1287-1293, 1990.

17. *Wang, W.-J., Cheng, G.-F., Dinger, B.G. and Fidone, S.J.* Effects of hypoxia on cyclic nucleotide formation in rabbit carotid body in vitro. Neurosci. Lett. 105: 164-168, 1989.

18. *Wang, W.-J., Cheng, G.-F., Yoshizaki, K., Dinger, B. and Fidone, S.* The role of cyclic AMP in chemoreception in the rabbit carotid body. Brain Res. 540: 96-104, 1991.

19. *Wang, Z.Z., Stensaas, L.J., de Vente, J., Dinger, B. and Fidone, S.J.* Immunocytochemical localization of cAMP and cGMP in cells of the rat carotid body following natural and pharmacological stimulation. Histochem. 96: 523-530, 1991.

20. *Weiner, N.* Control of the biosynthesis of adrenal catecholamines by the adrenal medulla. In: H. Blaschko, G. Sayers and D. Smith (eds) Adrenal Gland, vol. 6, sect. 7, Endocrinology. Handbook of Physiology. American Physiological Society, Washington, D.C., 1975, pp. 357-366.

21. *Williams, T. and Jew, J.* Monoamine connections in sympathetic ganglia. In: L.-G. Elfvin (ed.) Autonomic Ganglia. John Wiley and Sons, New York, 1983, pp. 235-264.

5 NEURONAL AND NEUROENDOCRINE MARKERS IN THE HUMAN CAROTID BODY IN HEALTH AND DISEASE

J. -O. Habeck* and W. Kummer

Pathologisches Institut
Städtische Kliniken
Chemnitz, FRG

INTRODUCTION

The human carotid body is subjected to both hypertrophy and proliferation of various cell types in the course of general diseases such as arterial hypertension and states of chronic hypoxemia. Exact identification of the cell types involved in these structural changes is often impossible in routinely stained preparations. Thus, the present light- and electronmicroscopical study was aimed to elucidate the value of some immunohistochemical markers generally used in neuropathology for cellular subtyping in the human carotid body.

MATERIAL AND METHODS

Carotid bodies from controls and patients suffering from chronic hypoxemia were obtained during routine autopsy and fixed by immersion in buffered 2% paraformaldehyde/15% saturated picric acid. Paraffin and cryostat sections were then subjected to single- or double-labelling immunofluorescence according to routine techniques (see Kummer and Habeck 1991). Primary antisera were directed against glial fibrillary acidic protein (GFAP; rabbit antiserum from Dakopatts, Glostrup, Denmark), protein gene product 9.5 (PGP; rabbit antiserum from Ultraclone, Cambridge, UK, and synaptophysin (mouse monoclonal antibody SY38 from Boehringer Mannheim, FRG). GFAP-immunoreactivity (GFAP-IR was also investigated at ultrastructural level using a standard pre-embedding protocol as described elsewhere (Kummer and Habeck 1991).

RESULTS

Controls

Glial fibrillary acidic protein. Lightmicroscopically, GFAP-IR was detected in elongated cellular elements within nerve bundles entering the carotid body and ramifying between the glomic lobules. GFAP-IR cells also surrounded the interlobular blood vessels, particularly arteries, as a dense plexus. Elongated immunoreactive cells with slender processes were numerous within the glomic lobules. These cells were preferentially located at the periphery of the lobules and surrounded groups of GFAP-negative cells (Fig. 1a).
Pre-embedding immunhistochemistry at ultrastructural level revealed GFAP-IR

in nerve bundles and at perivascular location to be represented by Schwann cells (Fig. 2). Immunoreaction was restricted to those Schwann cells which did not form a myelin sheath, whereas myelin-forming Schwann cells were devoid of GFAP-IR.

Immunoreactive cells within glomic lobules were represented by Schwann cells and sustentacular cells, which enveloped the granule-containing glomus cells with thin cytoplasmic processes.

Protein gene product 9.5 and synaptophysin. Both antisera revealed a large number of immunoreactive cells of oval or polygonal shape within the glomic lobules, some of them giving rise to coarse processes (Figs. 1a and 3). Double-labelling immunohistochemistry for PGP- and synaptophysin-IR revealed coexistence of both immunoreactivities in more than 95 % of labelled cells. A few cells displaying only one of the immunoreactivities were located singly between those cells exhibiting both markers. Double-immunofluorescence for GFAP- and synaptophysin-IR demonstrated, that synaptophysin-IR cells were covered by thin processes immunoreactive for GFAP (Fig. 1).

In addition to the immunoreactive cells within the glomic lobules, the PGP-antiserum labelled an extensive network of nerve fibers within the lobules as well as in the nerve fiber bundles and perivascular plexuses outside the lobules. Synaptophysin-antibodies did not label the intraglomic innervation to such an extent (Fig. 1b): Only perivascular varicosities exhibited synaptophysin-IR, whereas nerve bundles and the intralobular plexus lacked immunoreaction.

Chronic Hypoxemia

Hematoxylin-eosin stained sections were used to analyse the general morphology of the carotid bodies. The observations were consistent with previous reports (e. g. Habeck 1986), and, therefore, will not be reported here in detail. Such routinely stained sections served to select regions of the organ containing enlarged lobules or lobules with concentric proliferation of elongated cells for further immunohistochemical analysis.

Concentric proliferations of elongated cells consisted of GFAP-IR cells and were penetrated by numerous PGP-IR nerve fibers. Synaptophysin-IR was not present among the proliferating cells. The concentric proliferations engulfed a centre of polygonal to oval PGP-/synaptophysin-IR cells. They were less numerous than in normal glomic lobules.

DISCUSSION

As expected from the neuroectodermal origin of the carotid body (Le Douarin et al. 1972), proteins considered to have a potential value as neuroendocrine markers were detected in the human carotid body by means of immunohistochemistry.

Glomus cells share with neurons both a soluble cytoplasmic protein, i. e. PGP (Thompson et al. 1983), and a vesicle membrane antigen, i. e. synaptophysin (Wiedenmann and Franke 1985), thus underlining their "paraneuronal" character (Fujita et al. 1988). Entirely selective demonstration of glomus cells, however, was achieved by neither of these markers, since antisera directed against them also labelled neuronal elements of the carotid body. In case of PGP, the dense network of immunolabelled axons within glomic

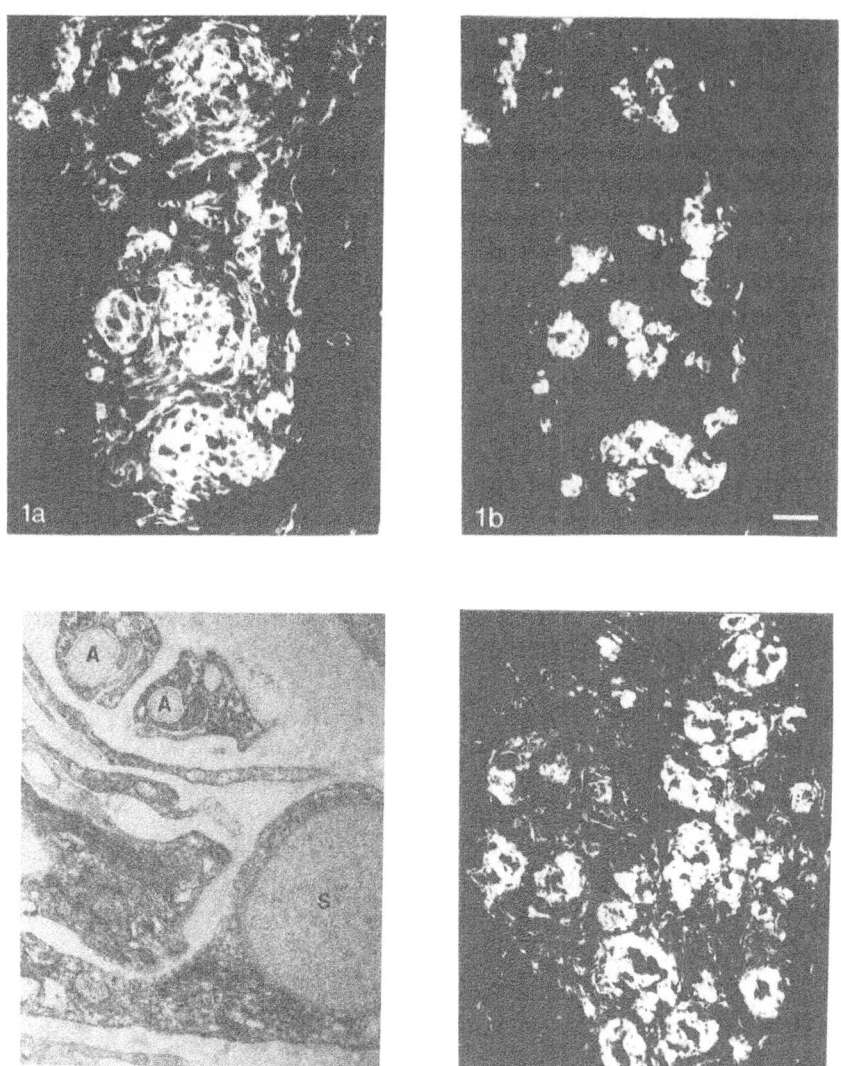

Fig.1.Double-labelling immunofluorescence demonstrating (a) GFAP- and (b) synaptophysin-IR in a glomic lobule. Elongated GFAP-IR cells both surround nests of glomus cells, which are highlighted by the synaptophysin-antibody, and form a plexus between them. Bar = 50 μm.

Fig. 2. GFAP-IR at ultrastructural level. The cytoplasm of non-myelin forming Schwann cells is filled with electron dense reaction product, whereas Schwann cell nuclei (S) and axons (A) are not labelled. Bar = 1 μm.

Fig. 3. PGP-immunofluorescence. Glomus cells a well as nerve fibers are labelled by the antibody. Bar = 50 μm.

lobules made a distinction of glomus cells from nerve fibers almost impossible. In contrast, synaptophysin-IR was detectable only in axon terminals (preferentially perivascular), but not in preterminal axons. Thus, immunolabelling within glomic lobules was primarily due to immunoreactive glomus cells.

Accordingly, antibodies to synaptophysin should be preferred to anti-PGP when selective labelling of glomus cells is desired. On the other hand, PGP-antisera (combined with synaptophysin-immunolabelling) may be useful to demonstrate the overall degree of glomic innervation: Those structures immunoreactive to PGP but immunonegative to synaptophysin represent the axons penetrating the glomic lobules. Further differentiation of intraglomic axons can be achieved by use of antisera towards neurofilaments and some neuropeptides, as reported in another chapter of this volume (Kummer and Habeck).

Sustentacular cells were found to share immunoreactivity to an intermediate filament, GFAP, with non-myelin forming Schwann cells. This finding corroborates earlier observations on the rat carotid body, where another protein expressed by Schwann cells, i. e. S-100, was immunohistochemically demonstrated in sustentacular cells (Kondo et al. 1982). At this meeting, immunoreactivity to S-100 has also been reported for sustentacular cells of the human carotid body (Habeck, Pallot and Abramovici, this volume). These findings underline the similarities between sustentacular cells and non-myelin forming Schwann cells, which have been pointed out before on the basis of their general ultrastructure (e. g. Bock 1982). There is so far no known distinctive cytological criterium between sustentacular cells and non-myelin forming Schwann cells, and they in fact may represent the same cell type.

Our findings on carotid bodies obtained from patients suffering from chronic hypoxemia showed that the markers evaluated in normal carotid bodies are also applicable to pathologically altered tissue. Thus, the cellular composition of concentric proliferations of glomic lobules could be easily and quickly analyzed without performing the tedious procedure of ultrastructural investigation. The results obtained are in full agreement with an electron microscopic study by Jago et al. (1984) who also classified the spindle-shaped cells in the periphery of such lobules as sustentacular/Schwann cells.

In conclusion, immunohistochemistry using antisera against GEAP, PGP, and synaptophysin not only provides more details on the cytological features of carotid body cell types but is also a useful tool for rapid and reliable analysis of proliferative events in the carotid body.

REFERENCES

Böck, P., 1982, "The paraganglia", In: Handbuch der mikroskopischen Anatomie des Menschen, Band VI, 8. Teil, Springer, Heidelberg.

Fujita, T., Kanno, T, and Kobayashi, S., 1988, "The paraneuron", Springer, Tokyo.

Habeck, J.-O., 1986, Morphological findings at the carotid bodies of humans suffering from different types of systemic hypertension or severe lung diseases, Anat. Anz., 162:17-27.

Jago, R., Smith, P., and Heath, D., 1984, Electron microscopy of carotid body hyperplasia, Arch. Pathol. Lab. Med., 108:717-722.

Kondo, H., Iwanaga, T., and Nakajima, T., 1982, Immunocytochemical study on the localization of neuronspecific enolase and S-100 protein in the carotid body of rats, Cell Tissue Res., 227:291-295.

Kummer, W., and Habeck, J.-O., 1991, Substance P- and calcitonin gene-related peptide-like immunoreactivities in the human carotid body studied at light and electron microscopical level, Brain Res., 554:286-292.

Le Douarin, N. M., Le Lièvre, C., and Fontaine, J., 1972, Recherches expérimentales sur l'origine embryologique du corps carotidien chez le Oiseaux, C. R. Acad. Sci. Paris, 275:583-586.

Thompson, R. J., Doran, J. F., Jackson, P., Dhillon, A. P., and Rode. J., 1983, A new marker for vertebrate neurons and neuroendocrine cells, Brain Res., 278:224-228.

Wiedenmann, B., and Franke, W. W., 1985, Identification and localization of synaptophysin, an integral membrane glycoprotein of Mr 38,000 characteristic of presynaptic vesicles, Cell, 41:1017-1028.

THE EFFECTS OF CHRONIC HYPOXAEMIA UPON THE STRUCTURE OF THE HUMAN CAROTID BODY

*J.-O. Habeck, **D. J. Pallot and ***A.Abramovici

*Inst. Pathology, Städtische Kliniken, Chemnitz, Germany; **Dept. Anatomy, Univ. Leicester, Leicester, England; ***Dept. Pathology, Tel Aviv Univ. Medical School, Tel Aviv, Israel

INTRODUCTION

Studies of the structure of the human carotid body in cases of chronic hypoxaemia have provided differing results. Some authors report an increase in the amount of glomic tissue caused predominantly by a proliferation of chief cells (Arias-Stella and Valcarcel, 1976; Janzer and Schneider, 1977), whereas others found enlarged carotid bodies accompanied by a proliferation of sustentacular and Schwann cells (Habeck, 1986; Heath et al., 1982; Smith et al., 1982). It may be that these differences are related to the aetiology of the hypoxaemia and perhaps its duration.

In cases of chronic hypoxaemia due to lung and/or airway disease, it is usually impossible to give exact information about the duration of the lowered arterial oxygen pressure, as these diseases begin slowly. On the other hand, in patients with congenital heart diseases associated with hypoxaemia such precise statements with regard to duration of hypoxia may be given.

The aim of this study was to obtain more information about the changing histological picture of human carotid bodies in chronic hypoxaemia and to examine the hypothesis that the response of the human carotid body to hypoxaemia is a two stage phenomenon. Initially the response is a proliferation and hypertrophy of the Type I cells. Continued exposure to the hypoxic stimulus lead to a gradual diminution in the number of Type I cells and their replacement by elongated cells which are probably represent a mixture of Type II and Schwann cells.

We have studied the carotid bodies from two cases of cyanotic heart disease with very different survival times. The findings from these cases will be compared with those of in other chronically hypoxaemic patients.

MATERIAL AND METHODS

The main data of cases with chronic hypoxaemia are documented in Table 1. In the 14 year old girl (Case l) with heart septal defects hypoxaemia was known since birth. In the 59 year old man (Case 5) with a univentricular heart, hypoxaemia was clinically documented over a period of 23 years. A longer time is supposed, as this malformation was congenital.

Table 1. Age, main disease and degree of spindle cell proliferation (SCP) within carotid bodies in cases with chronic hypoxaemia.

No.	Age (years)	Main disease	SCP in carotid body (- = no; + = moderate; ++ = extensive)
1	14	ventricular and atrial septal defect	-
2	46	partial resection of the lungs	+
3	56	chronic interstitial pneumonia	-
4	59	chronic interstitial pneumonia	+
5	59	univentricular heart	++
6	61	anthracosilicosis and emphysema	-
7	63	anthracosilicosis and emphysema	++
8	64	emphysema	++
9	72	interstitial fibrosis of the lungs	++
10	77	emphysema	++
11	82	interstitial fibrosis of the lungs	-
12	88	emphysema	+

Twelve age-matched cases in whom there was no evidence, either clinically or at postmortem, of neither pulmonary and/or congenital heart diseases nor arterial hypertension, served as controls.

The carotid bodies were removed during postmortems within 3 and 48 hours (mean: 13 hours) postmortem, fixed in Zamboni's fluid (20 cases) or in 70% ethanol (4 cases) and embedded in paraffin wax. Five μm thick sections were either stained with H & E or used for demonstrating S 100 and neurofilament protein immunoreactivity using a standard PAP technique.

The following primary antisera were applied: (1) rabbit anti-cow S 100 (Dakopatts, Glostrup, Denmark; lot 129, working dilution 1:200); (2) rabbit anti-neurofilament 200 KD (Sigma, St. Louis, USA; lot 69 F 4804, working dilution 1:100) (3) 3'diaminobenzidinetetrahydrochloride served as chromogen. Negative controls consisted of sections in which incubation with the primary antiserum was ommitted; in all cases such controls failed to exhibit any reaction product.

RESULTS

In the 14 year old girl with cyanotic heart disease, the carotid bodies were not enlarged but the diameter of glomic lobules was increased; there was no change in the number or arrangement of the sustentacular cells (figure 1). In marked contrast, the 59 year old man with a univentricular heart, showed significant enlargement of the carotid body and also the individual lobules as compared to age-matched controls. In this case two different histological pictures of glomic lobules were observed. Some were composed predominantly of Type I cells which were loosely arranged and embedded in a blood-filled capillary network; in others proliferated elongated cells formed concentric whorls around a small number of centrally situated Type I cells.

The other 10 cases with clinically established chronic hypoxaemia exhibited a variety of histological pictures (see also Table 1). Thus cases 3 and 11 had essentially normal carotid bodies with neither lobular nor whole organ enlargement. In case 6 the carotid bodies as well as the glomic lobules were enlarged and the increase in the size of the lobules was due to Type I cell hyperplasia/hypertrophy exclusively. Cases 2, 4 and 12 showed a moderate increase in elongated cells, whilst in cases 7, , 9 and 10 a marked increase of circular arranged spindle cells was observed (Figure 2).

In cases with increased number of elongated cells there was an increase in the number of S 100 - immunoreactive cells (Figure 3a) as well as numerous neurofilament-immunoreactive nerve fibres (Figure 3b) associated with the S 100 positive cells were seen. Sometimes the histological picture looks similar to a neuroma.

Neither an enlargement of the glomic lobules or the whole organ nor an increase in number of elongated cells was observed in the controls.

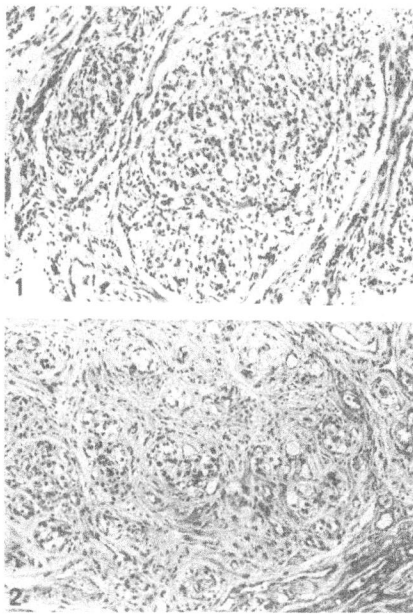

Fig. 1. Increased diameter of glomic lobule due to hyperplasia/hypertrophy of Type I cells (Case 1). 25x. H&E.

Fig. 2. Marked increase of circular arranged spindle cells in a glomic lobule (Case 10).25x. H&E.

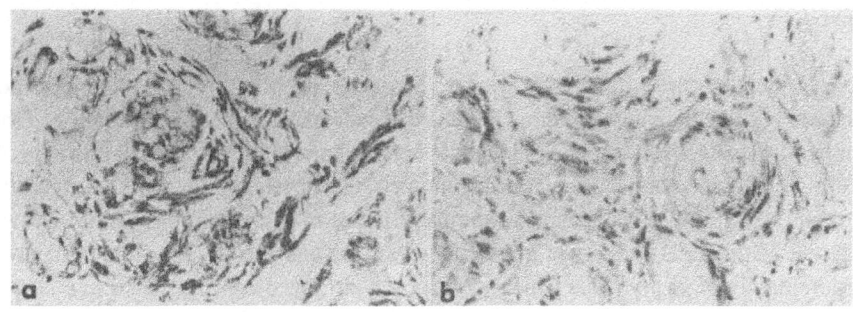

Fig. 3. (a) Increased number of S 100 immunoreactive cells (Case 9) and (b) of neurofilament immunoreactive nerve fibres (Case 8) in carotid bodies of chronically hypoxaemic humans. 40x.

DISCUSSION

The data reported here suggests that the response of the human carotid body to chronic hypoxia is more complex than has been suggested by some authors (for review see Pallot, 1987) and provide support for the hypothesis described by Pallot (1987). Initially there is a hyperplasia of the Type I cells which may be sufficient to cause enlargement of the organ or may simply lead to an increase in the size of individual lobules. Increasing duration, and perhaps severity, of hypoxia leads to further enlargement of the organ and to an increase in the number of elongated cells surrounding the islands of Type I cells. This process can result in some lobules being reduced to but a few Type I cells surrounded by a thick "capsule" of elongated Type II and Schwann cells. In some cases the increase in elongated cells is also associated with an increase in the number of nerve profiles; under these conditions some areas of the carotid body resemble a neuroma. It is tempting to postulate that the increase in the number of axons is brought about by the fact that the loss of Type I cells, normally innervated by these axons, leads to their sprouting in an attempt to find a targed organ. Whilst we are able to describe the changes from Type I cell hyperplasia to Type II/Schwann cell hyperplasia, we have, to date, no explanation for the mechanisms which result in these changes. Whether the response of the carotid body to hypoxia due to lowered inspired oxygen tension is different to that caused by lung and heart disease is unknown; we are however investigating this in animal models of lung disease.

The hypothesis above may only apply to lowlanders who are hypoxaemic due to pulmonary and cardiovascular diseases, as in highlanders adapted to live at low oxygen pressure only Type I cell proliferation has been reported (Arias-Stella and Valcarcel, 1976). This could support the assumption that in hypoxaemic lowlanders Type I cells die before Type II and Schwann's cell proliferation takes place. It could be that in highlanders the Type I cells are more adapted to low oxygen levels.

ACKNOWLEDGEMENT

We are grateful to Ms J. MacWilliam for expert technical assistance and the Wellcome Trust for financial support.

REFERENCES

Arias-Stella, J., and Valcarcel, J., 1976, Chief cell hyperplasia in the human carotid body at high altitudes, Human Pathology, 7: 361.

Habeck, J.-O., 1986, Morphological findings at the carotid bodies of humans suffering from different types of systemic hypertension or severe lung diseases, Anat. Anz., 162: 17.

Heath, D., Smith, P., and Jago, R., 1982, Hyperplasia of the carotid body, J. Pathol., 138: 115.

Janzer, R. C., and Schneider, J., 1977, The influence of chronically hypoxemic states on human carotid body structure and cardiac hypertrophy, Virchow's Arch. A Path. Anat. and Histol., 376: 75.

Pallot, D. J., 1987, The mammalian carotid body, Adv. Anat. Embryol. Cell Biol., 102:1.

Smith, P., Jago, R., and Heath, n., 1982, Anatomical variation and quantitative histology of the normal and enlarged carotid body, J. Pathol., 137: 287.

7 DOPAMINERGIC AND PEPTIDERGIC SENSORY INNERVATION OF THE RAT CAROTID BODY: ORGANIZATION AND DEVELOPMENT

D. M. Katz [1,2], J. C. W. Finley[2] and J. Polak[1]

[1]Dept. of Neurosciences, Case Western Reserve University School of Medicine, 2119 Abington Road and [2]Dept of Medicine, University Hospitals of Cleveland, Cleveland, OH 44106

INTRODUCTION

Despite widespread interest in the neural control of respiration, cellular and molecular mechanisms that mediate afferent transmission from peripheral chemoreceptors to brainstem respiratory neurons remain poorly understood. Recent studies, however, implicate specific classes of neuroactive substances as mediators and modulators of the hypoxic ventilatory response at the level of the primary afferent fibers that innervate the carotid body. This chapter focuses on expression and development of transmitter properties in carotid body afferent neurons located in the sensory ganglia of the glossopharyngeal nerve. These cells are the afferent link between the carotid body and brainstem and therefore play a pivotal role in translating hypoxic stimulation into an effective cardiorespiratory response.

BACKGROUND

Primary sensory neurons, located in the petrosal (PG) and jugular (JG) ganglia of the glossopharyngeal nerve, are the principal source of afferent innervation to the carotid body. These cells project peripherally in the carotid sinus branch of the glossopharyngeal nerve and terminate either in synaptic contact with Type I glomus cells or as free endings (McDonald & Mitchell, 1975; Kondo, 1976; Hansen, 1985; Chen, et al, 1986; Kondo and Yamamoto, 1988; Kummer, et al, 1989). The central projections of carotid body afferents terminate primarily in the brainstem nucleus tractus solitarius (Claps & Torrealba, 1988; Finley and Katz, 1992).

Several lines of evidence indicate that the sensory innervation of the carotid body, rather than being homogeneous, is composed of morphologically and neurochemically distinct subsets of neurons. Ultrastructural studies, for example, revealed at least three modes of afferent termination within the carotid body; small boutons apposed to glomus cells, large calyceal endings on glomus cells and free nerve endings (McDonald & Mitchell, 1975; Kondo, 1976; Hansen, 1985; Chen, et al., 1986; Kondo and Yamamoto, 1988; Kummer, et al., 1989). One possibility is that these morphologically distinct types of endings correspond to functionally distinct populations of afferent fibers (see below), though such correlations remain to be defined. Second, as described by MacDonald and Mitchell (1975) and Kondo (1976), many afferent endings make reciprocal synapses within the carotid body, that is, endings that are both pre- and post- synaptic to glomus cells. This finding raises the possibility that some afferent terminals in the carotid body may not only be receptive elements, but capable of transmitter release as well (McDonald & Mitchell, 1975; Kondo, 1976). The implication of these observations is that carotid body afferents may not simply be passive relays of chemosensory stimulation to the brainstem but actually play a role in modulating transmission and/or transduction in the carotid body.

NEUROCHEMICAL PROPERTIES OF CAROTID BODY AFFERENTS

Primary sensory neurons, including those in the petrosal and jugular ganglia, display an impressive diversity of neurotransmitter/neuromodulator properties (Katz & Karten, 1980; Katz, et al., 1983; Helke & Hill, 1988). However, only one phenotype expressed by cranial sensory neurons has been shown to be selectively associated with cells that innervate the carotid body (Katz & Black, 1986). We previously found that a large subpopulation of PG neurons in the adult rat express functional catecholaminergic properties, including catalytically active tyrosine hydroxylase (TH), the rate-limiting enzyme in catecholamine biosynthesis and catecholamine histofluorescence (Katz et al., 1983). 80-90% of the catecholaminergic PG neurons project peripherally in the carotid sinus nerve to the carotid body (Katz & Black, 1986), indicating a striking correlation between expression of this unique sensory transmitter phenotype and peripheral target innervation. A similar pattern of carotid body innervation by TH-containing PG neurons has recently been suggested in guinea pig (Kummer, et al., 1990). Catecholaminergic PG neurons are also immunoreactive for dopa decarboxylase, the dopamine-synthesizing enzyme (Figure 1; Finley, et al., in press) but do not exhibit immunoreactivity for dopamine beta-hydroxylase (Katz, et al., 1983), the enzyme responsible for converting dopamine to norepinephrine. These biochemical and immunocytochemical data indicate, therefore, that the TH-containing carotid body afferent neurons in the PG are dopaminergic (DA).

To better understand the role of dopaminergic afferents in carotid body innervation, we examined the ultrastructure of TH-containing sensory terminals using immuno-electron microscopy. These studies demonstrated that TH-immunoreactive afferent fibers are in direct synaptic contact with carotid body glomus cells (Figure 2; Finley, et al., in press). These contacts persist following removal of sympathetic fibers to the carotid body, thereby confirming their sensory origin. Both calyceal and bouton-like terminals on glomus cells are TH-immunoreactive. Interestingly, some of these synaptic contacts exhibit morphologic features of reciprocal synapses as originally described by McDonald and Mitchell (1975) and Kondo (1976). These data raise the possibility, therefore, that some afferents are capable of releasing dopamine peripherally in the carotid body. This possibility is consistent with studies by Fidone and colleagues who showed peripheral release of dopamine following carotid sinus nerve stimulation (Almaraz & Fidone, 1986) and dopamine synthesis in carotid sinus nerve fibers in the cat (Fidone, S., personal

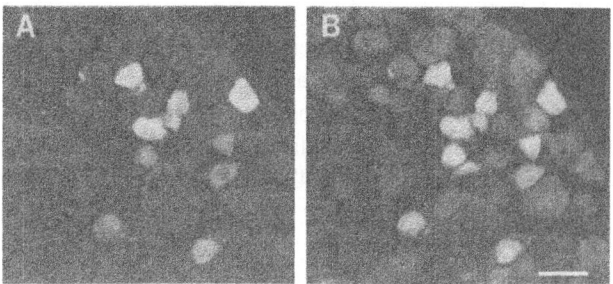

Figure 1. Fluorescence photomicrographs showing co-localization of TH (A) and DDC- (B) immunoreactivity in adult rat PG sensory neurons. Cells were double-stained with primary antisera directed against TH and DDC and secondary antisera coupled to rhodamine (TH) and fluorescein isothiocyanate (DDC). Scale bar equals 50 microns.

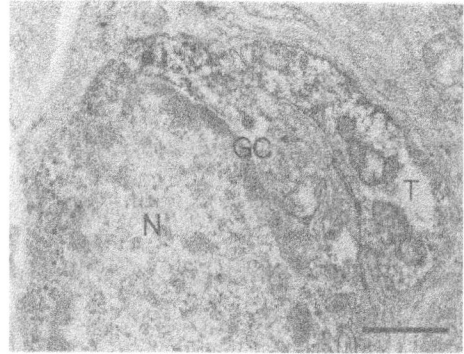

Figure 2. Electron micrograph showing a TH-immunoreactive afferent terminal (T) in contact with an immunoreactive glomus cell (GC). N indicates glomus cell nucleus. Scale bar equals 1 micron.

communication). We speculate that dopaminergic sensory terminals that are presynaptic to glomus cells could provide a morphologic substrate for the "efferent" inhibitory pathway to the carotid body originally proposed by Neil and O'Regan (1971). These workers showed that stimulation of the cut peripheral end of the CSN inhibits chemosensory responses to carotid body stimulation in cats. Subsequently, Lahiri and co-workers demonstrated a dopaminergic component in this inhibitory pathway (1984). It is conceivable, therefore, that dopaminergic inhibition in these experiments was mediated by stimulation-evoked release of dopamine from peripheral afferent terminals, rather than from true efferent axons. Under physiologic conditions, dopamine release from afferent terminals could have a local modulatory effect, analogous to that proposed for dendritic dopamine release in the substantia nigra (Nieoullon, et al, 1977). Unfortunately, few data are available on efferent inhibition in the rat carotid body, making comparisons between these physiologic studies and our morphologic data difficult. However TH-positive PG neurons (Katz, unpublished observations) and peri-glomic calyces (Stensaas and Fidone, personal communication) are also found in the cat, making it likely that similar synaptic interactions exist in both species.

Although the vast majority of catecholaminergic afferents in the PG project to the carotid body, these neurons represent only 42% of all carotid body afferents (Finley, et al , in press). This finding indicates that other, non-catecholaminergic afferents must also contribute to the sensory innervation of the carotid body. Indeed, diverse peptide immunoreactivities have been described in sensory fibers in the rat and cat carotid body, including Substance P (Lundberg, et al , 1979; Cuello &McQueen, 1980; Hess, 1981; Chen, et al , 1986; Kummer, 1988; Prabhakar, et al , 1989), enkephalin (Lundberg, et al, 1979) and calcitonin gene-related peptide (Kummer, 1987; Kondo & Yamamoto, 1988; Kummer, 1988; Kummer, et al , 1989). To begin defining the origin(s) of peptidergic fibers, we examined expression of Substance P immunoreactivity in sensory neurons retrogradely labeled following injections of dye tracers into the vascularly isolated rat carotid body in situ (Finley, et al , in press). As previously shown in our laboratory, tracer microinjections in this preparation selectively label sensory neurons with axons in the carotid body (Finley & Katz, 1992). In some experiments, tissue sections containing retrogradely labeled sensory neurons were double-stained for SP and TH to enable us to compare the size and location of dopaminergic and peptidergic afferents. These studies demonstrated that approximately 10% of carotid body afferents exhibit Substance P immunoreactivity. These cells are distinct from dopaminergic sensory neurons that innervate the carotid body, as we found no colocalization of Substance P and TH within individual cells. Moreover, SP cells differed markedly in both size and location from dopaminergic afferents. Whereas DA cells are small and topographically segregated in the distal PG, SP cells are medium- to large-sized and located in the proximal PG and JG. These morphologic distinctions between dopaminergic and peptidergic carotid body afferents raise the possibility that each subset of neurons plays distinct roles in chemoreception. This possibility is consistent with electron microscopic observations on Substance P-containing terminals in the rat carotid body (Chen, et al , 1986; Kummer, et al , 1989). Unlike the dopaminergic terminals described here, most Substance P fibers do not appear to synapse directly with glomus cells. Rather, these afferent terminate as free endings among connective tissue or vascular elements in the carotid body parenchyma (Chen, et al , 1986; Kummer, et al , 1989). The few SP terminals observed in contact with glomus cells do not appear to exhibit specializations typical of synaptic contacts (Chen, et al , 1986; Kummer, et al , 1989).

TRANSMITTER DEVELOPMENT IN CHEMORECEPTOR AFFERENTS

Immunocytochemical methods have been used to define the ontogeny of catecholaminergic and peptidergic properties in petrosal and jugular ganglion neurons (Ayer-LeLievre & Seiger, 1984; Katz & Erb, 1990). Surprisingly, although dopamine- and SP-containing cells comprise two distinct subpopulations, both phenotypes appear to develop concurrently. Specifically, TH-(Katz & Erb, 1990) and SP-(Ayer-LeLiever and Seiger, 1984) immunoreactive neurons become detectable in the petrosal and jugular ganglia on embryonic day (E) 16. 5 in the rat. These observations suggest that similar mechanisms may underlie neurochemical development in these two populations of cells.

Studies in our laboratory have begun to address cellular and molecular mechanisms that regulate development of dopaminergic properties in PG neurons (Katz and Erb, in preparation). Given the striking correlation between TH expression and innervation of the carotid body, we asked whether target-derived influences might play a role in dopaminergic developent in the PG. To address this possibility, studies in vivo examined the temporal correlation between TH expression in the PG and the development of connections between PG neurons and the carotid body. Using neurofilament protein immunocytochemistry we found that the CSN pathway between the PG and carotid body develops between E13. 5 and 16. 5. These data are consistent with electron microscopic studies of Kondo (1975) who found that nerve fibers first appear in the carotid body anlage around E13. 5. By E16. 5 the carotid body is well differentiated and invaded by numerous CSN fibers. In other words, by the time TH first appears in PG neurons, projections to the target carotid body have already developed, consistent with the posibility that neuron-target interactions play a role in regulating ganglion cell development at this stage. Interestingly, the fact that the CSN and carotid body develop simultaneously between E13.5 and 16.5 suggest that reciprocal interactions between PG afferents and the carotid body may be important for development of both the sensory neurons and their target cells.

To more directly examine how the carotid body might influence PG neuron development, ganglion cell maturation was monitored in a tissue culture model system. In these experiments, E16.5 PG were grown in explant culture in the presence or absence of the carotid body and analyzed for total neuronal survival, neurite outgrowth and expression of TH. These studies revealed a marked trophic effect of the carotid body on PG explants; coculture resulted in a two-fold increase in total neuronal survival compared to controls. In addition, PG explants in coculture exhibited abundant neurite outgrowth that extended in a fasciculated bundle towards the carotid body. Finally, cocultures contained an increased number of TH-positive PG neurons that was proportional to the overall increase in neuronal survival.

These stimulatory effects of the carotid body on PG neuron development in culture are consistent with the idea that maturation of chemoreceptor afferents in fetal animals is regulated in part by interactions with peripheral target tissues. This possibility is supported by our previous finding that connections with the periphery play a role in regulating TH levels in chemoafferent neurons in postnatal animals (Katz & Black, 1986). We hypothesize, therefore, that in addition to its role as a chemosensor, the carotid body is also a trophic regulator of chemoafferent development. Elucidating the role of neuron-target interactions in chemosensory transmitter development should further our understanding of mechanisms that underly functional maturation of peripheral chemoreflexes.

ACKNOWLEDGEMENT

This work was supported by NIH grant HL-42131 (DMK) and the American Lung Association and Parker B. Francis Foundation (JCWF).

REFERENCES

Almaraz, L. , and Fidone, S ., 1986, Carotid sinus nerve C-fibers release catecholamines from the cat carotid body, Neurosci. Letts. , 67:153-8.

Ayer-LeLievre, C .S. , and Sieger, A ., 1984, Development of substance P- immunoreactive neurons in cranial sensory ganglia of the rat, Int. J. Devl. Neurosci. , 2: 451-463.

Chen, I. -L., Yates, R.D., and Hansen, J.T., 1986, Substance P-like immunoreactivity in rat and cat carotid bodies: Light and electron microscopic studies, Histol. Histopath., 203-212.

Chen, I. -L. , Yates, R. D. , and Hansen, J. T. , 1984, Localization of substance P and dopamine-beta-hydroxylase-like immunoreactive (SP-I and DBH-I) in the rat and cat carotid bodies, Anat. Rec., 208: 29A-30A.

Claps, A. , and Torrealba, F. , 1988, The carotid body connections: A WGA-HRP study in the cat, Brain Res., 455: 123-133.

Cuello, A. C., and McQueen, D. S., 1980, Substance P: A carotid body peptide, Neurosci. Lett., 17: 215-219.

Finley, J.C.W., Polak, J. and Katz, D.M., Transmitter diversity in carotid body afferent neurons: dopaminergic and peptidergic phenotypes, Neurosci., In Press.

Finley, J.C.W., and Katz, D.M., 1992, The central organization of carotid body afferent projections to the brainstem of the rat, Brain Res., 572: 108-116.

Hansen, J. T., 1985, Ultrastructure of the primate carotid body: a morphometric study of the glomus cells and nerve endings in the monkey (Macaca fascicularis), J. Neurocytol., 14:13-32.

Helke, C.J., and Hill, K.M., 1988, Immunohistochemical study of neuropeptides in vagal and glossopharyngeal afferent neurons in the rat, Neurosci., 26:539-551.

Hess, A. , 1981, On the origin of subtance P-positive fibers in the rat carotid body, Anat. Rec. , 199: 114A.

Katz, D.M. and Black, I.B. , 1986, Expression and regulation of catecholaminergic traits in primary sensory neurons: Relationship to target innervation in vivo, J. Neurosci., 6: 983-989.

Katz, D.M., and Erb, M., 1990, Developmental regulation of tyrosine hydroxylase expression in primary sensory neurons of the rat, Dev. Biol., 137:233-242.

Katz, D.M., and Erb., M., Target regulation of petrosal ganglion cell development, in preparation.

Katz, D.M., and Karten, H.J., 1980, Substance P in the vagal sensory ganglia: Localization in cell bodies and pericellular arborizations, J. Comp. Neurol. , 193:549-64.

Katz, D.M., Markey, K.A., Goldstein, M., and Black, I.B., 1983, Expression of catecholaminergic characteristics by primary sensory neurons in the normal adult rat in vivo, Proc. Natl. Acad. Sci. Usa 80:3526-3530.

Kondo, H. , 1975, A light and electron microscopic study on the embryonic development of the rat carotid body, Am J. Anat., 144:275-294.

Kondo, H., 1976, Innervation of the carotid body of the adult rat, Cell Tiss. Res., 173:1-15.

Kondo, H., and Yamamoto, M., 1988, Occurrence, ontogeny, ultrastructure and some plasticity of CGRP (calcitonin gene-related peptide) -immunoreactive nerves in the carotid body of rats, Brain Res., 473: 283-293.

Kummer, W., 1987, Calcitonin gene-related peptide-immunoreactive nerve fibers in carotid body and in carotid sinus. Exper. Brain Res. Series 16:84-88.

Kummer, W., 1988, Retrograde neuronal labelling and double-staining immunohistochemistry of tachykinin-and calcitonin gene-related peptide-immunoreactive pathways in the carotid sinus nerve of the guinea pig, J. Autonom. Nerv. Syst., 23:131-141.

Kumer, W., Fischer, A., and Heym, C., 1989, Ultrastructure of calcitonin gene-related peptide- and substance P-like immunoreactive nerve fibres in the carotid body and carotid sinus of the guinea pig, Histochem., 92:433-439.

Kumer, W., Gibbins, I.L., Stefan, P., and Kapoor, V., 1990, Catecholamines and catecholamine-synthesizing enzymes in guinea-pig sensory ganglia, Cell Tiss. Res., 261:595-606.

Lahiri, S., Smatresk, N., Pokorski, M., Barnard, P., Mokashi, A., and McGregor, K.H., 1984 Dopaminergic efferent inhibition of carotid body chemoreceptors in chronically hypoxic cats, Am. J. Physiol., 247:R24-R28.

Lundberg, J.M., Hokfelt, T., Fahrenkrug, J., Nilsson, O., and Terenius, L., 1979, Peptides in the cat carotid body (glomus caroticum): VIP-, enkephalin-, and substance P-like immunoreactivity, Acta Physiol. Scand., 107:279-281.

Lundberg, J.M., Hokfelt, T., Nilsson, G., Terenius, L., Rehfeld, J., Elde, R., and Said, S., 1978, Peptide neurons in the vagus, splanchnic and sciatic nerves, Acta Physiol. Scand., 104:449-501.

Neil, E., and O'Regan, R.G., 1971, The effects of electrical stimulation of the distal end of the cut sinus and aortic nerves on peripheral arterial chemoreceptor activity in the cat, J. Physiol., (Lond.) 215:15-32.

Nieoullon, A., Cheramy, A., and Glowinski, J., 1977 Release of dopamine in vivo from cat substantia nigra, Nature 226:375-377.

Prabhakar, N.R., Landis, S.C., Kumar G.K., Mullikin-Kilpatrick, D., Cherniack, N.S., and Leeman, S., 1989, Substance P and neurokinin A in the cat carotid body: localization, exogenous effects and changes in content in response to arterial pO2, Brain Res., 481:205-214.

8

EFFECTS OF CELL-FREE PERFUSION AND ALMITRINE BISMESYLATE ON THE ULTRASTRUCTURE OF TYPE-I CELL MITOCHONDRIA IN THE CAT CAROTID BODY

*M. Kennedy, S. Giles, R.G. O'Regan, S. Feely and *Y. Evrard*

Department of Physiology and Histology, University College, Earlsfort Terrace, Dublin 2, Ireland and *Institut de Recherches Internationales Servier, Paris, France

INTRODUCTION

The vascularly-isolated carotid body perfused with cell-free solution has been shown to exhibit Type 1 cell mitochondrial disruption when the perfusion is carried out at constant inflow pressure and apparent mitochondrial "normality" with perfusion at constant inflow volumes (O'Regan et al, 1987). Functional deterioration in these preparations has been observed (Joels and Neil, 1968). The mere structural integrity of an organ does not imply perfect functioning of that organ. However, when both physiological and ultrastructural features are in agreement certain deductions may be made. It is proposed that mitochondria have an important role in the chemotransduction process. A quantitative analysis was performed to be more precise about the nature of the mitochondrial changes observed during cell-free perfusion of the vascularly-isolated carotid body.

Almitrine bismesylate causes marked chemoexcitation of peripheral chemoreceptors (Laubie and Schmitt, 1980, O'Regan et al., 1983). When administered acutely it has no effect on Type-1 cell ultrastructure (Kennedy et al., 1990a) while when administered chronically mitochondria exhibit some changes (Kennedy et al, 1990b). Based on the inhibitory effect of light on the carotid chemoexcitatory response to almitrine it has been proposed that this agent forms a light-dissociable complex with carotid body mitochondria, and that this in altering mitochondrial function, stimulates chemoreceptors (Maxwell, Mottershead and Nye, 1987). Joels and Neil (1962) showed that light reversed the chemoexcitatory effect of high partial pressures of carbon monoxide; this gas forms a light-dissociable complex with cytochrome a_3. Could almitrine react similarly with the mitochondrial electron transport chain? Conflicting reports exist about almitrine's effect on mitochondria, high levels of almitrine increase and maintain cerebral mitochondrial functional states, with increases in states 3 and 4 of mitochondrial metabolism being noted (Lepagnol and Weinachter, 1988) indicating a hypermetabolic type response. However, high concentrations of almitrine seem to inhibit mitochondria in rat liver (Mottershead, Nye and Quir, 1988) with decreases in O_2 consumption in state 3 of between 75% and

90%. Almitrine did not seem to act via cytochrome c or aa$_3$ although almitrine has an affinity for a component associated with electron transport along the mitochondrial chain. Presumably this is related to its mechanism of action. Agents which affect mitochondrial function are potent stimulators of carotid chemoreceptor activity.

In this present work, carotid bodies were perfused 1) with blood, 2) with physiological saline at constant inflow pressure, 3) with physiological saline at constant inflow volumes and a stereological analysis performed on Type-1 cell mitochondria from these groups. An analysis was also performed on carotid bodies treated with almitrine over 6 weeks.

METHODS

Adult cats were anaesthetized with pentobarbitone sodium (induction 42-48mg/kg i.p.; maintenance, 6-12mg, i.v.). Systemic arterial blood pressure was continuously monitored via a femoral arterial cannula which also enabled blood-gas analysis to be monitored. The temperature of the animal was maintained at 37° using a homeothermic electric blanket. Both carotid bodies were perfused with Ringer-Locke solution using techniques described elsewhere (Neil and O'Regan), 1971). Only the common carotid, external carotid and the arterial supply to the glomus remained patent. Cannulae were inserted into the lingual and dorsal muscular arteries. One carotid body was perfused naturally with blood and served as a control, while the other carotid body was perfused with Ringer-Locke solution (P0$_2$ 18-21kPa, PH 7.38-7.42, 37°C) at constant inflow pressure (100-130 mm Hg) or at constant inflow volume (100-200 µl/min). After cell-free perfusion both organs were fixed by perfusion with Karnovsky's solution for a period of 20 min upon which each carotid body was removed and placed in fixative.

The carotid bodies were post-fixed for 1 h in a 1% osmium tetroxide solution in 0.1M cacodylate buffer (PH 7.3, 4°C). After washing in cacodylate buffer, the tissue was dehydrated through graded alcohols and embedded in Epon. Ultrathin sections (60, nm) were stained with uranyl acetate and lead citrate, and viewed with a Phillips 201 electron microscope. For stereological examination, electron micrographs of 30 type 1 cells were obtained randomly at different planes of section, from each carotid body, and printed at a final magnification of 30,000. Analysis was carried out using a MOP AMO2 Kontron Ltd. Image .Analyser System facilitated by computer tracings on a digitizer board. Taking mitochondrial to be ellipsoidal, the shape coefficient (ß) of mitochondria in each set of micrographs (Weibel, 1969) was calculated. This together with the number of mitochondria pen unit area (N$_A$) and the fraction of cytoplasm occupied by mitochondria (V$_V$) allowed the number of mitochondria per unit volume of cytoplasm (N$_V$) and the mean mitochondrial volume (MMV) to be calculated (Steer, 1981).

An analysis of variance (ANOVA) was used in order to study whether there were any significant differences between control blood perfused organs (CONT), organs perfused with saline at constant inflow pressure (CONP) and carotid bodies perfused with saline at constant inflow volume (CONV), for periods of perfusion averaging 30 minutes. Where the F value indicated that the null hypothesis of no significant difference among the means of the 3 groups was rejected, Tukeys multiple comparisons procedure was used to see where these differences lay. The confidence limits were set at the 95% value and the critical differences between the means (D) calculated for each pair of subgroups CONT-CONP; CONT-CONV; CONP-CONV). Where probability 'p' was found to be less than 0.05 the result was judged to be statistically significant.

The same system of perfusion fixation and analysis was used for rat carotid bodies treated with almitrine bismesylate for 6 weeks. The qualitative analysis of this study has been described elsewhere (Kennedy et al., 1990). The quantitative analysis of these carotid bodies is presented here. N_A, V_V, N_V and MMV were calculated (Steer 1981) and a students t-test was used to compare values from control and almitrine treated organs with P values of ≤ 0.05 considered to be significant.

RESULTS

Results are presented in tabular form. The key to the symbols used in these tables is given below.

KEY TO SYMBOLS USED IN TABLES

CONT	=	Carotid bodies perfused with blood
CONP	=	Carotid bodies perfused with saline at constant inflow pressure
F	=	Result from ANOVA for comparison of group means
D	=	Critical differences derived from Tukey test. This 'D' was calculated when the ANOVA 'F' was significant
P	=	Probability of statistical significance where confidence level is set at 95%, i.e. 0.05
S	=	Statistical significance of result
***	=	Statistically significant result
NS	=	Non-significant result

$$NA = \frac{\text{NO. OF MITOCHONDRIA}}{\mu M^2 \text{ CYTOPLASM}}$$

$$V_V = \frac{\text{FRACTION OF CYTOPLASM}}{\text{OCCUPIED BY MITOCHONDRIA}}$$

$$N_V = \frac{\text{NO. OF MITOCHONDRIA}}{\mu M^3 \text{ CYTOPLASM}} \qquad N_V = \frac{N_A^{\frac{3}{2}}}{V_V^{\frac{1}{2}}} \quad X \quad \frac{1}{\beta}$$

β = SHAPE COEFFICIENT

$$\text{MEAN MITOCHONDRIAL VOLUME (MMV)} = \frac{V_V}{N_V}$$

TABLE 1. Comparison of the mean number of mitochondria per unit area (N_A) between experimental groups.

F = 4.28   P = 0.031

	No. of Animals	Mean N_A (μm) $\pm$ SEM	Difference Between Means D	Significance S
CONT	8	1.225 ± 0.113		
CONP	6	0.725 ± 0.130		
CONV	6	1.054 ± 0.130		
CONT-CONP			0.501	***
CONT-CONV			0.171	NS
CONP-CONV			-0.329	NS

A statistically significant decrease in the number of mitochondria per unit area was seen in the type 1 cells of carotid bodies perfused with saline at constant inflow pressure.

TABLE 2. Comparison of the mean volume fractions occupied by mitochondria (Vv) between experimental groups.

F = 4.88   P = 0.021

	No. of Animals	Mean Vv $\pm$ SEM	Difference Between Means D	Significance S
CONT	8	0.094 ± 0.001		
CONP	6	0.090 ± 0.005		
CONV	6	0.072 ± 0.005		
CONT-CONP			0.004	NS
CONT-CONV			0.022	***
CONP-CONV			0.018	NS

Table 2 indicates that there is a slight, but non-significant drop in the volume fraction occupied by the mitochondria in the constant pressure perfused group as compared with controls. By contrast, there is a significant decrease in volume fraction occupied in the constant volume perfused group.

TABLE 3. *Comparison of the mean numbers of mitochondria in a unit volume of cytoplasm (Nv) between the three groups.*

$$F = 4.92 \qquad P = 0.021$$

	No. of Animals	Mean Nv (μm^3) ± SEM	Difference Between Means D	Significance S
CONT	8	2.913 ± 0.326		
CONP	6	1.415 ± 0.337		
CONV	6	2.681 ± 0.337		
CONT-CONP			1.498	***
CONT-CONV			0.232	NS
CONP-CONV			-1.266	NS

There is a significant decrease in the number of mitochondria per unit volume for the constant pressure perfused group. This implies that the number of mitochondria decreased while the individual volumes increased (as indicated by the Vv comparison).

TABLE 4. *Comparison of the mean volumes occupied by a single mitochondria (MMV) between the experimental groups.*

$$F = 4.99 \qquad P = 0.02$$

	No. of Animals	Mean MMV (μm^3) ± SEM	Difference Between Means D	Significance S
CONT	8	0.034 ± 0.015		
CONP	6	0.096 ± 0.017		
CONV	6	0.030 ± 0.017		
CONT-CONP			-0.062	***
CONT-CONV			0.004	NS
CONP-CONV			0.065	***

TABLE 5. *Stereological Analysis of Chronic Almitrine Study.*

	N_A $\mu m^2 \pm$ S.E.M.	Vv $\pm$S.E.M.	Nv μm^3 $\pm$ S.E.M.	MMV μm^3 $\pm$ S.E.M.
CONTROL (n = 5)	1.454 $\pm$ 0.10	0.094 $\pm$0.003	3.771 $\pm$0.710	0.030 $\pm$0.006
ALMITRINE (n = 5)	0.969$\pm$0.10 ***	0.078 $\pm$0.004 ***	2.293 $\pm$0.320 NS	0.037 $\pm$0.005 NS

*** = $p \leq 0.05$, significant
NS = nonsignificant

The volume occupied by a single mitochondria in the constant pressure group is significantly higher than either the control or the constant volume perfused group.

Stereological analysis confirms qualitative observations that the type 1 cell mitochondria in almitrine treated organs were fewer in number, and occupied less volume in the cytoplasm, while having a slightly larger individual volume compared to controls.

DISCUSSION

The constant pressure-perfused group showed a significant decrease in the number of mitochondria per unit area (N_A) while at the same time a significant increase in the mean mitochondrial volumes. The volume fraction decrease was slight indicating that the decrease in numbers was accompanied by a proportionally larger increase in size. It may be postulated that the remaining mitochondria actually enlarged in a compensatory attempt to maintain oxidative metabolism at a maximum. As regards the possibility of mitochondrial fusion, the ratio of the particle axes indicating the shape of the mitochondria, in the constant pressure perfused Type 1 cells was different from the control mitochondria implying 'rounder' mitochondria. Relatively less membrane is thus required to enclose fewer but larger mitochondria, this fact would also contribute to the total mitochondrial volume fraction being relatively unchanged (0.094 +/- 0.005, control, and 0.090 +/- 0.005 constant pressure perfused). The size difference in mitochondria is consistent with a 280% increase (0.034 μm^3, control, to 0.096 μm^3, constant pressure perfused).

The changes were particularly striking in experiments where a carotid body on each side was fixed by perfusion under exactly the same conditions fixative solution, time, pressure etc.) and yet the saline perfused organ showed changes even after short periods of perfusion at constant inflow pressure while the contralateral blood perfused organ demonstrated a normal ultrastructural appearance.

Alterations in mitochondrial function and structure have been observed in rat liver mitochondria (Kimberg, Loud and Wiener, 1968). Following cortisone treatment mitochondria are larger and fewer in number. The average mitochondrial volume is increased four-fold in the peripheral and midzonal regions of the liver, with a proportional decrease in the number of mitochondria per cell and the proportions of the total cytoplasmic volume occupied by mitochondria remain relatively unaltered. The treated mitochondria are also defective.

The constant volume perfused organs showed a decrease in numbers of mitochondria per unit area and volume of cytoplasm, but this did not achieve significance, it did however account for a significant decrease in the volume of cytoplasm occupied by mitochondria. Thus a true decrease is demonstrated here without any associated swelling. Their shape remained similar to control mitochondria. Mitochondria are known to undergo finely tuned volume changes and the maintenance of volume homeostasis is important, as excessive swelling of the matrix impairs those mitochondrial functions essential for cell survival.

The mechanisms affecting the constant pressure group, as they are linked to perfusion with saline solution, are offset somewhat by the maintenance of constant flow. It thus would seem that perfusion of the carotid body in vivo is more "physiological" under constant volume conditions, while constant pressure conditions cause pronounced damage to the mitochondria.

O'Regan has asserted (1979) "Paradoxically, it appears that the chemoreceptor responses of in vitro carotid body preparations are similar to blood perfused organs and superior to glomi perfused with saline solutions". Verna, Roumy and Leitner (1981) report normal chemoreceptor functioning and ultrastructure of the rabbit carotid body after 3-6 hours of superfusion, and there were no features of anoxia even at the centre of the carotid body. While these authors did not do a stereological analysis of their results, the 'intactness' of the superfused preparation is in contrast to the in vivo perfused cat carotid body. Verna et al were looking at rabbit carotid bodies, and their shortest ultrastructural observations were at 3 hours. It would be interesting, in the light of the present study, to see if, in the short term (15-60 min) the superfused organ undergoes mitochondrial changes which later adapt and therefore appear more 'normal'. Certainly, constant flow mechanisms seem beneficial in the superfused organ.

A key role for mitochondria has been postulated in the mechanism of action of almitrine (Maxwell, Mottershead and Nye, 1987; Mottershead, Nye and Quirk, 1988; Lepagnol and Weinachter, 1988). Agents which affect mitochondrial function are potent stimulators of carotid chemoreceptor activity. It is difficult to set a hard and fast guideline to relating mitochondrial structure and state of functioning. The appearance of the mitochondria in the acute studies reflected normally metabolising units (Kennedy et al, 1990a) with tubular cristae similar to the mitochondrial structure described by others (Lea and Hollenberg, 1989). Functional changes can be presumed to precede ultrastructural changes. The qualitative results of chronic studies (Kennedy et al, 1990b) would bear this out in that mitochondrial changes can be observed and may indicate an increase above normal in metabolism and thus a specificity of action of almitrine on these organelles. Chronic almitrine administration induced a decrease in numerical density of mitochondria in the type 1 cells. This decrease was statistically significant and accompanied by a decrease in volume fraction, again significant. The mitochondria in the almitrine treated group occupy less space, relatively, in the cytoplasm. There was a slight increase in the volume of the average mitochondrion (which does not attain significance). This correlates well with the qualitative observations of variability of effects in the almitrine-organs. An approach, other than ultrastructural, is needed to identify almitrine's true mechanism of action. The mitochondria indeed seem of direct relevance, but biochemical studies would yield clearer results on the actual pathways involved at a molecular level.

The results of this study support the idea of a specific sensor located in the mitochondria which in turn relies on the perfusing fluid for its adequate stimulus. In the absence of the latter, swelling and dissolution of the mitochondria takes place. The lack of a metabolic drive from the mitochondria affects the glomus cell functioning. The slower rate of deterioration in the volume-perfused organ points to another sensor in the carotid body - that of flow. It is postulated that mitochondrial energy production is an important element in the "sensing" process and that there may exist another factor, a "flow" factor on Type 1 cells or on the endothelial cell membrane which responds to either a substrate present in blood, a factor present on the red blood cell, or dissolved oxygen.

REFERENCES

Joels, N. and Neil, E. (1962), The excitation mechanism of the carotid body. Br.Med.Bull., 19:21.

Kennedy, M., Ennis, S., O'Regan, R.G. and Evrard, Y. (1990), Almitrine bismesylate and the carotid body: an ultrastructural study in: Arterial Chemoreception, C. Eyzaguirre, S.J. Fidone, R.S. Fitzgerald, S. Lahiri and D.M. McDonald, eds., Springer-Verlag, New York.

Kennedy, M., Lane, H., O'Regan, R.G. and Evrard, Y. (1990). Ultrastructure of type-1 cell mitochondria in the rat carotid body following chronic oral administration of almitrine bismesylate in: "Chemoreceptors and Chemoreceptor Reflexes", H. Acker, A. Trzebski and R.G. O'Regan, eds, Plenum Publishing Co. Ltd., London, New York, Washington.

Kimberg, D.V., Loud, A.V. and Wiener, J. (1968). Cortisone-induced alterations in mitochondrial function and structure. J.Cell. Biol., 37:63.

Laubie, M. and Schmitt, H. (1980). Long-lasting hyperventilation induced by almitrine: evidence for a specific effect on carotid and thoracic chemoreceptors. Europ. J. Pharmacol., 61 (2):125.

Lea, P. J . and Hollenberg M. J . (1989). Mitochondrial structure revealed by high-resolution scanning electron microscopy. Am. J. Anat., 184:245.

Lepagnol, J . M . and Weinachter, S . N . (1988). In vitro and in vivo effects of almitrine on oxidative phosphorylation in mitochondria isolated from rat brain. J. Physiol. (Lond.), 396: 92p.

Maxwell, D.L., Mottershead, J.P. and Nye, P.C. (1987). Light reduces carotid body chemoreceptor discharge in the cat after infusion of almitrine bismesylate. J.Physiol. (Lond.) 382: 58p.

Mottershead, J.P. Nye, P.C.G. and Quirk, P.G. (1988)The effects of almitrine on electron transport in mitochondria isolated from rat liver. J. Physiol. (Lond.), 396: 91p.

Neil, E. and O'Regan, R.G. (1971). The effects of electrical stimulation of the cut sinus and aortic nerves on peripheral arterial chemoreceptor activity in the cat. J.Physiol. (Lond.), 215: 15.

O'Regan, R.G. (1979). Responses of the chemoreceptors of the cat carotid body perfused with cell-free solutions. Ir. J. Med. Sci., 148 (3):78

O'Regan, R.G., Kennedy, M., Cottell, D., and Feely, S. Ultrastructural studies of the cat carotid body perfused for short periods with physiological saline solutions in: "Chemoreception in Respiratory Control", J. A. Ribeiro and D. J. Pallot eds., Croom Helm., London, Sydney.

O'Regan, R.G. and Majcherczyk, S. (1983). Control of peripheral chemoreceptors by efferent nerves in: "Physiology of the Peripheral Arterial Chemoreceptors", H. Acker and R.G. O'Regan eds., Elsevier Biomedical Press: Amsterdam, New York, Oxford.

Shirahata, M., Andronikou, S. and Lahiri, S. (1987). Differential effects of oligomycin on carotid chemoreceptor responses to O_2 and CO_2 in the cat. J. Appl. Physiol., 65 (5): 2084.

Steer, M.W. (1981).Understanding Cell Structure. Cambridge University Press: London.

Verna A., Roumy, M. and Leitner, L.M. (1982). Ultrastructural features of the carotid body after in vitro experiments: Correlation with physiological results, J. Neurocytol., 10: 659.

Weibel, E.R. (1969). Stereological principals for morphometry in electron microscopic cytology. Int. Rev.Cytol., 26: 235.

9 MULTI-UNIT COMPARTMENTATION OF THE CAROTID BODY CHEMORECEPTOR BY PERINEURIAL CELL SHEATHS: IMMUNO-HISTOCHEMISTRY AND FREEZE-FRACTURE STUDY

H. Kondo and M. Yamamoto

Department of Anatomy, School of Medicine
Tohoku University
Sendai 980, Japan

SUMMARY

The existence and extent of the perineurial cell envelope in the carotid body was clearly demonstrated by the immunoreactivity for nerve growth factor receptor (NGFR) and the freeze-fracture analysis. Consequently the chemoreceptor is regarded as multiple units, each of which consists of certain numbers of chief and sustentacular cells and nerves covered as a whole by a sheath of the perineurial cells.

INTRODUCTION

Axon-Schwann cell fascicles in the peripheral nerves are surrounded concentrically by the perineurium which is composed of flattened perineurial cells and intercalated collagen fibrils (Akert and Sandri, 1976; Gamble and Eames, 1964; Reale et al., 1975). The perineurium is presumed to play roles in the anatomical organization of the nerve fascicle grouping, in the endoneurial fluid homeostasis and in maintenance of tenile strength (Malmgren and Olsson, 1980; Shinowara et al., 1982). In the carotid body chempreceptor, the afferent sinus nerve is divided into many fine axon-Schwann cell fascicles and each of them penetrates the parenchymal target cell groups composed of chemoreceptive chief cells and enclosing sustentacular cells. However, no detailed information has been available concerning how the perineurium extend and terminate in the peripheral targets including the carotid body, mainly because of the lack of specific histological markers for the perineurial cells. Recently we have recognized the consistent occurrence of immunoreactivity for nerve growth factor receptor (NGFR) in the terminal perineurial cells of the normal peripheral nerves (Yamamoto et al., 1991). Using this NGFR-like immunoreactivity as a marker for the perineurial cell, as well as the freeze-fracture analysis, the present study was attempted to disclose the extent and arrangement of the perineurial cells in the carotid body chemoreceptor.

Neurobiology and Cell Physiology of Chemoreception
Edited by P.G. Data *et al.*, Plenum Press. New York. 1993

MATERIALS AND METHODS

For immunohistochemistry, young albino rats weighing 150-200 gm were fixed under pentobarbital anesthesia by transcardiac perfusion with 4 % paraformaldehyde in 0.1 M phosphate buffer, pH 7.4 for 10 min. The carotid bodies were excised from perfused rats and immersed in the same fixative for further 4 hrs. After rinse with phosphate buffer, tissue blocks were immersed in 30 % buffered sucrose overnight at 4° C, and 15 um thick sections were made on a cryostat. The sections were mounted on glass slides and incubated with 2 ug/ml of mouse monoclonal antibody 192-IgG directed against the rat NGFR (a kind gift from Dr. E. Johnson, Washington Univ.) in the phosphate buffered saline (PBS) overnight at room temperature. The specificity of this antibody has already been detailed (Yan and Johnson, 1988). The sections were then incubated for 45 min. at room temperature with biotinylated anti-mouse IgG horse serum (Vector Lab. Inc., Burlingame, CA) in PBS which contained 2% of normal rat serum to eliminate non-specific adsorption of the anti-mouse IgG to the rat tissues. The sections were further incubated with peroxidase-conjugated streptavidin (Dakopatts A/S, Glostrup, Denmark) at a dilution of 1 : 600 in PBS at room temperature for 45 min. The sites of antigen-antibody reaction were made visible by 3',3'-diaminobenzidine tetrahydrochloride (DAB) at room temprature. for immunoelectron microscopy, the tissue sections were further postfixed with 0.5 % OsO_4 in 0.1 M cacodylate buffer, pH 7.4 for 10 min., after completion of the DAB reaction procedure, and they were then embedded in Epon. Ultrathin sections were made and examined under electron microscope.

For immunoreaction control, the incubation with the 192-IgG antibody was omitted or the 192-IgG was replaced by purified non-immune mouse IgG in the whole imunohistochemical procedure.

For freeze-fracture analysis, the carotid bodies were extirpated from rats fixed by transcardiac perfusion as described above and immersed for 2 hrs in 3 % glutaraldehyde buffered with 0.1 M sodium cacodylate, pH 7.4, at 4° C. The tissues were then soaked for 24 hrs in a 40 % glycerol mixed with 3 % buffered glutaraldehyde at 4° C. they were frozen in liquid nitrogen and fractured and shadowed in a freeze-fracture device (FD-2S, EIKO Enginerering Co. Ltd., Japan). Tissue replicas were cleaned in bleach, washed in ethanol and distilled water, mounted on grids with a single hole (1 x 2 mm) which had been covered by formvar films, and were examined with an electron microscope.

RESULTS AND DISCUSSION

In immunolight microcopy, intense immunoreactivity for NGFR was detected in many nerve fibers of variable thickness and it was also observed in the parenchymal cell groups in forms of honeycombs representing the contour of individual cells (Fig. 1).

In immunoelectron microscopy, NGFR-immunoreactive material was localized along most portions of apposed membranes between chief cells and sustentacular cells, and those between axons and sustentacular cells (Fig. 2). Portions of the plasma membrane of chief and sustentacular cells covered by the basal lamina were largely immunonegative. The immunoreactive material was deposited intermittently along the apposed membranes of two adjacent chief cells and no immunoreactive material was localized at the synaptic junctions on the chief cells. In axon-Schwann cell fascicles NGFR-immunoreactive material was deposited lightly along the apposed membranes between axons and Schwann cells. No immunoreaction material was deposited along the plasma membrane of Schwann cells exposed to the interstitial space.

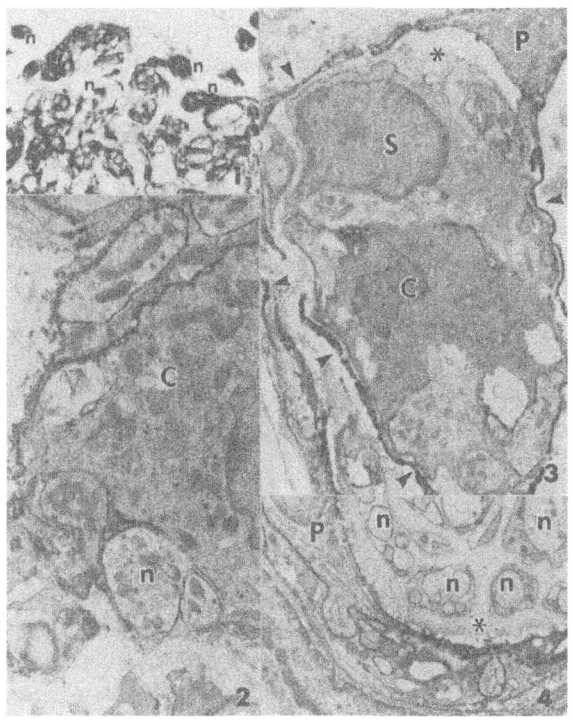

Fig. 1. Immunolight micrograph showing NGFR-immunoreactive nerve fibers (n) and the honeycomb-appearence of NGFR-immunoreactivity in the parenchymal cell groups of the rat carotid body. X 330

Fig. 2-4. Immunoelectron micrographs of the carotid body. NGFRimmunoreaction is localized along the apposed plasma membranes between chief (C) and sustentacular (S) cells, and those between axons (n) and sustentacular or Schwann cells. Note the dense immunoreaction along the membrane of the perineurial cells (P) which extend their processes (arrowheads) and enclose the parenchymal cell groups and axon-Schwann cell fascicles with collagen-filled interstitial space (asterisks) interevening. X 18,000 (Fig. 2), 8,300 (Fig. 3), 12,000 (Fig. 4)

The most intense immunoreaction for NGFR was observed along the plasma membranes of flattened cells enclosing concentrically axon-schwann cell fascicles and parenchymal cells with distinct interstitial space intervening (Figs. 3, 4). The immunoreactive flattened cells was characterized by an oblong nucleus, paucity of perikaryal cytoplasm and long cytoplasmic processes. Adjacent processes were directly apposed to each other with focal junctional specializations. The cytoplam was relatively poor in organelles, but a few cisterns of rough-endoplasmic reticulum were often seen in the perikaryal cytoplasm. Pinocytotic vesicle were well-developed beneath the plasma membrane and the basal lamina was poorly developed. Some of the pinocytotic vesicles were filled with the immunoreactive material. These histological features make it possible to identify the immunoreactive cells as the perineurial cells.

In freeze-fracture analysis, the characteristic features of the chief cells and sustentacular cells have already been describes in detail in our previous study (Kondo, 1981). Besides the two major cell types of the parenchymal cell groups, the perineurial cells were identified at close proximity to the parenchymal cells as well as nerve fascicles. The perineurial cells had a widespread sheet-like extension and surrounded the entire surface of the parenchymal cell groups with a narrow spaces containing collagen fibers intervening (Fig. 5). The fracture faces of the plasma membrane of the perineurial cells showed coarse furrows of variable depth, width and length, running singly or in bundles in all directions. These furrows were interpreted as those pressed by collagen fibers in the interstitial space. Tight junctions characterized by meshworks of numerous ridges and grooves were formed between two adacent perineurial cells, but they were rather short and simple and confined to circumscribed areas. No gap junctions were found on the perineurial cells. Pits of pinocytotic vesicles were scattered randomly over the entire surface of the perineurial cells. These freeze-fracture features are quite similar to those of the perineurial cells of other peripheral nerves reported previously by Reale et al. (1975).

In hitherto published electron microscopy of the carotid body as well as the present one, flattened cells can be discerned to be arranged in parallel to the surface of the parenchymal cell groups (Kondo, 1971). However, few attention has been paid to such cells in the previous studies and they have not been identified as representing a distinct cell entity. The present immunohistochemical and freeze-fracture study clarifies for the first time that the perineurial cells, contiguous to those of nerves innervating the carotid body, surround the parenchymal cell groups, and that the cells exhibit NGFR-immunoreactivity intensely. Consequently the carotid body chemoreceptor is regarded as multiple units, each of which consists of certain num bers of chief and sustentacular cells and innervating nerves covered as a whole by a sheath of the perineurial cells. Since the tight junctions between the perineurial cells are confined to circumscribed areas, it is unlikely that the space inside the perineurial cell sheath is completely isolated from the outside. The functional significance of this multi-unit cellular organization by the perineurial cells remains to be elucidated in relation to more clear understanding of the chemoreception mechanism.

With regard to the appearance of NGFR-immunoreactivity in non-neuronal Schwann cells, Taniuchi et al. (1988) have suggested a possibility that NGFR retains and concentrates NGF on the surface of Schwann cells, resultly in supply of trophic support and haptotactic gaidance for regenerating axons. Whether or not this suggestion can be applicable to NGFR in the perineurial cells also remains to be elucidated.

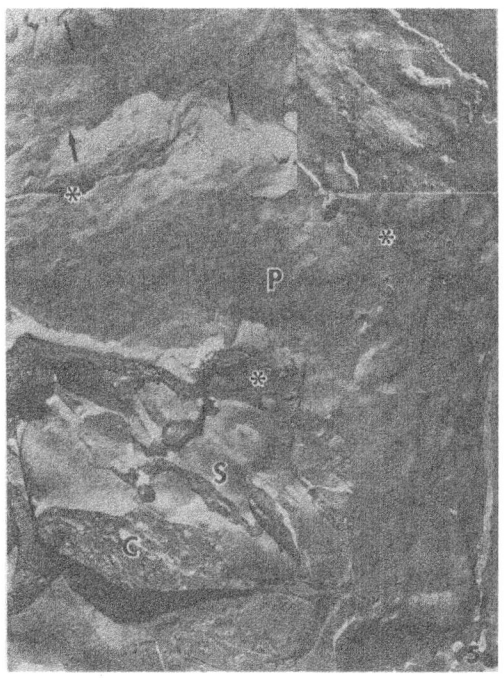

Fig.5. *Freeze-fracture replica image of the carotid body. Sheet-like cytoplasmic extensions of the perineurial cells (P) enveloping the chief (C) and sustentacular (S) cell clusters. Arrows indicate furrows running in random direction on the fractured plasma membrane of the perineurial cells. Note narrow spaces (asterisks) filled with collagen fibers intervening between the perineurial cell and the parenchymal cell cluster or between two adjaxent perineurial cells. Inset shows circumscribed tight junctions between two adjacent perineurial cells. X 14,500; Inset, 72,000*

ACKNOWLEDGEMENT

The authors wish to thank Dr. E. M. Johnson, Jr. and Ms. P. Osborne, Dept. pharmacol. Washington Univ. St. Louis, U.S.A. for supplying the antiserum for NGFR.antiserum for NGFR. We also thank Mr. H. Iwasa and Mr. S. Yamazaki for their technichal assistance. This study wag supported in part by a grant from Takefu Psychiatric Hospital, Takefu, Fukui Prefecture, Japan.

REFERENCES

Akert, K. and Sandri, C., 1976, The fine structure of the perineurial endothelium, Cell Tiss. Res., 165: 281.

Gamble, H.J. and Eames, R.A., 1964, An electron microscopic study of the connective tissues of human peripheral nerve, J. Anat., 98: 655.

Kondo, H., 1971, An electron microscopic study on innervation of the carotid body of guinea pig, J. Ultrastr. Res., 37: 544.

Kondo, H., 1981, Evidences for the secretion of chief cells in the rat carotid body, in: "Arterial chemoreceptors", C. Belmonte et al., ed,. pp. 45, Leicester University Press, Great Britain.

Malmgren, L.T. and Olsson, Y., 1980, Difference between the peripheral and central nervous system permeability to sodium fluorescein, J. Comp. Neurol., 191: 103.

Reale, E., Luciano. L. and Spitznas, M., 1975, Freeze-fracture faces of the perineurial sheath of the rabbit sciatic nerve, J. Neurocytol., 4: 261.

Shinowara, N.L., Michel, M.E. and Rapport, S.I., 1982, Morphological correlates of permeability in the frog perineurium: vesicles and "transcellular channels", Cell Tiss. Res., 227: 11.

Taniuchi, M., Clark, H.B., Schweitzer, J.B. and Johnson, E.M., Jr., 1988, Expression of nerve growth factor receptors by Schwann cell of axotomized peripheral nerves: Ultrastructural location, suppression by axonal contact, and binding properties, J. Neurosci., 8: 664.

Yan, Q. and Johnson, E.M., 1988, An immunohistochemical study of the nerve growth factor receptor in developing rats, J. Neurosci., 8: 3481.

Yamamoto, M., Amano, O. and Kondo, H., 1991, Immunoreactivity for nerve growth factor receptor in the perineurial cells of the peripheral nerves: normal appearance and response to nerve section, Neurosci. Res., Supple. 14: S140.

10 LIGHT- AND ELECTRONMICROSCOPICAL IMMUNOHISTOCHEMICAL INVESTIGATION OF THE INNERVATION OF THE HUMAN CAROTID BODY

W. Kummer and *J.-O. Habeck*

*Institute for Anatomy and Cell Biology
University of Heidelberg
Heidelberg, FRG

INTRODUCTION

The innervation of the carotid body has been extensively studied in several laboratory animals commonly used in chemoreceptor research. These studies revealed some features common to all species investigated - e. g. the presence of calcitonin gene-related peptide-immunoreactive (CGRP-IR) fibers of sensory origin (Kummer 1987, 1988; Kondo and Yamamoto 1988) but also some clear cut species differences - e. g. the presence of neuropeptide Y (NPY) containing sensory fibers in the rat (Czyczyk-Krzeska et al. 1991) but not in the guinea pig (Kummer 1990). Thus, data obtained in laboratory animals cannot be extrapolated to other species like man, and a detailed immunohistochemical study on the innervation of the human carotid body was initiated.

MATERIAL AND METHODS

Carotid bodies from patients without chronic respiratory diseases or arterial hypertension were obtained during routine autopsy at postmortem intervals of 3 to 12 h and fixed by immersion in buffered 2% paraformaldehyde/15% saturated picric acid. Cryostat sections cut at 14 μm were used for single- or double-labelling immunofluorescence as described before (Kummer 1990). Frozen sections of 40 μm thickness were immunohistochemically processed utilizing an indirect peroxidase method, dehydrated and flat embedded in epoxy resin. Ultrathin sections cut from these preparations were used for electronmicroscopical immunohistochemistry(for technical details see Kummer et al. 1989).

Primary antisera used in this study were directed against aromatic-L-amino-acid-decarboxylase (= AADC; polyclonal rabbit antiserum; Eugene Tech International, Allendale, NJ, USA), CGRP (polyclonal rabbit antiserum; Peninsula, St. Helens, U.K.), dopamine-β-hydroxylase (= DβH; polyclonal rabbit antiserum; Eugene Tech International) neurofilament$_{160KD}$ (= NF; mouse monoclonal antibody; Boehringer Mannheim, FRG), NPY (polyclonal rabbit antiserum, Amersham Buchler, Braunschweig, FRG), substance P (= SP; rat monoclonal antibody; Dunn, Asbach, FRG, and polyclonal rabbit antiserum kindly provided by Dr. R. Murphy, Melbourne, Australia), tyrosine hydroxylase (= TH; mouse monoclonal antibody; Boehringer Mannheim, FRG, and polyclonal rabbit antiserum; Eugene Tech International), and vasoactive intestinal peptide (= VIP; polyclonal rabbit antiserum; Immuno Nuclear Corporation, Stillwater, MN, USA).

RESULT

General structure. The specific tissue of the human carotid body is arranged to lobules separated by connective tissue septa carrying the supplying blood vessels. Thus, a clear distinction between the innervation of glomic lobules and that of large supplying blood vessels can be made, which is not that evident in small laboratory animals.

Perivascular innervation. The interlobular arteries (and, to a lesser extent, the veins) were innervated by two neurochemically distinct types of axons. 1) Most frequent were axons immunoreactive to TH, AADC, DβH, and NPY. 2) Another subset of perivascular nerve fibers displayed coexistence of SP- and CGRP-IR. Immunoelectronmicroscopy revealed that both types of fibers were unmyelinated and contained numerous vesicles in axonal swellings only partly covered by Schwann cells.

Innervation of glomic lobules. Three kinds of axons differing neurochemically were identified within glomic lobules.

1) CGRP/SP-IR fibers. Axons of this type were much more frequent within the glomic lobules than around interlobular blood vessels. Ultrastructurally, they were identified as thin unmyelinated axons (median diameter: 0.28 μm). CGRP/SP-IR axons usually run together with non-reactive axons within the same Schwann cell sheath. Reaching the sustentacular cell/glomus cell complexes, CGRP/SP-IR axons developed vesicle containing axonal swellings which sometimes came into direct contact to glomus cells. Membrane specializations did not become evident between glomus cells and this type of axons (Fig. 1).

2) TH/NF-IR fibers. This kind of axon formed a very dense network within the glomic lobules and was also observed in branches of the sinus nerve. At lightmicroscopical level, TH/NF-IR fibers appeared much thicker than any other fiber population within the human carotid body. This was confirmed by morphometric measurements at ultrastructural level: Unmyelinated TH/NF-IR axons within terminal branches of the sinus nerve exhibited a median diameter of 0.69 μm, thus being significantly larger than CGRP/SP-IR axons (Kolmogoroff-Smirnoff test). Schwann cells covering an unmyelinated TH/NF-IR axon usually did not envelope an additional axon, thus establishing a 1:1 ratio of axon to Schwann cell. In addition to these unmyelinated TH/NF-IR fibers, myelinated axons with the same neurochemical characteristics were present, measuring about 2.5 μm in diameter (including myelin sheath).

In contrast to CGRP/SP-IR axones, TH/NF-IR terminals containing many mitochondria and vesicles made extensive specialized contacts with glomus cells (Fig. 2).

3) NPY/VIP-IR fibers. This population was least numerous within glomic lobules. Varicose fibers of this type preferentially surrounded the nests of glomus cells instead of penetrating them.

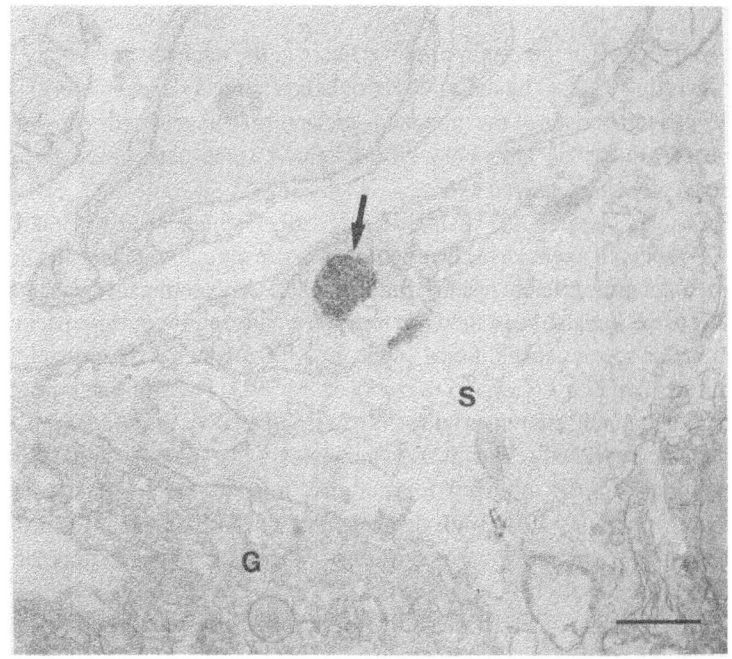

Fig. 1. A small SP-IR axon terminal (arrow) *filled with electron dense reaction product is enveloped by cytoplasmic processes of a sustentacular cell (S) without being in direct contact to a glomus cell (G). Bar = 1 μm.*

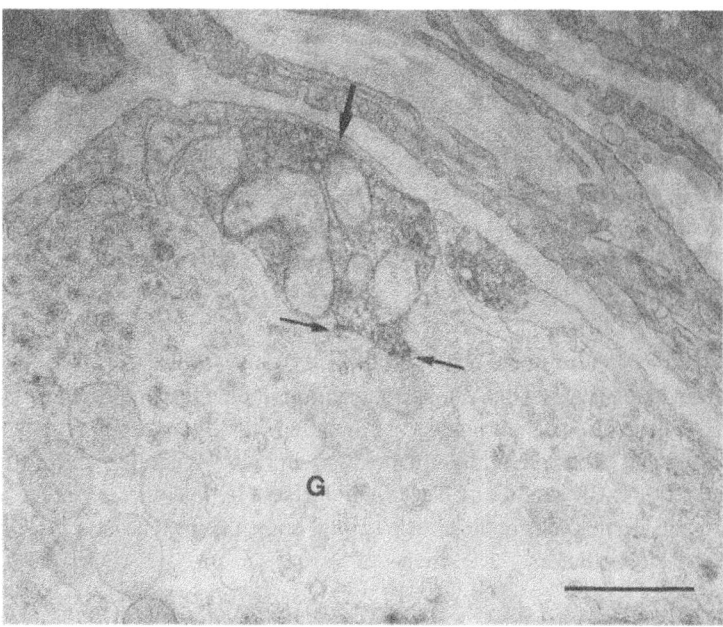

Fig.2. This TH-IR axon ending (arrow) *containing numerous vesicles and mitochondria displays specialized membrane thickenings* (small arrows) *at its contact site with a glomus cells (G). Bar = 1 μm.*

DISCUSSION

It is obvious that the origin of nerve fibers in the human carotid body cannot be directly investigated, e. g. by means of retrograde tracing or degeneration experiments. Nevertheless, comparison of the immunohistochemical data reported here with previous experimental studies on laboratory animal allows some clues as to the source of neurochemically characterized axons.

Catecholaminergic (i. e. TH/DβH-IR) nerve fibers simultaneously containing NPY-IR, and NPY/VIP-IR axons have been also identified in the guinea pig carotid body where they are of autonomic efferent origin (Kummer 1990). A predominant sympathetic origin of NPY-IR axons has also been shown in the rat (Kondo et al. 1986; Kummer et al. 1989), although in this species an additional supply of NPY-producing sensory neurons to the carotid bifurcation exists (Czyczyk-Krzeska et al. 1991). These findings are strongly in favour of an autonomic efferent origin of TH/AADC/DβH/NPY- and NPY/VIP-IR fibers in the human carotid body. The clear topographical separation of glomic lobules from their supplying arteries allowed for the first time to define separate targets of catecholaminergic/NPY-IR and noncatecholaminergic/NPY-IR axons within the carotid body: the former representing vasomotor fibers whereas the latter address the elements of the glomic lobules itself.

A sensory origin of CGRP- and SP-IR fibers in the carotid body has been proven experimentally in guinea pig (Kummer 1987, 1988) and rat (Kondo and Yamamoto 1988; Czyczy-Krzeska et al. 1991). Since fibers of this neurochemical type are a general feature of the vasculature, doubt has been raised as to a relationship of CGRP/SP-IR axons to the specific cellular elements and functions of the carotid body (Chen et al. 1986; Kummer 1988). The present data show, that indeed a purely perivascular component is involved in the CGRP/SP-IR innervation of the carotid body. The vast majority of such axons, however, is intimately attached to the glomic lobules and obviously does not belong to the general sensory vascular innervation. In agreement with previous ultrastructural studies on cat, rat, and guinea pig (Chen et al. 1986; Kondo and Yamamoto 1988; Kummer et al. 1989), CGRP- and SP-IR nerve terminals make only unconspicuous contacts with glomus cells and originate from unmyelinated axons. The axonal diameters measured in this study correspond to a calculated conduction velocity of 0.5 m/s, according to the formula provided by Gasser (1955). Taken together, they are likely to represent the C-fiber class of chemoreceptor afferents.

TH-IR but DβH-negative axons innervating the carotid body have been traced to sensory neurons located in the petrosal ganglion (Katz and Black 1986; Kummer et al. 1990). In the human carotid body, this set of neurons is obviously represented by the TH/NF-IR axon population. Their mode of termination within the carotid body yet has not been investigated ultrastructurally in laboratory animals. According to the presence of myelinated TH/NF-IR axons, and because of the large immunoreactive nerve terminals contacting the glomus cells in the human carotid body, these axons have to be classified as A-fiber chemoreceptors.

REFERENCES

Chen, I-L., Yates, R. D., and Hansen, J. T., 1986, Substance P-like immunoreactivity in rat and cat carotid bodies: Light and electron microscopic studies, Histol. Histopathol., 1:203-212.

Czyczyk-Krzeska, M.F., Bayliss, D.A., Lawson, E.E., and Millhorn, D.E., 1991, Expression of messenger RNAs for peptides and tyrosine hydroxylase in primary sensory neurons that innervate arterial baroreceptors and chemoreceptors, Neurosci. Lett., 129:98-102.

Gasser, H.S., 1955, Properties of dorsal root unmedullated fibers on the two sides of the ganglion, J. Gen. Physiol., 38:709-728.

Katz, D.M., and Black, I.B., 1986, Expression and regulation of catecholaminergic traits in primary sensory neurons: relationship to target innervation in vivo. J. Neurosci., 6:983-989.

Kondo, H., and Yamamoto, M., 1988, Occurrence, ontogeny, ultrastructure and some plasticity of CGRP (calcitonin gene related peptide)-immunoreactive nerves in the carotid body of rats, Brain Res., 473:283-293.

Kondo, H., Kuramoto, H., and Fujita, T., 1986, Neuropeptide tyrosine-like immunoreactive nerve fibres in the carotid body chemoreceptor of rats Brain Res., 372:353-356.

Kummer, W., 1987, Calcitonin gene-related peptide-immunoreactive nerve in carotid body and carotid sinus, in: "Histochemistry and cell biology of of autonomic neurons and paraganglia", C. Heym, ed., Exp. Brain Res. Series, 16:78-83, Springer, Heidelberg.

Kummer, W., 1988, Retrograde neuronal labelling and double-staining immunohistochemistry of tachykinin and calcitonin gene-related peptide-immunoreactive pathways in the carotid sinus nerve of the guinea pig, J. Autonom. Nerv. Syst., 23:131-141.

Kummer, W., 1990, Three types of neurochemically defined autonomic fibres innervate the carotid baroreceptor and chemoreceptor regions in the guinea pig, Anat. Embryol., 181:477-489.

Kummer, W., Fischer, A., and Heym C., 1989, Ultrastructure of calcitonin gene-related peptide- and substance P-like immunoreactive nerve fibres in the carotid body and carotid sinus of the guinea pig, Histochemistry, 92:433-439.

Kummer, W., Gibbins, I. L., and Heym, C., 1989, Peptidergic innervation of arterial chemoreceptors, Arch. Histol. Cytol., 52 (Suppl.):361-364.

Kummer, W., Gibbins, I. L., Stefan, P., and Kapoor, V., 1990, Catecholamines and catecholamine synthesizing enzymes in guinea-pig sensory ganglia, Cell Tissue Res., 261:595-606.

11 SEROTONIN (5-HYDROXYTRYPTAMINE) EXPRESSION IN PULMONARY NEURO-ENDOCRINE CELLS (NE) AND A NETUMOR CELL LINE

C. Newman, D. Wang and E. Cutz

Department of Pathology, The Research Institute
The Hospital For Sick Children and University of Toronto
Toronto, Ontario

INTRODUCTION

Pulmonary neuroepithelial bodies (NEB) are clusters of innervated NE cells scattered throughout the airway mucosa and are thought to function as hypoxia sensitive chemoreceptors (Lauweryns et al., 1977, Lauweryns and Cokelaere, 1973). NEB may play an important role during neonatal adaptation as well as in infant disease states including Sudden Infant Death Syndrome (Perrin et al., 1991). Because of their small number and diffuse distribution within the lung, NEB are difficult to study. A cell line derived from human lung small cell carcinoma (H69) exhibits many features of NE differentiation and hence may be a useful in vitro model for studies on pulmonary NE regulation at the cellular and molecular level. Although the amine mechanism is an important feature of normal and neoplastic NE cells, it has not been fully characterized.

In this study 5-HT content, localization and the presence of the enzyme tryptophan hydroxylase (TPH- the rate limiting enzyme in the synthesis of 5-HT), has been examined in H69 cells and compared with 5-HT content in NEB cells isolated from rabbit fetal lungs (Cutz et al., 1985). The presence of high concentrations of 5-HT in both the normal and neoplastic NEB may give further insight into the functional role of these cells relevant to pulmonary physiology and pathology.

MATERIALS AND METHODS

Cell Preparation

H69 cells. The human male SCLC cell line, H69, was obtained from American Type Culture Collection (Rockville, MD). Cells were maintained in alpha medium with 20% fetal calf serum (Gibco, Grand Island, NY). Flasks containing the cells were incubated at 37°C in a 5% CO_2 incubator.

Fetal Rabbit NEB. Detailed methods for the isolation and culture of NEB cells from fetal lungs have been previously reported (Cutz et al., 1985). For a single experiment, two litters (8-12 fetuses each) of New Zealand white rabbit fetuses of 24 days gestation were required to obtain a sufficient number of cells. The cell dissociation procedure for lung tissue consisted of a mechanical and enzymatic step followed by gradient centrifugation. The single-cell fraction of enriched NEB cells was resuspended in alpha-MEM with 0.5% fetal bovine serum (plus supplements) and seeded (1×10^6) into Lab-Tech slide culture chambers for morphologic studies and in Petri dishes for biochemical analysis. Cultures were maintained in a 5% CO_2 incubator (37°C) for 3 days before analysis.

Immunostaining

Slides with monolayer cultures of lung cells were fixed for 45 minutes with 10% neutral-buffered formalin. H69 cells were spun into a pellet, fixed overnight in 10% neutral-buffered formalin, and embedded in paraffin. Both cell preparations were immunostained using indirect immunoperoxidase methods. A monoclonal antibody against 5-HT (Sera Lab UK) was used in a dilution of 1:100 followed by avidin-biotin complex reagent (Vectastain), peroxidase visualization and counterstain with hematoxylin.

Biochemical Analysis of Cell Cultures

The content of 5-HT was measured in both NEB cell cultures and H69 tumor line by HPLC with electrochemical detection (Waters #460) set at a potential of 0.46V against a KCl electrode. The $HClO_4$ extracts of both preparations were homogenized, centrifuged and injected directly into the HPLC. An 0.1M acetate mobile phase (flow rate 0.5 ml/min) with 10-15% methanol was used with a Bondapak reverse-phase C18 column. Protein content was assessed by the Bio-Rad method (Bio-Rad, CA) and 5-HT concentration was expressed as mean pg/mg protein.

Northern Blot Analysis

Total RNA was prepared, according to the method of Chomezynski and Sacchi (1987), by homogenizing fetal rabbit lung and H69 cell pellet in guanidinium thiocyanante followed by phenol-chloroform extraction under acidic conditions, centrifugation and isopropanol precipitation. Total RNA was subjected to electrophoresis and detected by hybridization with TPH complementary RNA (cRNA) probe. The full length rabbit TPH cDNA clone (gift from Dr. R. C. Eisensmith) was cut with EcoR1 and the 700 bp fragment was further subcloned in the pSP72 vector. The template was linearized by ClaI and the cRNA probe was transcribed with SP6 RNA polymerase. The probe was separated from free nucleotide by Sephadex G-50 column. Hybridization reaction was preformed using standard protocols (Maniatis et al., 1989) and filters exposed 24-48 hours at -70 °C to Kodak XAR film with intensifying screen.

RESULTS

The cultures of rabbit lung cells, immunostained for 5-HT showed numerous immunoreactive NEB cells (Fig 1). The overall shape of the cells varied form oval to elongated, often with short neurite-like processes.

Paraffin sections of H69 cell pellets immunostained for 5-HT showed positive immunoreactivity in almost all cells although occasional clusters of 5-HT negative cells were observed (Fig. 2). In 5-HT positive cells, the staining was evenly localized throughout the cytoplasm.

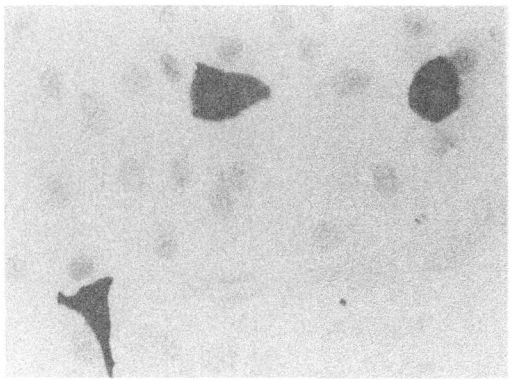

Fig. 1. A 3-day culture of enriched rabbit fetal lung cells showing NEB cells immunostained for 5-HT (400x).

Both cultures of fetal rabbit NEB and H69 cells contained significant amounts of 5-HT as measured by HPLC with electrochemical detection. In both cases, the 5-HT peaks were clearly separated from the other peaks. No significant amounts of 5-HT metabolites were detected except for a small 5-hydroxyindole acetic acid (5-HIAA) peak (Fig 3). The mean concentration of 5-HT in cultures of NEB cells was 51 ± 3.4 pg/mg protein and in H69 cells was 365 ± 12.9 pg/mg protein.

Northern blot analysis of total RNA from cultures of rabbit fetal NEB cells showed a prominent band of 1.9 kb in lane 2 indicating the presence of mRNA for TPH in these cells. Likewise, RNA extraction from H69 cells showed an even more intense TPH mRNA band in the same position (lane 1) indicating that these cells also expressed the message for TPH (Fig 4). As a negative control, rabbit adrenal gland showed no band (lane 3).

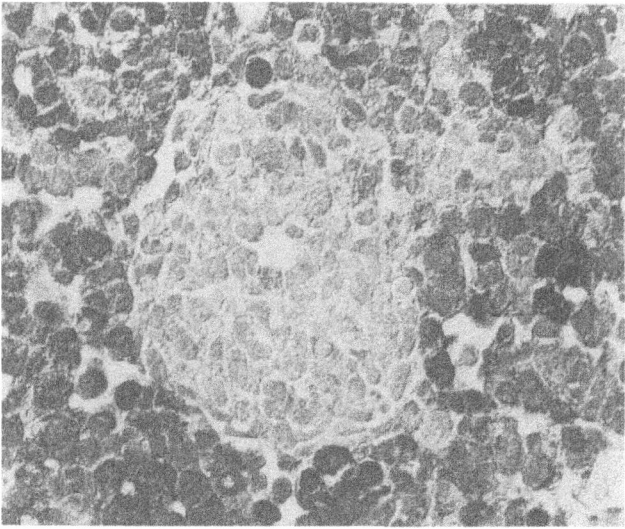

Fig. 2. Paraffin section of a cell pellet of H69 cells immunostained for 5-HT with darkly stained positive cells. An island of 5-HT negative cells is shown in the center (450x).

75

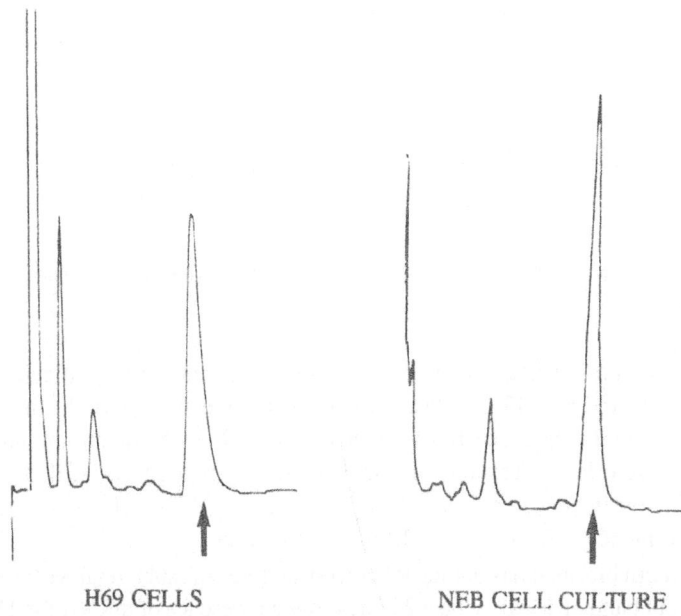

Fig. 3. Chromatograms of H69 and rabbit fetal lung cultures showing 5-HT peak indicated by arrows.

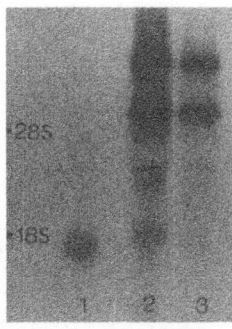

Fig. 4. Northern blot analysis: the 1.9kb band around the 18s position indicate expression of tryptophan hydroxylase mRNA in H69 cells (lane 1) and in fetal rabbit NEB (lane 2). As a negative control,rabbit adrenal gland was used (lane 3).

DISCUSSION

Presented here are both qualitative as well as quantitative data on the presence of 5-HT in NEB cultures as well as the SCLC tumor line (H69), a malignant counterpart of normal NEB cells. The demonstration of mRNA for the rate-limiting enzyme TPH in both cell preparations indicates that these cells synthesize 5-HT.

Although the precise role of 5-HT in NEB cells is unknown, Lauweryns et al., (1977; Lauweryns and Cokelaere, 1973) suggested that 5-HT released from these cells may be a mediator of hypoxia-induced pulmonary vasoconstriction, influencing local blood circulation in the lung. Another possibility is that 5-HT could be functioning as a neurotransmitter important in the chemoreceptive process. The role of dopamine and perhaps also 5-HT in chemoreception by carotid body chief cells is well known (Fishman et al., 1985).

The low numbers and diffuse distribution of NEB cells within lung tissue make them difficult to study. The findings documented here demonstrate that normal as well as neoplastic NEB cells contain 5-HT. Tumor cell lines such as H69 may provide a useful model to study the amine function(s) at the cellular and molecular level.

ACKNOWLEDGEMENTS

This work was supported by grants from Medical Research Council of Canada (PG-42) and NICHD (1 ROlHD22713). Ch.N. was a recipient of Dr. Sidney Segal studentship of Canadian Foundation on Study of Infant Death.

REFERENCES

Chomezynski, P., and Sacchi, N., 1987, Single-step method of RNA isolation by acid guanidinium thiocynate-phenol-chloroform extraction, Anal. Biochem., 162: 152.

Cutz, E., Yeger, H., Wong, V., Bienkowski, E., and Chan, W., 1985, In vitro characteristics of pulmonary neuroendocrine cells isolated from rabbit fetal lung, Lab Invest., 53: 672.

Fishman, C. M., Greene, W. L., and Platika, D., 1985, Oxygen chemoreception by carotid body in cells in culture, Proc. Nad. Acad. Sci., 82: 1448.

Lauweryns, J. M., Cockelaere, M., Deleersnyder, M., and Liebens, M., 1977, Intrapulmonary neuro-epithelial bodies in newborn rabbits, Cell. Tiss. Res. 182: 425.

Lauweryns, J.M., and Cokelaere, M.C., 1973, Hypoxia-sensitive neuro-epithelial bodies. Intrapulmonary secretory neuro-receptors, modulated by the CNS? Z Zellforsch 145: 521.

Maniatis, T., Fritsch, E.M., and Sambrook, J., 1989, in: "Molecular Cloning: A Laboratory Manual," Cold Spring Harbor Laboratory Press, Cold Spring Harbor, NY.

Perrin, D.G., McDonald, T.J., and Cutz, E., 1991, Hyperplasia of bombesin immunoreactive pulmonary neuro-endocrine cells and neuro-epithelial bodies in sudden infant death syndrome, <u>Ped. Path.</u>, 11: 431.

12 EFFECTS OF HYPOXIA ON CULTURED CHEMORECEPTORS OF THE RAT CAROTID BODY: DNA SYNTHESIS AND MITOTIC ACTIVITY IN GLOMUS CELLS

C.A. Nurse and C. Vollmer

Department, of Biology, McMaster University
Hamilton, Ontario, Canada. L8S 4K1

INTRODUCTION

Enlargement of the carotid body, an arterial chemosensory organ, is known to occur under natural conditions of chronic hypoxia or during hypoxic lung disease (Edwards et al 1971a,b; Dhillon et al 1984). For example, natives or animals living at high altitude (e.g. the Peruvian Andes) have enlarged carotid bodies and a high incidence of carotid body tumors (Edwards et al 1971b). This enlargement is due to both hyperplasia and hypertrophy of various cell types, including endothelial cells of the vasculature and the arterial chemoreceptors (Dhillon et al 1984; Bee et al 1986), which recent evidence indicates are the glomus or type 1 cells (Lopez-Barneo et al 1988; Biscoe and Duchen 1990; Stea and Nurse1991a,b). These cells are considered members of the sympathoadrenal sublineage of neural crest derivatives, and show several ultrastructural and biochemical similarities to other neuroendocrine cells of this lineage, i.e. adrenalmedullary chromaffin cells and the small intensely fluorescent (SIF) cells in sympathetic ganglia (Kobayashi 1971; McDonald and Blewett 1981; Doupe et al 1985). The mechanisms by which a simple environmental stimulus, i.e. hypoxia, triggers hyperplasia or cell division in the neuronal-like glomus cells, especially in the adult carotid body, are of general neurobiological interest since the inability of mature neurons to divide severely limits the repair capacity of the nervous system. Understanding these mechanisms is difficult in vivo since hypoxia causes a number of other cardiovascular effects that cannot be easily separated or controlled. In this study we overcame some of these limitations by using an established in vitro system consisting solely of dispersed rat carotid body cells, in which the chemoreceptor glomus cells survive for up to several months (Nurse 1990; Stea and Nurse 1991a). We therefore asked whether hypoxia can stimulate DNA synthesis and mitotic activity in glomus cells, grown in this simple culture system containing relatively few cell types. To identify glomus cells undergoing DNA synthesis and mitosis we use double-label immunofluorescence to detect colocalization of the cytoplasmic marker tyrosine hydroxylase (TH), and nuclear incorporation of the thymidine analog, bromodeoxyuridine (BrdU).

METHODS

Details of the procedures for the culture of glomus cells following combined enzymatic and mechanical dissociation of the rat carotid body have been described elsewhere (Nurse 1990; Stea and Nurse 1991a). In this study for each experimental series carotid bodies from 12 rat pups, 6-9 days old, were dissociated and the resulting cell suspension was plated into central wells (ca. 1 cm diameter) of 6 modified 35mm culture dishes. Initially all the cultures were grown at 37°C in a humidified atmosphere of 20% $O_2/5\%$ CO_2 (normoxia) in F-12 growth medium supplemented with 10 % fetal calf serum and other additives (Nurse 1990; Stea and Nurse 1991a). After approximately 2 days in vitro the cultures were fed and 3 of them were transferred to a similar incubator equilibrated with 5% $O_2/5\%$ CO_2 (hypoxia); the remaining 3 cultures were returned to the original (control) normoxic incubator. In most cases the experiment was terminated ca. 60 hr later, though during the final 8 to 20 hr bromodeoxyuridine (BrdU; 10μm) was present in the culture medium. Both groups of normoxic and hypoxic cultures were then fixed in absolute methanol at 4°C for 10 min and treated with 0.1M glycine for 5 min at room temperature. Following 3x3 min rinses with 0.1M phosphate buffer, the cultures were exposed to 2N HCl for 10 min at 37 °C to denature the DNA. The acid was then neutralized during 2x 5 min treatments with 0.1M sodium borate at pH 8.5 and the cultures washed 3x (3 min each) with phosphate buffer before a 1 hr exposure at 37°C to a mixture of rabbit anti-TH antiserum (1:1000; Nurse 1990) and mouse anti-BrdU monoclonal antibody (6 μg/ml; Boehringer Mannheim). After 3 rinses in phosphate buffer the cultures were incubated for 30 min at room temperature in PBS-GS (see Nurse 1990) containing Texas-red conjugated goat anti-mouse IgG (1:400) and FITC-conjugated goat anti-rabbit IgG (1:50). The immuno-stained cultures were visualized and scored with the aid of a Zeiss IM35 inverted phase contrast microscope equipped with epi-illumination plus rhodamine and fluorescein filter sets.

RESULTS

In these cultures glomus cells grow in discrete clusters or islands that are easily recognizable under phase contrast microscopy (Fig.1A; see also Nurse 1990; Stea and Nurse 1991a). Following TH-immunostaining almost all cells in the cluster appear as TH+ (Fig. 1B) and in addition, singly-isolated glomus cells as well as doublets of glomus cells (e.g. Fig. 1E) are readily revealed. Usually in living cultures, clusters of 3 or more glomus cells can be reliably identified though our level of confidence decreases for the singles and doubles. Since the carotid body is a tiny organ the yield of glomus cells is generally low; this, together with the broad variability in cluster size and distribution, can result in quite variable numbers of surviving glomus (TH+) cells among the dishes, even after careful attempts to apply a uniform volume of the cell suspension to each culture. Routine comparisons of control (normoxic) and hypoxic cultures after similar exposure times revealed no obvious differences when viewed under phase contrast microscopy. However, as described below quantitative differences were unmasked under the two conditions when uptake of the thymidine analog, bromodeoxyuridine (BrdU), by TH+ glomus cells was assayed by immunofluorescence.

DNA synthesis in glomus cells during normoxia and hypoxia

In cultures grown under normoxic (20% O_2) conditions and exposed to a 8-20 hr pulse of BrdU, a proportion of the TH+ glomus cells was always double-labelled. This proportion was relatively constant among sister cultures (ca.20% for a 20 hr exposure) and showed far less variability than the actual number of TH+ cells among the cultures.

Since BrdU is a thymidine analog, it is taken up in the nucleus during the S phase of the cell cycle when DNA synthesis occurs. Thus, double labelled cells show a green (fluorescein) cytoplasmic fluorescence and a red (Texas red) nuclear fluorescence corresponding to TH and BrdU localization respectively. An example of a glomus cell cluster containing 5 clearly visible TH+/BrdU+ cells is shown in Fig.1A,B,C. Following exposure to hypoxia (5% O_2) there was a significant increase in the proportion of TH+ cells that were double-labelled (i.e. TH+/BrdU+). This result has been confirmed in 7 separate experiments and data from one experiment are shown in Fig.2. The bins on the right of Fig.2 show the mean number of TH+ and TH+/BrdU+ glomus cells in triplicate cultures exposed to normoxic (N) or hypoxic (H) atmospheres for ca. 60hr; the bins on the left compare the frequency of TH+/BrdU+ or 'cycling' glomus cells under the two conditions. The relatively high proportion of cycling glomus cells under normoxic conditions (ca. 20% in Fig.2) may be partly due to the activity of potential growth factors in the fetal calf serum present in the culture medium.

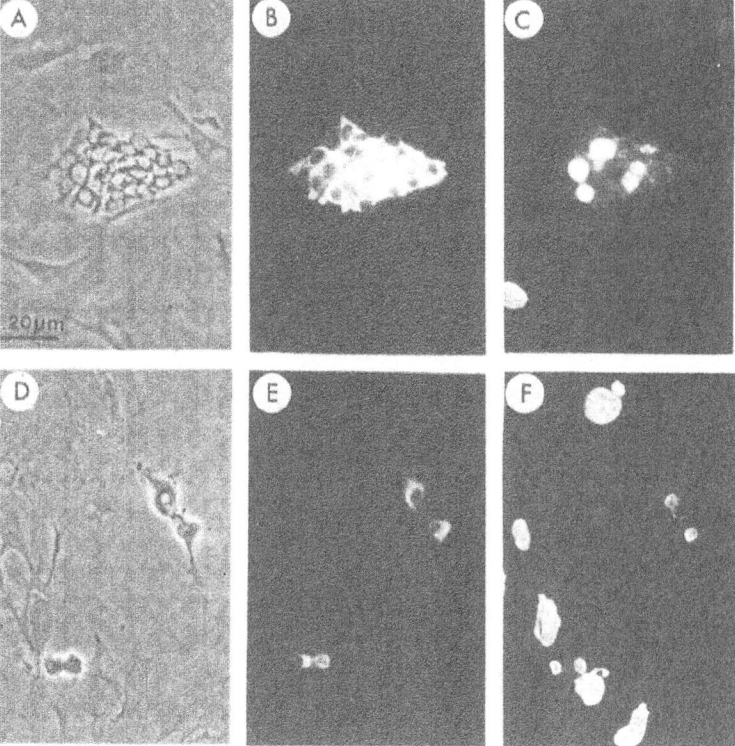

Fig.1.TH and BrdU immunoreactivity in carotid body cultures. The same field containing a glomus cell cluster is shown in the phase contrast micrograph (A), and after exposure to reveal TH+ and BrdU+ immunofluorescence with a fluorescein (B) and Texas red (C) filter set respectively; 5 cells in the cluster are clearly double-labelled representing glomus cells that have undergone DNA synthesis during the final 20hr when BrdU was present. A field from a different culture is shown in the same sequence in D,E,F; the TH+ doublets in E (upper and lower fluorescent pairs) are also BrdU+ as shown in F. The nuclei of other cells in the field have also taken up BrdU in F.

Does hypoxia stimulate mitosis in glomus cells?

The increased percentage of TH+/BrdU+ glomus cells seen under hypoxic conditions indicates an increase in DNA synthesis but does not directly address whether there is an increase in mitosis or hyperplasia in glomus cells. Because of the large variability in the number of TH+ glomus cells among the dishes on plating, a small expansion of this cell population is difficult to demonstrate under the present conditions. However, we have seen several indications that glomus cells are capable of division in culture. First, we have observed a rare example of a single TH+ cell in anaphase, on the periphery of a large cell cluster (not shown). Second, we routinely see profiles where pairs of glomus cells seem to be undergoing cytokinesis (e.g. Fig. 1D,E,F). Analysis of immunostained cultures often revealed some fused TH+ doublets where both nuclei were BrdU+ (Fig.1E,F), and others where both TH+/BrdU+ cells of the pair were separated by tens of microns but remained tethered by a thin string of cytoplasm (not shown). In addition some doublets appeared as adjacent distinct TH+ cells where one or both were negative for BrdU-immunoreactivity. In several experiments the proportion of double-labelled cells in the first two bins (i.e. singles and doubles) was higher in the hypoxic group and this is expected if mitotic activity was increased. We consider these results as preliminary evidence that hypoxia also stimulates mitosis in cultured glomus cells, though this point requires validation.

DISCUSSION

In this study we provide evidence that hypoxia stimulates DNA synthesis in glomus cells when grown in dispersed cell culture. This was indicated by increased incorporation

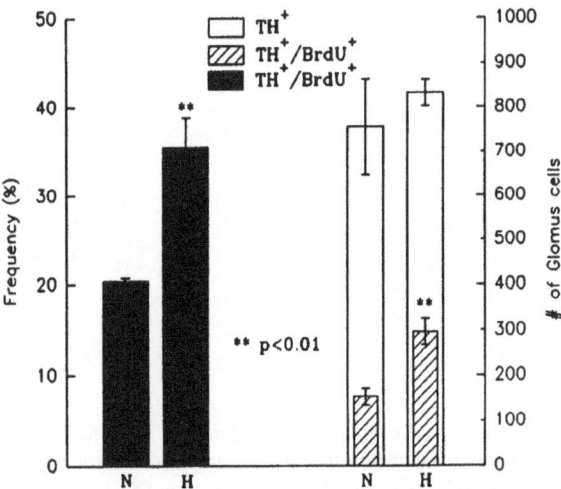

Fig.2. Effects of normoxia and hypoxia on BrdU uptake by TH$^+$ glomus cells in carotid body cultures. Right hand bins indicate the number of TH$^+$ and double labelled TH$^+$/BrdU$^+$ glomus cells grown under normoxic (N; 20% O$_2$) and hypoxic (H; 5% O$_2$ for 60 hr) conditions; bars represent mean ($\pm$ S.E.M.) of triplicate cultures. Left hand bins indicate the mean frequency ($\pm$ S.E.M.) of TH$^+$/BrdU$^+$ glomus cells in the total TH$^+$ population.

of the thymidine analog, BrdU, into TH+ glomus cells even after only ca. 60hr of hypoxia (5% O_2). Since this is a necessary (though not sufficient) condition for increased mitotic activity, our results suggest that the observed hyperplasia of glomus cells in the carotid body in vivo (Dhillon et al 1984; Bee et al 1986), following prolonged exposure of humans and animals to hypoxia is likely due to factors endogenous to the carotid body and does not require a functioning afferent pathway. Since factors present in serum may also stimulate DNA synthesis and mitosis in these cultured glomus cells, the effect of hypoxia alone may well be greater than that obtained in this study. We have recently succeeded in culturing glomus cells in chemically defined media for at least 5 days and this should allow us to address this point directly.

The simplest explanation for these findings is that the observed increase in DNA synthesis is mediated via a direct interaction between the hypoxic stimulus and the chemoreceptor glomus cells. Low PO_2 is known to suppress the outward K^+ current in glomus cells (Lopez-Barneo et al 1988; Stea and Nurse 1991a) as well as elicit a rise in intracellular calcium (Biscoe and Duchen 1990). These and /or other physiological responses that accompany hypoxia in the carotid body (e.g. changes in cAMP or pH) may well trigger the increase in DNA synthesis. However, since our cultures are not pure we cannot as yet exclude a role for the other surviving cell types including sustentacular cells, fibroblasts, endothelial cells. The attractive feature of the present system is that it should allow the underlying mechanisms responsible for the hypoxia-induced hyperplasia (and hypertrophy) of carotid body cells to be studied under controlled condidons.

ACKNOWLEDGEMENTS

This work was supported by a grant from the NIH ROl HL 43412.

REFERENCES

Bee D., Pallot D.J., Barer G.R. (1986) Division of type 1 and endothelial cells in the hypoxic rat carotid body. Acta anat. 126:226 229.

Biscoe T.J., Duchen M.R. (1990) Responses of type 1 cells dissociated fom the rabbit carotid body to hypoxia. J. Physiol. 428:39-59.

Dhillon D.P., Barer G.R., Walsh M. (1984) The enlarged carotid body of the chronically hypoxic and chronically hypoxic and hypercapnic rat: A morphometric analysis. Quart. J. Ept. Physiol. 69:301-317.

Doupe A.J., Patterson P.H., Landis S.C. (1985) Small intensely fluorescent cells in culture. Role of glucocorticoid and growth factors in their development and interconversions with other neural crest derivatives. J. Neurosci. 5:2143-2160.

Edwards C., Heath D., Harris P. (1971a) The carotid body in emphysema and left ventricular hypertrophy. J. Pathol. 104:1-13.

Edwards C., Heath D., Harris P., Castillo Y., Kruger H., Arias-Stella J. , (1971b) The carotid body in animals at high altitude. J. Pathol. 104:231-238.

Kobayashi S., (1971) Comparative cytological studies of the carotid body I. Demonstration of monoamine-storing cells by correlated chromaffin reaction and fluorescence histochemistry. <u>Arch. Histol. Jpn.</u> 33:319-339.

Lopez-Barneo J., Lopez-Lopez J.R., Urena J., Gonzalez C. (1988) Chemotransduction in the carotid body: K^+ current modulated by PO_2 in type 1 chemoreceptor cells. <u>Science</u> 241:580-582.

McDonald D.M., Blewett R.W. (1981) Location and size of carotid body-like organs (paraganglia) revealed in rats by the permeability of blood vessels to Evans blue dye. <u>J. Neurocytol.</u> 10:607-643.

Nurse C.A. (1990) Carbonic anhydrase and neuronal enzymes in cultured glomus cells of the carotid body of the rat. <u>Cell Tissue Res</u>, 261:65-71.

Stea A, Nurse CA (1991a) Whole-cell and perforated-patch recordings from O_2-sensitive rat carotid body cells grown in short and long-term cultures. <u>Pfluegers Arch.</u> 418:93-101.

Stea A., Nurse C.A. (1991b) Contrasting effects of Hepes vs HCO_3-buffered media on whole-cell currents in cultured chemoreceptors of the rat carotid body. <u>Neurosci. Lett.</u> 132:239-242.

13 LOCALIZATION OF DOPAMINE D2 RECEPTOR mRNA IN THE RABBIT CAROTID BODY AND PETROSAL GANGLION BY in situ HYBRIDIZATION

A. Schamel and A. Verna

Laboratoire de Cytologie, Université de Bordeaux II and URA CNRS
n° 339, Talence, France

INTRODUCTION

The carotid body (CB) of mammals is known to contain large amounts of catecholamines and it has been demonstrated that dopamine (DA) is predominant in this structure among several species (rat, rabbit, human ..., see Fidone et al., 1983). There is no doubt that DA is localized to the glomus (type I) cells but the role of DA in chemoreception processes is controversial. Both inhibitory and excitatory effects on the chemoafferent activity have been reported, depending on the dose, species and experimental technique used. Pharmacological studies with specific agonists and antagonists have established that DA-receptors exist in the CB (Zapata et al., 1983), some of them belonging to the D2 subtype (McQueen and Mir, 1984; Mir et al., 1984) which are not linked to adenylate cyclase activation (Kebabian and Calne, 1979). To better understand the role of DA in chemoreception, it appears to be essential to define the distribution and the precise localization of DA receptors in the CB. Until now, only indirect data have been obtained on this issue. Binding studies, using tritiated spiroperidol or domperidone (Dinger et al., 1981; Mir et al., 1984) have suggested that most of DA receptors in the CB are linked to the afferent innervation but the location of the other receptors remains unknown (glomus cells? blood vessels? sympathetic nerve fibers?). However, some data have shown that DA can modify the resting electrical parameters of glomus cells, suggesting that DA-receptors (autoreceptors) are present on these cells (Goldman and Eyzaguirre, 1984; Benot and Lopez-Barneo, 1990).

The recent cloning of the D2 dopamine receptor (D2R) cDNA (Bunzow et al., 1988) now makes it possible to identify and localize cells which express this gene by the in situ hybridization method (isH). We have therefore used this method in the present study to determine if glomus cells express the D2R gene. Since previous studies suggested that most DA receptors are linked to the CB sensory innervation (coming from the petrosal ganglion via the sinus nerve), we also used isH to identify petrosal neurons expressing the D2R gene. Using this method, we have been able to demonstrate the presence of D2R mRNA in glomus cells as well as in some petrosal neurons.

Neurobiology and Cell Physiology of Chemoreception
Edited by P.G. Data *et al.*, Plenum Press, New York, 1993

MATERIAL AND METHODS

New-Zealand female white rabbits weighing 2 to 3 kg were anaesthetized with sodium pentobarbital (30 mg/kg) via the ear vein. Fixative was then perfused through the common carotid artery for 5 min after which the ipsilateral carotid bifurcation and petrosal ganglion were removed and dissected out.

Tissue processing

Cryostat sections. Formaldehyde 1% in phosphate buffer was used as fixative for the initial perfusion. After dissection, tissues were immersed in the same fixative for 1 h, cryoprotected (30% sucrose, overnight) and frozen in isopentane cooled with liquid nitrogen. Cryostat sections were made, mounted on gelatin-coated glass slides, dried and stored at - 80°C.

Vibratome sections. A mixture of 4% formaldehyde and 0.1% glutaraldehyde in phosphate buffer was used as fixative. After perfusion (5 min) and immersion (1 h) in this fixative, 100 µm-thick sections were made with a vibratome, washed in 4 x SSC (standard sodium citrate) buffer at 4°C and processed immediately.

Probe preparation and hybridization procedure

The D2R probe was a mixture of three synthetic oligodeoxynucleotides synthesized by the phophoramidite procedure (Le Moine et al., 1990). These oligonucleotides were chosen in regions of the D2R cDNA that had no homology with other sequences of the same family and were labelled by terminal deoxynucleotidyl-transferase with (^{35}S)thiodATP. Theses probes equally recognized the two D2R isoforms generated from the same gene by alternative splicing (Dal Toso et al., 1989; Giros et al., 1989). After labelling, the probes were precipitated with cold absolute ethanol-3 M sodium acetate, dried and resuspended in the hybridization buffer (final concentration: 3 pg/ml).

Cryostat sections were warmed to room temperature and 35 ml of probe containing buffer were deposited on the slides. The slides were incubated overnight at 40°C in airtight boxes containing moistened filter paper. After incubation, slides were washed with 4 x SSC, rinsed in decreasing concentrations of SSC and dehydrated. The sections were then coated with Ilford K5 emulsion, left in the dark for 2-4 weeks at 4°C, developed and stained with toluidine blue.

Vibratome sections were incubated overnight under gentle agitation, at 40°C, in the hybridization solution containing the probes (5 mg/ml). At the end of the incubation period, sections were washed, dehydrated and embedded in araldite. After polymerization, semi-thin (2 µm) sections were made from the vibratome sections and processed for autoradiography, as above (exposure time: 2 months).

Quantification. The labelling intensity was estimated by measuring the silver grain density with a Biocom 200 image analyser enabling the quantification of the number of silver grains over a defined structure and a measure of the surface of that structure. Grain density of the background was measured over regions taken at random around the structures of interest (glomus cells and petrosal neurons).

Controls. Some sections were treated with hybridization solution containing no probe or unrelated probes (somatostatin, vasopressin). Others were hybridized in normal hybridization solution to which increasing amounts of related or unrelated unlabelled oligonucleotides were added (Le Moine et al., 1990). Brain sections were also performed at the level of the striatum and used as positive controls.

RESULTS

Carotid body

Cryostat sections showed, clearly, a positive labelling over glomus and sustentacular cell clusters. However, the poor resolution of the method did not permit unequivocal conclusions as to whether the labelling was restricted to glomus cells or concerned both types of cells. Furthermore, silver grain density of the background was relatively high, particularly over collagen bundles.

Semi-thin sections (made from vibratome sections) gave clearer pictures, due to a reduced background level. Weak labelling was observed over glomus cells (fig. 1). The silver grain density was variable but always 2 or 3 times higher than over the surrounding connective tissue. Sustentacular cells were not labelled. Silver grains over glomus cells seemed to be uniformly distributed over most of the cells. However, some cells showed a preferential localization of silver grains over the cytoplasm periphery and/or over the nucleus margin. The other cell types of the CB (connective, vascular, Schwann cells) were not labelled.

Control sections incubated in medium containing no probe or unrelated probes did not show any labelling. Adjacent sections treated with medium to which increasing concentrations of unlabelled probes were added showed that labelling disappeared for oligonucleotide concentrations 20 times higher than that of the labelled probe. On the other hand, the addition of unlabelled and unrelated oligonucleotides (concentration 20 times higher than that of the probe) did not modify the labelling of glomus cells.

Finally, positive controls were obtained using cryostat sections through the rabbit brain, at the level of the striatum. After the hybridization procedure, these sections were exposed on X-ray films (Kodak X-Omat) for 72 h. After development, an intense autoradiographic reaction was observed on the films, showing a similar distribution as that observed by Le Moine et al. (1990) on sections throughout the rat brain.

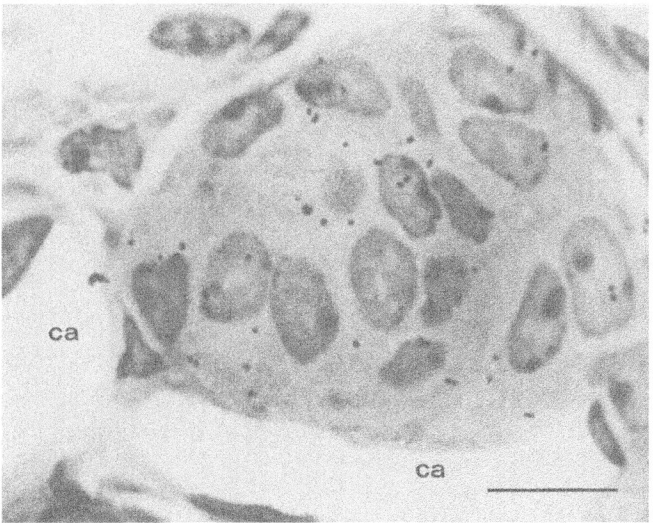

Fig. 1. Detection of D2R mRNA in the rabbit carotid body. Weak labelling is visible over a group of glomus cells. Ca: capillary; Bar: 5 μm.

Petrosal ganglion

Crysotat sections through the petrosal ganglion showed many labelled neurons (fig. 2). The silver grain density over these neurons was higher than that observed over glomus cells. Most labelled neurons were of relatively small size (20 to 30 μm) and were intensely stained by toluidine blue. These neurons seemed to be dispersed throughout the ganglion. In contrast, a few labelled neurons were of larger size (40 to 50 μm). Silver grains, over labelled neurons, were mostly associated with the cytoplasm, with no apparent local accumulation. Some neurons were visible on two consecutive adjacent sections: both sections showed similarly distributed silver grains. At last, many large neurons and a few small ones, were unlabelled.

Control sections gave similar results to those obtained with CBs and thus confirmed the specificity of labelling.

DISCUSSION

It has previously been shown by Dinger et al. (1981) that high affinity dopaminergic receptors are present in the rabbit CB. This result was obtained by binding experiments using ^{3}H-spiroperidol in normal and denervated CBs. Since deafferentation reduced specific binding by 64%, it was supposed that most DA receptors were associated with terminals of the sinus nerve. Dinger et al. (1981) were unable to determine the precise localization of the remaining receptors from their data, but suggested that these receptors could be associated with vascular elements, sympathetic nerve endings or the glomus cells themselves (autoreceptors).

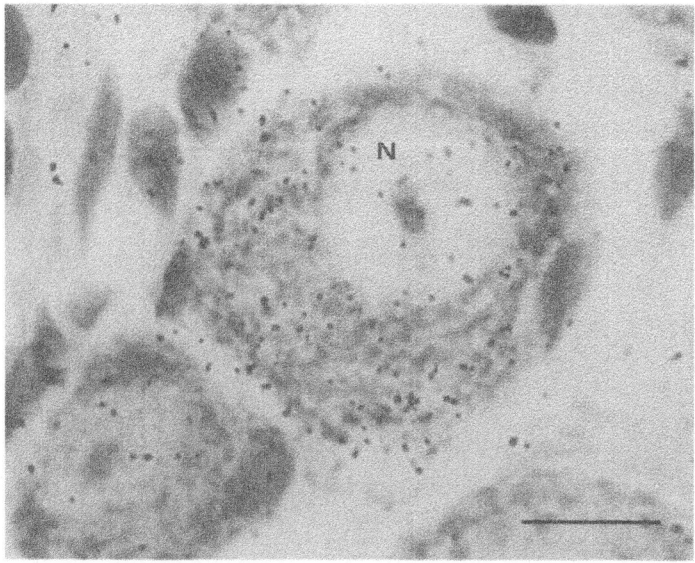

Fig. 2. Detection of D2R mRNA in the rabbit petrosal ganglion. A remarkable accumulation of silver grains is visible over a small-sized neuron. N: nucleus. Bar: 10 μm.

Using selective antagonists, Zapata et al. (1983) concluded from a combined pharmacological and electrophysiological study, that DA-induced chemosensory inhibition, in the cat, is mediated by D2R. This was confirmed by neuropharmacological and binding studies by Mir et al. (1984) who showed, using a specific ligand (^{3}H-domperidone), that D2R are present in the rabbit CB and that chronic de-afferentation results in a 32% decrease in specific ^{3}H-domperidone binding. Mir et al. (1984) concluded that the majority of D2 sites may be present on cells (type I or type II) rather on nerve terminals.

Our results after isH with D2R probes, are, therefore, consistent with the above suggestions concerning the presence of D2R in the CB and enable us to conclude that D2R mRNA is present in glomus cells but not in the other cell types.

Since the results obtained by Dinger et al. (1981) and Mir et al. (1984) suggested that part of the CB D2R are linked to the sinus nerve terminals, we expected to find petrosal neurons expressing the D2R gene and, indeed, we found many labelled petrosal neurons. However, our results must be considered as preliminary ones, since the neurons we observed were not identified as chemoafferent or baroafferent in function. It is well known that the petrosal ganglion contains both types of neurons and morphometric studies, in the cat, have shown that it is not possible to differentiate these two types by their size (Claps and Torrealba, 1988; Torrealba and Claps, 1988). The fact that the labelled neurons we observed were preferentially of small size is not sufficient, therefore, to conclude that these neurons were chemoafferent in function.

It has been shown by immunocytochemistry that primary afferent neurons in the dorsal root ganglia and petrosal ganglion (in the rat) contain tyrosine hydroxylase (TH). Since these neurons lacked dopamine β-hydroxylase, they were supposed to be dopaminergic (Price and Mudge, 1983; Katz et al., 1983). Retrograde tracing combined with immunocytochemistry revealed the CB as the target of TH-positive petrosal neurons (Katz and Black, 1986). If there are, in fact. dopaminergic sensory neurons in the petrosal ganglion, our observations showing the presence of D2R mRNA in petrosal cell bodies might be interpreted as indicating the possible presence of autoreceptors on these neurons, exactly as is the case in glomus cells. However, it has been recently established by Kummer et al. (1990) that TH-positive petrosal neurons (in the guinea-pig) lack immunoreactivity for DA as well as for DOPA decarboxylase and do not display glyoxylic acid-induced fluorescence. Kummer et al. (1990) concluded that the presence of DA in sensory neurons of the petrosal ganglion has to be questioned. In this context, the presence of D2R mRNA may not be related to autoreceptors but to receptors located at the level of terminals in contact with glomus cells and activated by DA released by these cells. However, further studies combining isH and retrograde labelling (for example) will be necessary to determine with certainty if the labelled neurons we observed here were actually chemoafferent or not.

ACKNOWLEDGEMENTS

The authors wish to gratefully acknowledge Pr. B. Bloch and his colleagues for their invaluable help during the course of this work and Dr. T. Durkin for his careful revision of the manuscript.

REFERENCES

Benot, A. R. and Lopez-Barneo J., 1990, Feedback inhibition of Ca^{2+} currents by dopamine in glomus cells of the carotid body. Eur. J. Neurosci.: 2, 809.

Bunzow J. R., Van Tol H.H.M., Grandy D.K., Albert P., Salon J., Christie M., Machida C.A., Neve K.A. and Civelli O., 1988. Cloning and expression of rat D2 dopamine receptor cDNA. Nature, 336: 783.

Claps A. and Torrealba F., 1988, The carotid body connections: a WGA-HRP study in the cat. Brain Res., 455: 123.

Dal Toso R., Sommer B., Ewart M., Herb A., Pritchett D. B., Bach A., Shivers B. D. and Seeburg P. H., 1989, The dopamine receptor: two molecular forms generated by alternative splicing. Eur. molec. Biol. Org. J., 8: 4025.

Dinger B., Gonzalez C., Yoshizaki K. and Fidone S., 1981. ^{3}H-spiroperidol binding in normal and denervated carotid bodies. Neurosci. Lett., 21: 51.

Fidone S., Stensaas L. and Zapata P., 1983, Sites of synthesis, storage, release and recognition of biogenic amines in carotid bodies. In: Physiology of the peripheral arterial chemoreceptors, H. Acker and R.G. O'Regan, eds., Elsevier, Amsterdam, pp 21-44.

Giros B., Sokoloff P., Martres M.P., Riou J.F., Emorine L.F. and Schwartz J.C., 1989, Alternative splicing directs the expression of two D2 dopamine receptor isoforms. Nature, 342: 923.

Goldman, W.F. and Eyzaguirre C., 1984, The effect of dopamine on glomus cell membranes in the rabbit. Brain Res., 321: 337.

Katz D. M. and Black I. B., 1986, Expression and regulation of catecholaminergic traits in primary sensory neurons: relationship to target innervation in vivo. J. Neurosci., 6: 983.

Katz D.M., Markey K.A., Goldstein M. and Black I.B., 1983, Expression of catecholaminergic characteristics by primary sensory neurons in the normal adult rat in vivo. Proc. Natl. Acad. Sci. USA. 80: 3526.

Kebabian J. W. and Calne D. B., 1979, Multiple receptors for dopamine. Nature. 277: 93.

Kummer W., Gibins I. L., Stefan P. and Kapoor V., 1990, Catecholamines and catecholamine-synthesizing enzymes in guinea-pig sensory ganglia. Cell Tissue Res., 261: 595.

Le Moine, C., Normand E., Guitteny A. F., Fouque B., Teoule R. and Bloch. B., 1990, Dopamine receptor gene expression by enkephalin neurons in rat forebrain. Proc. Natl. Acad. Sci. USA, 87: 230.

McQueen D. S. and Mir A. K., 1984, Changes in carotid body amine levels and effects of dopamine on respiration in rats treated neonatally with capsaicin. Br. J. Pharmacol., 83: 909.

Mir, A. K., McQueen D. S., Pallot D. J. and Nahorski S. R., 1984. Direct biochemical and neuropharmacological identification of dopamine D2-receptors in the rabbit carotid body. Brain Res., 291: 273.

Price J. and Muldge A. W., 1983, A subpopulation of rat dorsal root ganglion neurons is catecholaminergic. Nature, 301: 241.

Torrealba F. and Claps A., 1988, The carotid sinus connections: a WGA-HRP study in the cat. Brain Res., 455: 134.

Zapata P., Serani A. and Lavados M., 1983. Inhibition in carotid body chemoreceptors mediated by D-2 dopaminoceptors: antagonism by benzamides. Neurosci. Lett., 42: 179.

14 NORADRENERGIC GLOMUS CELLS IN THE CAROTID BODY: AN AUTORADIOGRAPHIC AND IMMUNOCYTOCHEMICAL STUDY IN THE RABBIT AND RAT

A. Verna[1], A. Schamel[1] and J.-M. Pequignot[2]

[1]: Laboratoire de Cytologie, Université de Bordeaux II and URA CNRS n° 339, Talence, France
[2]: Laboratoire de Physiologie A, Faculté de Médecine Grange-Blanche and URA CNRS n° 1195, Lyon, France

INTRODUCTION

It is well known that the rat and rabbit carotid body (CB) contains more dopamine (DA) than norepinephrine (NE) (Hellström and Koslow, 1975; Hansen and Christie, 1981; Mir et al., 1982; Leitner et al., 1986) and it has been supposed for a long time that the storage sites for DA and NE are the glomus cells and the postganglionnic sympathetic innervation, respectively. However, biochemical results have shown that sympathectomy does not induce a total disappearance of the CB NE content in the rabbit (Fidone and Gonzalez, 1982; Leitner et al., 1986) as well as in the rat (Pequignot et al., 1986). These results suggest that part of the CB NE is probably stored in the glomus cells. However, several possibilities must be considered: are there different kinds of glomus cells (i.e. dopaminergic/noradrenergic) or do each glomus cell contains a mixture of DA and NE? If so, is the DA/NE ratio variable from cell to cell or is it the same for all cells?

To obtain some answers to these questions, we used autoradiography afer ^{3}H-NE administration to know if some glomus cells are able to take up exogenous NE; we also used immunocytochemistry to localize dopamine β-hydroxylase (DBH) and NE, in order to know if some glomus cells are able to synthesize and store NE. A morphometric study was undertaken to precise some morphological features of the cells revealed by the above procedures. Lastly, the effects of chronic hypoxia on the number of NE-containing glomus cells were investigated in the rat.

MATERIAL AND METHODS

Autoradiography after ^{3}H-NE administration

This part of the study was carried out on New Zealand female white rabbits weighing 2 to 3 kg. Animals were anaesthetized with sodium pentobarbital (30 mg/kg). Fifty μCi of ^{3}H-NE (Saclay, 30 Ci/mM) were diluted in Krebs-Henseleit medium (K-H) to obtain a 10^{-7} M NE concentration. This medium was perfused through the common carotid artery for 30 min at a flow rate of 100 ml/min. Blood flow was then reestablished for 1 hour after which the CB was fixed by perfusion (2% glutaraldehyde, 0.5% formaldehyde in cacodylate buffer).

Neurobiology and Cell Physiology of Chemoreception
Edited by P.G. Data *et al.*, Plenum Press, New York, 1993

Tissue processing. After a 5 min perfusion, samples were fixed by immersion in the same fixative for 1 h at 4°C. The tissue was then post-fixed, dehydrated and embedded in araldite. Ultrathin sections (60 nm thick) were made, contrasted with uranyl acetate and lead citrate, coated with a thin layer of carbon, and with Ilford L 4 nuclear emulsion. Autoradiographs were exposed at 4°C for one month, developed in Kodak Microdol and observed in the electron microscope. Semi-thin sections were also made and coated with Ilford K 5 emulsion. After being exposed for 8-12 days at 4°C, the autoradiographs were developed in Kodak D19 and stained with toluidine blue before observation in the light microscope.

Quantitative study. Dense cored vesicles were studied on electron micrographs, at a final magnification of x 42500. Instead of measuring the external diameter of the vesicles, we measured the dense core diameters because they are less sensitive to fixation artefacts than the vesicle membrane (especially osmotic artefacts) and because the better contrast of dense cores makes it easier to use a computerized image analysis system (Biocom 200). The labelling intensity was evaluated by counting the number of silver grains per unit of cytoplasmic area (measured with the image analyser) on the micrographs.

Immunocytochemistry

Dopamine β-hydroxylase. Carotid bodies were perfused through the common carotid arteries with 4% paraformaldehyde and 0, 2% glutaraldehyde in phosphate buffer for 10 min, immersed in the same fixative for 3 h at 4°C, cryoprotected (sucrose 10% 1 h; 30%: overnight) and frozen. Lastly, 10-20 μm thick sections were made in a cryostat. Cryostat floating sections were incubated in DBH antiserum (rabbit polyclonal antibody against DBH from bovine adrenal medulla, 1/1000) and processed according to the PAP procedure. Some sections were then post-fixed with 1% osmium tetroxide, dehydrated in ethanols and embedded in araldite. Semi-thin and ultrathin sections were observed without any staining.

Norepinephrine . Carotid bifurcations were perfused with 50 ml of K-H medium and rapidly removed. CBs were frozen with isopentane precooled with liquid nitrogen, serial sections were cut in a cryostat, mounted on glass slides and immersed immediately in fixative (2.5% glutaraldehyde in cacodylate buffer).

Other carotid bifurcations were perfused first with 50 ml K-H medium and then with fixative containing 2.5% glutaraldehyde for 10 min. CBs were dissected out and immersed in the same fixative for 1h. Specimens were then rinsed with buffer and 50 μm thick sections were obtained with a vibratome.

Cryostat and vibratome sections were rinsed and treated with sodium borohydride (0.1 M) for 30 min to reduce double bounds between glutaraldehyde, NE and tissue proteins (Geffard et al., 1986). After several washes, sections were processed for immunocytochemistry using a glutaraldehyde-conjugated NE-antibody (polyclonal, 1/5000) and the PAP procedure. Vibratome sections were then post-fixed with 1% osmium tetroxide, dehydrated and embedded in araldite. Semi-thin and ultrathin sections were observed without any staining.

Controls. The procedures for raising antibodies and the specificity tests have been previously described by Geffard et al. (1986) and Tillet et al. (1987). Positive controls for DBH and NE consisted of rabbit adrenal gland. Negative controls consisted of omitting the primary or secondary antibody or replacing it by non-immune rabbit serum. In each case, all cellular immunoreactivity was abolished, although a faint background staining persisted, especially on collagen bundles.

Effect of chronic hypoxia

This part of the study was carried out on male Sprague-Dawley rats (about 200 g). Eight rats were kept for 15 days in a special chamber ventilated with an hypoxic gas mixture (10% O_2 in N_2) while the CO_2 concentration was maintained at the basal level of room air. Eight other rats were kept in the same room, breathing normal air. At the end of the experimental period, all animals were anaesthetized by an intraperitoneal injection of sodium pentobarbital (30 mg/kg) and perfused with fixative through the heart. Cryostat sections were made at the level of the carotid bifurcation and processed according to the above procedure for NE localization.

RESULTS

Autoradiography after ^{3}H-NE (rabbit CB)

Light microscopy. Semi-thin sections showed some immunolabelled type I glomus cells (Pl. 1, A) which were either isolated or mixed with unlabelled cells in a same cluster. The number of these cells was variable from one CB to another but always very low (about 0.5 to 1% of the total number of glomus cells). After simultaneous administration of ^{3}H-NE and desipramine, labelled glomus cells were no longer observed.

High resolution autoradiography. Most of labelled glomus cells differed from the unlabelled ones by the presence of larger DCVs, similar to those of the adrenal medulla. However, very small DCVs were also present (Pl. 1, B). The nucleus of some labelled cells showed a more irregular shape and a more electron dense chromatin than that of unlabelled cells. Many labelled cells were also characterized by a great number of glycogen particles.

The labelling intensity varied from cell to cell. A few weakly labelled glomus cells showed DCVs of medium size, as in unlabelled cells. Other weakly labelled glomus cells exhibited swollen DCVs of very large diameter (up to 400 nm), numerous lysosomes and/ or autophagic vacuoles. However, many intermediates were found between the typical unlabelled cells having DCVs of medium size and the few cells having very large DCVs.

Quantitative study. Dense core diameters, in labelled cells, were distributed according to a unimodal distribution showing a peak at about 115 µm. Compared to unlabelled cells, the major difference is the extension towards both large and small diameters in labelled cells (Fig. 1). The mean diameter of dense cores was significantly larger in labelled cells than in unlabelled ones: 127 ± 54 nm versus 113 ± 33 nm (mean $\pm$ SD; $P < 0.01$, Z-test for large samples).

The labelling intensity varied with the size of DCVs: it was maximum for cells having dense cores between 130 and 170 nm in mean diameter but decreased for cells with dense cores smaller than 130 nm or larger than 170 nm (the mean diameter of dense cores in 7 unlabelled cells taken at random was found to be distributed between 86 and 115 nm).

Immunocytochemical study (rabbit CB)

DBH-immunocytochemistry. A few DBH-immunoreactive glomus cells were observed with the light microscope (Pl. 1, C). Their number was variable from one CB to another but always low (less than 10 cells per 10 µm-thick section). The immunolabelling intensity was also very variable. At the ultrastructural level, immunopositive cells were characterized by very large DCVs (Pl. 1, D).

Norepinephrine immunocytochemistry. Strong immunoreaction was demonstrated in a few glomus cells (Pl. 1, E). Electron micrographs showed that labelled cells contained a high number of very large DCVs (Pl. 1, F). Immunostained cells were observed singly, or mixed with unreactive cells. The immunolabelling intensity varied from cell to cell. Cells showing different levels of immunostaining were occasionnally observed in a same cell cluster.

After intraperitoneal injection of dexamethasone (0.1 mg/kg/day for 1 week) the number of NE-immunopositive glomus cells was found to be increased, as compared to controls.

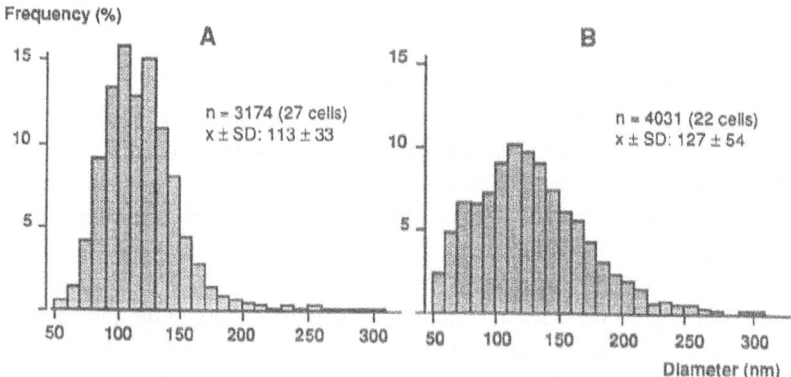

Fig. 1. Distribution of dense core diameters in unlabelled (A) and labelled (B) glomus cells after ^{3}H-NE administration in the rabbit.

Effects of chronic hypoxia (rat CB)

The CBs of hypoxic rats differed from controls by a larger size and the important development of blood vessels. The number of NE-immunopositive glomus cells in control rats was very low, as in the rabbit CB (0 to 5 cells per 10 μm-thick section). On the contrary, the number of NE-positive cells was dramatically increased in the CB of hypoxic rats (50 to 100 cells per section; Pl. 1, G, H).

Plate 1.
A and B: autoradiography after ^{3}H-NE (rabbit). A: semi-thin section showing labelled (arrows) and unlabelled (arrowheads) glomus cells. c: capillary; bar: 5 μm. B: electron micrograph showing labelled (left) and unlabelled glomus cells (right). Note that the size of the DCVs is larger and more variable in the labelled cell than in the unlabelled one. Bar: 1 μm.
C and D: DBH-immunocytochemistry (rabbit). C: semi-thin section (made through a cryostat section) showing immunopositive (arrows) and immunonegative (arrowheads) glomus cells. Bar: 5 μm. D: electron micrograph (through a cryostat section). Note that the immunopositive cell (right) contains DCVs larger than those of the immunonegative cell (left); N: nucleus (tangentially sectionned in the labelled cell). Bar: 1 μm.
E to H: NE-immunocytochemistry. E: cryostat section (rabbit CB) showing a small cluster of immunopositive glomus cells (arrows). Bar: 10 mm. F: electron micrograph (rabbit CB). Note that this NE-positive cell contains numerous large DCVs. N: nucleus. Bar: 2 μm.
G and H: effect of 2 weeks hypoxia in the rat. G: control CB. NE-immunopositive glomus cells are very few (arrows). Bar: 100 μm. H: hypoxic CB. Many cell clusters are made of NE-immunopositive glomus cells (arrows). Bar: 100 μm.

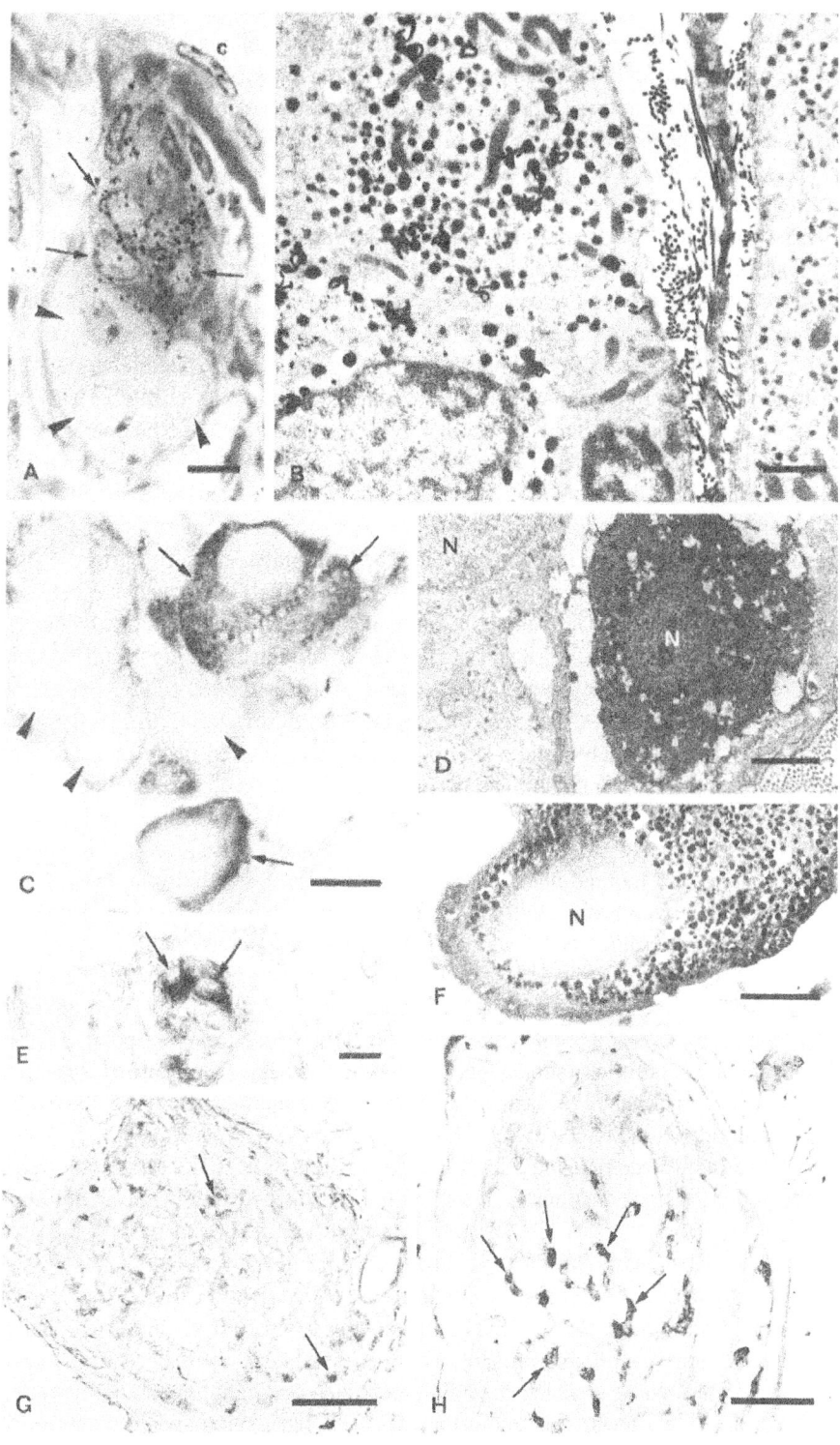

97

DISCUSSION

The results of the present study show that a few glomus cells, in the rabbit CB, differ from the other by the ability to take up exogenous NE and by the presence of larger dense-cored vesicles. Many of these large vesicles are similar to those of the adrenal medulla. This NE uptake phenomenon is specific since it occurs at low NE concentration (10^{-7} M) and is suppressed by desipramine. Since other glomus cells, also in small numbers, and also characterized by the presence of large dense cored vesicles, are immunopositive for DBH and for NE itself, it is tempting to speculate that there are, in the rabbit CB, two types of glomus cells: one type is rare and caracterized by the ability to take up, synthesize and store NE; the other type, more common, is represented by glomus cells having dense cored vesicles of medium size, and which do not synthesize NE. In other words, most of glomus cells in the rabbit are dopaminergic whereas a few ones are noradrenergic. This conclusion is consistent with the biochemical studies of Fidone and Gonzalez (1982) and Leitner et al. (1986) which have shown that a significant part of the NE content of the rabbit CB does not disappear after sympathectomy, and, therefore, is probably linked to glomus cells. In fact, the distinction of two types of glomus cells in the rabbit CB is too simple and our quantitative study shows that the problem is more complex. The distributions of dense core diameters in labelled/unlabelled cells, after [3]N-NE administration, are largely overlapping and it is possible to find all intermediates between cells with small (around 80 nm) and large (around 200 nm) mean dense core diameters. If we consider, now, the labelling intensity after ^{3}H-NE administration, it is again possible to find all intermediates between unlabelled and strongly labelled cells. It would be arbitrary, therefore, to distinguish only two types of cells. Similarly, our immunocytochemical studies, although not quantitative, showed many intermediates between negative and strongly immunopositive cells. It seems conceivable therefore, to suppose that noradrenergic glomus cells derive from dopaminergic ones by some unknown process, having both metabolic (NE uptake and synthesis ability; increased glycogen store) and morphologic consequences (increased variability in dense-cored vesicle size). This hypothesis would explain the different aspects of glomus cells in terms of functionnal states rather than in terms of different types of cells.

The immunocytochemical study showed that NE-storing glomus cells are few both in rabbit and in the rat CB. This is in agreement with the results of Chen et al. (1985) who found only a few DBH-immunoreactive glomus cells in the rat. From this point of view, the rat and rabbit CBs are very similar, but they differ from the cat CB since, in this species, NE is more abundant than DA (Armengaud et al., 1988) and most glomus cells exhibit DBH-immunoreactivity (Chen et al., 1985).

Furthermore, it seems that the dopaminergic/noradrenergic equilibrium is controlled by several factors since the number of NE-storing cells is increased by dexamethasone and by chronic hypoxia as well. These observations are in agreement with those of Hellström and Koslow (1976) who have shown that dexamethasone elicits an increase in the DA and NE content of the rat CB and whith those of Pequignot et al. (1987) who have demonstrated that chronic hypoxia leads to an increase in the rat CB NE (and DA) content.

The CB innervation also probably controls the DA/NE ratio since sinus nerve section increase, two weeks later, the NE content linked to glomus cells, in the rabbit (Leitner et al., 1986). It seems, therefore, that the NE content of glomus cells is controlled by humoral, environmental and nervous factors. However, some data suggest that it is more influenced by long-term than by acute changes.

What is the role, if any, of noradrenergic glomus cells? The effects of NE on the chemoafferent activity are controversial, probably because of species and experimental differences. However, the dominant effect of NE seems to be exitatory, possibly by an action on β-adrenergic receptors (Milsom and Sadig, 1983). Furthermore, Leitner et al. (1983) have shown that the CB sensitivity to hypoxia, which is abolished by reserpine treatment, can be restored not only by DA, but also by NE. At last, it has been demonstrated that nicotine evokes, in the rabbit, a preferential NE release from glomus cells (Gomez-Nino et al., 1990). These results suggest that NE from glomus cells could have some important function in the CB physiology but further studies will be necessary to specify this function.

REFERENCES

Armengaud, C., Leitner, L.-M., Malber, C.-H., Roumy, M., Ruckebusch, M. and Sutra, J. F., 1988, Comparison of the monoamine and catabolite content in the cat and rabbit carotid bodies. Neurosci. Lett., 85: 153.

Chen, I.-L., Hansen, J. and Yates, R. D., 1985, Dopamine-ß-hydroxylase-like immunoreactivity in the rat and cat carotid body; a light and electron microscopic study. J. Neurocytol., 14: 131.

Fidone S. and Gonzalez C.. 1982, Catecholamine synthesis in rabbit carotid body in vitro. J. Physiol.: 333, 69.

Geffard, M., Patel, S., Dulluc, J. and Rock, A.M., 1986, Specific detection of noradrenaline in the rat brain by using antibodies. Brain Res., 363: 395.

Gomez-Nino, A., Dinger, B., Gonzalez. C. and Fidone, S. J., 1990, Differential stimulus coupling to dopamine and norepinephrine stores in rabbit carotid body type I cells. Brain Res., 525: 160.

Hansen, J. T. and Christie. D. S., 1981, Rat carotid body catecholamines determined by high performance liquid chromatography with electrochemical detection. Life Sci., 29: 1791.

Hellström, S. and Koslow, S., 1975, Biogenic amines in rat carotid body of adultes and infant rats: a gas chromatographic-mass spectrometric assay. Acta Physiol. Scand., 93: 540.

Hellström, S. and Koslow, S., 1976, Effect of glucocorticoid treatment on catecholamine content and ultrastructure of adult rat carotid body. Brain Res.102,245.

Leitner, L.-M., Roumy, M. and Verna A., 1983, In vitro recording of chemoreceptor activity in catecholamine-depleted rabbit carotid bodies. Neuroscience, 10: 883.

Leitner, L.-M., Roumy, M., Ruckebusch, M. and Sutra, J. F., 1986, Monoamines and their catabolites in the rabbit carotid body; effects of reserpine, sympathectomy and carotid sinus nerve section. Pflügers Arch., 406: 552.

Milsom, W. K. and Sadig, T., 1983, Interaction between norepinephrine and hypoxia on carotid body chemoreception in rabbits. J. applied Physiol.: Resp. env. Ex. Physiol., 55: 1893.

Mir, A. K., Al-Neamy, K., Pallot, D. J. and Nahorski, S. R., 1982, Catecholamines in the carotid body of several mammalian species: effects of surgical and chemical sympathectomy. Brain Res., 252: 335.

Pequignot, J.-M., Cottet-Emard, J.-M., Dalmaz, Y., De Haut de Sigy, M. and Peyrin, L., 1986, Biochemical evidence for norepinephrine stores outside the sympathetic nerves in rat carotid body. Brain Res., 367: 238.

Pequignot, J.-M., Cottet-Emard, J.-M., Dalmaz, Y. and Peyrin, L., 1987, Dopamine and norepinephrine dynamics in rat carotid body during long-term hypoxia. J. Auton. Nerv. Syst., 21: 9.

Tillet, Y., Thibault, J. and Dubois, M. P., 1987, Immuno-cytochemical demonstration of the presence of catecholamine and serotonin neurons in the sheep olfactory bulb. Neurosci., 20: 1011.

15   THE MODULATION OF INTRACELLULAR pH IN CAROTID BODY GLOMUS CELLS BY EXTRACELLULAR pH AND pCO$_2$

K.J. Buckler, R.D. Vaughan-Jones, C. Peers,
D. Lagadic-Gossmann and P.C.G. Nye*

University Laboratory of Physiology
Parks Road, Oxford, OX1 3PT, England
*Department of Pharmacology, Worsley Medical and Dental Building
Leeds, LS2 9JT, England

INTRODUCTION

The effects of changes in blood pH upon afferent, carotid sinus nerve (CSN) discharge are well documented: a decreased pH$_o$ increases firing rate (Gray, 1968; Biscoe, Purves & Sampson; 1970). The mechanisms by which this is brought about, however, remain largely unknown. A number of observations suggest that the type -I, or glomus, cell is the primary site of pH chemoreception (McClodcey, 1968; Ridderstrale & Hanson, 1984; Rigual, Inequez, Carreres & Gonzalez, 1985; Rigual, Lopez-Lopez & Gonzalez, 1991; Rigual, Gonzalez, Gonzalez & Fidone, 1986; Lopez-Lopez et al., 1989; Peers 1990). One of the fundamental questions concerning pH chemoreception is whether these cells respond primarily to a change of extracellular pH or to a secondary change of intracellular pH. We have shown that type-I cells possess at least three pH$_i$-regulatory mechanism which serve, under conditions of constant pH$_o$, to maintain pH$_i$ at a fairly constant level of 7.2-7.3 in physiological media (Buckler, Vaughan-Jones, Peers & Nye, 1991a). This has prompted us to determine the extent to which pH$_i$ in the type-I cell is influenced by external pH and to consider whether changes of pH$_i$; are required for chemoreception of arterial pH.

METHODS

Type I cells were enzymatically isolated from the carotid bodies of 7-11 day old rat pups (Peers & O'Donnell, 1990). Only the most commonly found spherical, phase-bright cells of approximately 10 μm diameter, or clusters of such cells, were used in experiments.

HCO$_3$-buffered Tyrode contained nM NaCl 117, KCl 4.5, NaHCO$_3$ 23, MgCl$_2$ 1.0, CaCl$_2$ 2.5, glucose 11 and was equilibrated with 5% CO$_2$/95% air, pH at 37°C was 7.4 - 7.45. In other HCO$_3$ buffered Tyrode, NaHCO$_3$ was varied by substitution with NaCl.

Intracellular pH (pH$_i$) was measured using the dual emission fluoroprobe carboxy-SNARF-1. Cells were loaded with SNARF-1 by incubation in a 5μM solution of the acetoxymethyl ester (SNARF-1-AM) for 8 min. For further details regarding the use and calibration of this dye see Buckler & Vaughan-Jones 1990.

Neurobiology and Cell Physiology of Chemoreception
Edited by P.G. Data *et al.*, Plenum Press, New York, 1993

RESULTS

Fig.1A shows the response of pH_i in an isolated single type-I cell to simulated respiratory acidosis and alkalosis. The cells were initially superfused with Tyrode containing 23 mM HCO_3^- and equilibrated with 5% CO_2 in air ($pH_o = 7.4$). Upon changing to an identical Tyrode equilibrated with 10% CO_2 in air ($pH_o \sim 7.1$, simulated respiratory acidosis) there was a rapid decrease in pH_i to a new lower steady-state level. Upon return to the control solution equilibrated with 5% CO_2, pH_i rapidly returned to its previous control level. In response to a simulated respiratory alkalosis (same Tyrode equilibrated with 1% CO_2 in air, $pH_o \sim 8.1$) there was a rapid increase in pH_i to a higher steady-state level. This effect was also rapidly reversed upon return to 5% CO_2. These data clearly show that changes in pH_o produced by changing pCO_2, as in respiratory acidosis/alkalosis, greatly affect pH_i. The quantitative relationship between pH_i and pH_o is plotted in Fig. lD. It is remarkably steep (slope, 0.63) indicating that pH_i is particularly sensitive to changes of pH_o.

Fig. 1B shows the effect on pH_i of raising CO_2 from 5% to 10% but this time at a constant pH_o (achieved by simultaneously increasing $[HCO_3^-]$ in the Tyrode solution). There is a rapid initial acidosis associated with CO_2 entry into the cell. This fall of pH_i, however, is only transient; pH_i recovers to control levels. Conversely, decreasing CO_2 from 10% back to 5% produces only a transient intracellular alkalosis. Thus, in contrast to the dramatic effects of changes in pCO_2 at a constant $[HCO_3^-]_o$ (see above), steady-state pH_i was not significantly affected by isohydric changes of CO_2 (see also Buckler et al 1991b). These data suggest that steady-state pH_i is only affected by changes in pH_o and is independent of changes in pCO_2 per se. If so, then varying pH_o at a constant CO_2 should produce similar effects upon steady-state pH_i as change in CO_2 at constant .$[HCO_3^-]$.

Fig. 1C shows that lowering pH_o at constant CO_2, equivalent to a metabolic acidosis, causes a monotonic fall in pH_i and ,conversely, increasing pH_o, equivalent to a metabolic alkalosis, causes a monotonic rise in pH_i. These changes take much longer to reach a steady-state than during simulated respiratory acidosis or alkalosis (compare Fig 1A and 1C). The relationship between steady-state pH_i and pH_o for a simulated metabolic acidosis/ alkalosis is plotted in Fig. lD (broken line). Note that the relationship between pH_i and pH_o is again steep (slope 0.68) and almost identical to that found in respiratory acidosis/ alkalosis (Fig 1D solid line). Thus metabolic acidosis/alkalosis and simulated respiratory acidosis/alkalosis have near identical effects upon steady state pH_i.

DISCUSSION

Effects of pH_o on pH_i

The above data suggest that pH_o is the principal determinant of steady-state pH_i in the type-I carotid body cell, its effect being apparently independent of changes of pCO_2 and $[HCO_3^-]$. Thus raising pCO_2 only reduces steady-state pH_i if there is a simultaneous fall in pH_o. If pH_o is kept constant, then raising pCO_2 initially lowers pH_i, but the effect is only transient. A unit change in pH_o leads ultimately to a 0.6-0.7 unit change in steady-state pH_i, this change in pH_i is much greater than has been reported for other mammalian cell types (typically 0.2 - 0.35, Ellis & Thomas, 1976; Vaughan-Jones, 1986; Aickin & Thomas, 1977; Tolkovsky & Richards, 1987; Aickin, 1984;). Since a steady-state pH_i occurs only when any acidifying influences on the cell are balanced by acid extrusion, changes in pH_o could modify steady-state pH_i by altering acid efflux, intracellular acid

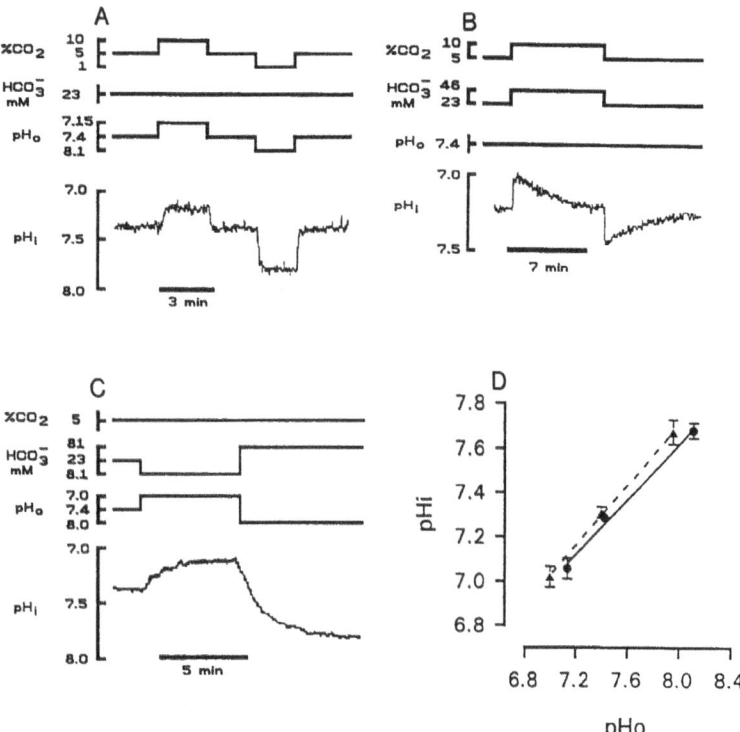

Fig.1. A) Effects of changing pCO_2 and pH_o at constant $[HCO_3^-]_o$ on pH_i in type-I cells. B) Effects of changing pCO_2 at constant pH_o (i.e isohydric change of pCO_2). Single cell, initially in a standard 5% CO_2, 23mM HCO_3^- buffered Tyrode (pH 7.4). Increasing pCO_2 at constant pH_o caused a rapid cellular acidification followed by recovery of pH_i back to its initial value. C) Effects of changes in pH_o caused by increasing or decreasing $[HCO_3^-]_o$ at constant CO_2 of 5%. Note that under these conditions, following a change in pH_o, pH_i takes longer to reach its new steady state than during a simulated respiratory acidosis, compare Fig. 1a. D) Effects of changing pH_o on pH_i: data are means $\pm$ standard error, line of best fit determined by linear regression. Circles: effects on steady-state pH_i of changing pH_o by altering pCO_2 at constant $[HCO_3^-]_o$: line of best fit (solid line) pH_i = $0.63.pH_o$ + 2.59, regression coefficient = 0.95, n=7. Triangles: effects on steady-state pH_i of changes in pH_o at constant CO_2: line of best fit (broken line) pH_i = $0.68.pH_o$ + 2.25, regression coefficient = 0.95, n=5.

production, acid influx or a combination of all three. Acid efflux has been shown to be modulated by changes of pH_o in other tissues. In the cardiac Purkinje fibre for instance the pH_i versus pH_o relationship in nominally CO_2-free solutions is determined by the inhibitory effect of extracellular H^+ ions upon Na-H exchange (Vaughan-Jones & Wu, 1990); this effect generates a roughly linear pH_i versus pH_o relationship with a slope of about $0.3\ pH_i/pH_o$. Since the pH_i versus pH_o relationship in the type I cell is much steeper than 0.3, this might reflect a greater inhibitory effect of low pH_o on Na-H exchange and/ or upon the Na-HCO_3, dependent acid extrusion mechanism. Low pH_o might also accelerate HCO_3^- efflux from the type-I cell. In an open CO_2-system, this would be formally equivalent to an increased acid equivalent influx. Accelerated HCO_3^- efflux could occur either via Cl-HCO_3 exchange which, even at normal pH_o, seems to mediate a small HCO_3^- efflux (Buckler et al. 1991a) or it could occur via HCO_3^- ions moving passively out of the cell through anion channels (cf. Stea & Nurse, 1989). Alternatively there may be other, as yet unidentified, pH_o sensitive acid influx mechanisms.

Does pH_i Play a Role in Chemotransduction?

The question of whether the first stage of pH chemoreception by the carotid body occurs at an extracellular or intracellular site has been the subject of much conjecture. It is instructive to compare the effects of changes in pH_o and CO_2 upon pH_i with their effects upon CSN discharge. We note that the steady state response of CSN discharge-rate is reported to be a unique function of extracellular pH, irrespective of how the latter is changed (Hornbein & Roos, 1963; Gray, 1968; Hornbein, 1968). We have now shown that the steady state pH_i in the type-1 cell also appears to be a unique function of pH_o. The speed of the response of CSN discharge to a respiratory acidosis is however faster than that to a metabolic acidosis (Gray 1968); the data we have presented illustrates that this is true of changes in pH_i also. Rapid CO_2-induced changes in pH_i are to be expected since type-1 cells are small (8-11 μm) and contain carbonic anhydrase (Ridderstrale & Hanson 1984, Buckler et al. 1991a). Where pCO_2 is changed **at constant pH_o** however changes in both pH_i and CSN discharge are transient (again cf Gray 1968). The response of CSN discharge to changes in pH_o and CO_2 therefore follows a pattern which more closely matches changes in pH_i than changes in CO_2, pH_o or $[HCO_3^-]_o$ alone. Our results therefore argue strongly for a central role of type-1 cell pH_i in the chemotransduction process for both respiratory and metabolic acidosis. This is supported further by a recent report that reducing pH_i in the rabbit carotid body promotes catecholamine release (Rigual et al 1991). It is possible, therefore, that the steep pH_i/pH_o relationship in the type-1 cell is of functional advantage, since it maximises the pH_i-change in response to an extracellular acid/base challenge, thus maximising the chemoreceptor's H^+_o-sensitivity .

The Next Step in Chemotransduction

The data presented here provide no direct information on the step in the process of pH-chemotransduction following a change of pH_i. Since changes in both blood pO_2 and pH modulate discharge of the same nerve fibre, chemotransduction of both variables must converge at some point. There is compelling evidence for a role of intracellular calcium in the chemotransduction of hypoxic stimuli in the type-I cell (Biscoe & Duchen, 1990) and indirect evidence for a role of intracellular calcium in "acid" chemoreception (Rigual et al., 1991). Recent research in our laboratory indicates that the same changes in pH_o and CO_2 that we have demonstrated here to lower pH_i, also increase $[Ca^{2+}]_i$ (Buckler & Vaughan-Jones 1992). A number of mechanisms could be responsible for such an elevation of

[Ca^{2+}]$_i$. Rocher et al., (1991) have proposed that a link between low pH$_i$ and elevated Ca^{2+}$_i$ in the rabbit carotid body is provided by stimulation of Na$_o$-dependent acid extrusion. This, it is proposed, will raise Na^+_i and hence, via transmembrane Na-Ca exchange, will elevate Ca^{2+}$_i$. However this model requires that Na-dependent acid extrusion be stimulated by low pH$_i$ when pH$_o$ is <u>also</u> low, a condition known to <u>inhibit</u> extrusion in other cells (e.g. Aronson, 1985; Vaughan-Jones & Wu, 1990). Alternatively, since Biscoe and Duchen (1990) have shown that type-1 cells contain releasable stores of Ca^{2+}, it is conceivable that low pH$_i$ could promote Ca^{2+} release or retard its re-uptake into these stores. Finally, low pH$_i$ might activate Ca^{2+}-influx via voltage-gated channels through the inhibition of a K$^+$ conductance and consequent membrane depolarization (Peers, 1990; Peers & Green, 1991).

ACKNOWLEDGEMENTS

This work was supported by the Wellcome trust (R.V-J, P.C.G.N, K.J.B & C.P) the British Heart Foundation (R.V-J, K.J.B) & the French Foreign Office (D.L-G).

REFERENCES

Aickin,C.C.,1984,Direct measurement of intracellular pH and buffering power in smooth muscle cells of Guinea - pig vas deferens. J.Physiol., 349:571.

Aickin, C.C. & Thomas, R.C., 1977, Microelectrode measurement of the intracellular pH and buffering power of mouse soleus muscle fibres. J. Physiol., 267:571.

Aronson, P.S., 1985, Kinetic properties of the plasma membrane Na$^+$-H$^+$ exchanger. Ann. Rev. Physiol., 47:545.

Biscoe, T.J. & Duchen, M.R., 1990, Responses of type I cells dissociated from the rabbit carotid body to hypoxia. J. Physiol., 428:39.

Biscoe, T.J., Purves, M.J. & Sampson, S.R., 1970, The frequency of nerve impulses in single carotid body chemoreceptor afferent fibres recorded in vivo with intact circulation. J. Physiol., 208:121.

Buckler, K.J. & Vaughan-Jones, R.D., 1990, Application of a new pH-sensitive fluoroprobe (carboxyl-SNARF-I) for intracellular pH-measurement in small isolated cells. Pflugers Archive., 417:234.

Buckler, K.J., Vaughan-Jones, R.D., Peers, C. & Nye, P.C.G., 1991a, Intracellular pH and its regulation in isolated type I carotid-body cells of the neonatal rat. J. Physiol., 436:107.

Buckler, K.J., Vaughan-Jones, R.D., Peers, C., Lagadic-Gossmann, D. & Nye, P.C.G., 1991b, Effects of extracellular pH, pCO$_2$ and HCO$_3^-$ on intracellular pH in isolated type I cells of the neonatal rat carotid body. J. Physiol., 444:703.

Buckler, K.J. & Vaughan-Jones, R.D., 1992, Effects of acidic stimuli on intracellular calcium in rat neonatal carotid body glomus cells. (Manuscript in preparation).

Ellis, D. & Thomas, R.C., 1976, Microelectrode measurement of the intracellular pH of mammalian heart cells. Nature, 262:224.

Gray, B.A., 1968, Responses of the perfused carotid body to changes in pH and pCO$_2$. Resp. Physiol., 4:229.

Hornbein, T.F. & Roos, A., 1963, Specificity of H$^+$ ion concentration as a carotid chemoreceptor stimulus. J. Appl. Physiol., 18:580.

Hornbein, T.F., 1968, The relation between stimulus to chemoreceptors and their response. In "Arterial chemoreceptors," R.W. Torrance, Ed., Blackwell, Oxford.

Lopez-Lopez, J., Gonzalez, C., Urena, J. & Lopez-Barneo, J., 1989, Low PO2 selectivity inhibiytis K channel activity in chemoreceptor cells of the mammalian carotid body. J. Ge. Physiol., 93:1001-1015.

McCloskey, D.I., 1968, In :"Arterial chemoreceptors," R.W. Torrance, Ed., Blackwell, Oxford: pp 279-296.

Peers, C., 1990, Selective effect of lowered extracellular pH on Ca^{2+}-dependent K$^+$ currents in type I cells isolated from the neonatal rat carotid body. J. Physiol., 422:381.

Peers, C. & Green, F.K., 1991, Inhibition of Ca^{2+}-activated K$^+$ currents by intracellular acidosis in isolated type-I cells of the neonatal rat carotid body. J. Physiol., 437:589.

Peers, C. & O'Donnell, J., 1990, Potassium currents recorded in type-I carotid body cells from the neonatal rat and their modulation by chemoexcitatory agents. Brain Res., 522:259.

Ridderstrale, Y. & Hanson, M.A., 1984, Histochemical localization of carbonic anhydrase in the cat carotid body. Ann. N. Y. Acad. Sci., 429:398.

Rigual, R., Gonzalez, E., Gonzalez, C. & Fidone,S., 1986, Synthesis and release of catecholamines by the cat carotid body in vitro: effects of hypoxic stimulation. Brain Res., 374:101.

Rigual, R., Iniguez, C., Carreres, J. & Gonzalez, C., 1985, Carbonic anhydrase in the carotid body and the carotid sinus nerve. Histochem., 82:577.

Rigual, R., Lopez-Lopez, J.R. & Gonzalez, C., 1991, Release of dopamine and chemoreceptor discharge induced by low pH and high pCO$_2$ stimulation of the cat carotid body. J. Physiol., 433:519.

Rocher, A., Obeso, A., Gonzalez, C. & Herreros, B., 1991, Ionic mechanisms for the transduction of acidic stimuli in rabbit carotid body glomus cells. J. Physiol., 433:533.

Stea, A. & Nurse, C.A., 1989, Chloride channels in cultured glomus cells of the rat carotid body. Am. J. Physiol., 257:C174.

Tolkovsky, A.M. & Richards, C.D., 1987, Na^+/H^+ Exchange is the major mechanism of pH regulation in cultured sympathetic neurones: measurements in single cell bodies and neurites using a fluorescent pH indicator. Neurosci., 22:1093.

Vaughan-Jones, R.D., 1986, An investigation of chloride-bicarbonate exchange in the sheep cardiac Purkinje fibre. J. Physiol., 379:377.

Vaughan-Jones, R.D. & Wu, Mei-Lin., 1990, Extracellular H^+ inactivation of $Na^+ -H^+$ exchange in the sheep cardiac Purkinje fibre. J. Physiol., 428:441.

16 EVIDENCE FOR GLUCOSE UPTAKE IN THE RABBIT CAROTID BODY

M.A. Delpiano

Max-Planck-Institut für Systemphysiologie
Rheinlanddamm 201, 4600 Dortmund 1, Germany

INTRODUCTION

The carotid body located in the bifurcation of the carotid arteries is able to detect gas changes in blood composition (PO_2, PCO_2/pH) and to transduce them into afferent nerve signal. The intimate mechanism responsible for this chemoreceptive response to PO_2 changes is not yet well understood. Various reports point to a chain of events initiated by disturbance of cell metabolism followed by transmitter release and depolarization of nerve endings (Mills and Jöbsis, 1972; Hayashida et al., 1980). It seems, however, that glycolysis also plays an essential role in chemoreception, because lowering PO_2 stimulates glycolysis in the cat carotid body in vitro (Delpiano and Acker, 1985). The aim of this article is to present evidence that this may not be the case as inhibition of glucose utilization by substrate deprivation or by uptake blocks almost totally glycolytic-dependent pH decrease but only partially chemoreception.

MATERIAL AND METHODS

Carotid bodies excised from anesthetized rabbits were superfused in a leucite chamber with a modified Locke's solution (in mM): NaCl 128, KCl 5.6, $CaCl_2$ 2.1, D-glucose 5.5, $NaHCO_3$ 10, HEPES 10 adjusted with sucrose to 300 mosml; constant pH (7.4) and PCO_2 (23 Torr). Solution was equilibrated with different gas mixtures (PO_2: 280 or 20 Torr) using two gas mixing pumps (Wösthoff, Germany). Chemosensory nerve activity was recorded from cut sinus nerve with two platinum electrodes lifted into paraffin oil as previously reported (Delpiano, 1985a). The extracellular pH (pHe) was recorded with pH sensitive microelectrodes. Briefly, double-barrelled pH sensitive electrodes with a diameter of about 2-3 mm consisting of a DC channel filled with Mg^{2+}-acetate (1 mol/l) and a pH channel filled with a proton cocktail (82500 Fluka, Germany) were manufactured similarly as described for ion-selective electrodes (Delpiano and Acker, 1989). Electrodes were connected via silver chlorided wires to a high impedance FET voltage follower (10^{15} Ohm) bound to a differential preamplifier which substracted DC signal electronically, so that voltage signal from pH channel was proportional to the $-\log_{10}$ of the proton concentration measured in the tissue. Calibration with different phosphate buffers was done before and after each experiment.

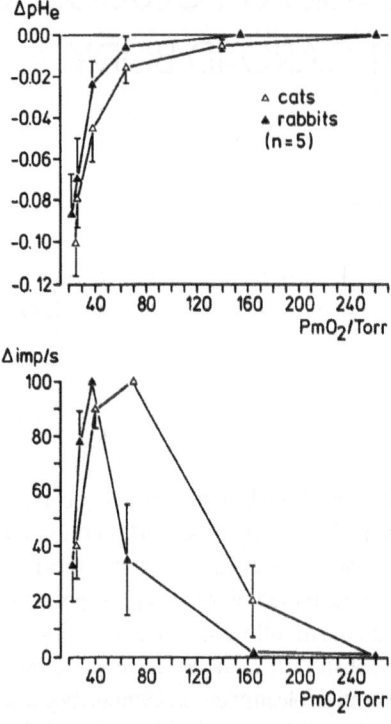

Fig. 1. Compared pH and nerve changes to hypoxia in cat and rabbit carotid bodies. Ordinate: units of pH difference (upper panel) and nerve discharge compared to control (lower panel). Abscissa: oxygen partial pressure in the superfusion medium.

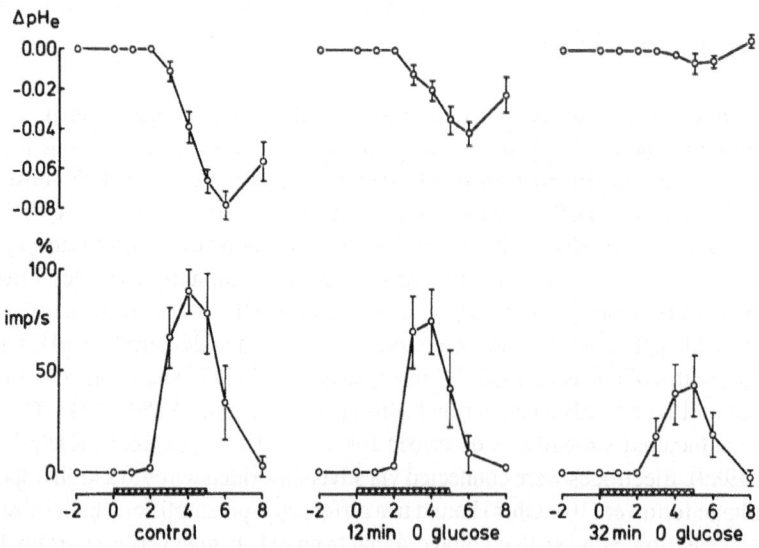

Fig. 2. Changes in extracellular pH (upper panel) and chemosensory nerve discharge (lower pannel) to hypoxia in glucose-free superfusion. Ordinate: units of pH difference for extracellular pH (pHe) (upper) and % of nerve discharge (lower) obtained from control values. Abscissa: time period during control and after 12 and 32 minutes superfusion without glucose in medium.

Locke's solutions with different substances were freshly prepared before each experiment. Chemicals used were obtained from Merck (Darmstadt, Germany) and were of analytical purity. Pyruvate (Na^+ salt) from Boehring (Mannheim, Germany), Phlorizin and Phloritin from Sigma (St. Louis, USA).

RESULTS AND DISCUSSION

In a previous paper Delpiano and Acker (1985a) reported that in the cat carotid body in vitro hypoxia stimulated tissue acidification. Since extracellular pH decrease and sensory nerve response were suppressed by different inhibitors of glycolysis (Delpiano and Acker, 1985a, b), it was assumed that hypoxia stimulated the glycolytic pathway probably by activating the enzyme lactate dehydrogenase (LDH). Therefore hypoxia could stimulate pH decrease by glycolytic-increased lactate production. To prove whether activation of glycolysis also occurs in carotid bodies of other animals and whether it is necessary for chemoreception, tissue pH (pH_e) and nerve excitation were recorded in rabbit carotid bodies. A shown in Fig. 1, in a comparative diagram, lowering PO_2 from about 260 to 24 Torr both pH and nerve response change in rabbit carotid bodies in the same mode as in cats. In these experiments it seems however, that pH and nerve discharge start in rabbits at lower PO_2 values. Independent of this apparent difference, that was not further analysed, in rabbits also inhibitors of glycolysis abolished pH changes and nerve discharge (not shown). Fig. 2 illustrates that, in glucose-free superfusion, pH and nerve response decrease in a time dependent manner but with different time courses. At 12 and 32 min after zero glucose, pH changes have been attenuated by about 62 and 94% respectively, compared to maximal pH amplitude during control. In contrast, nerve discharge was attenuated only 16 and 46% at the same timespan (Fig. 2). The persistence of the chemosensory discharge until pH changes have been almost abolished points to the question as whether activation of glycolysis is a prerequisite for or a consequence of the PO_2-sensing mechanism. From earlier pH measurements on cats (Delpiano and Acker, 1985a; Delpiano, 1987) it was shown that after long-time superfusion with zero glucose or 2-deoxyglucose, when both responses were markedly attenuated, addition of pyruvate into the medium restored nerve discharge to hypoxia but not pH. Such a restoring effect of pyruvate illustrated in Fig. 3 was inhibited competitively by malonate (Delpiano, 1987), which as known interferes pyruvate oxydation in the Krebs cycle by inhibiting succinate dehydrogenase (Webb, 1966). When pyruvate was removed from the medium the sensory nerve response exhausted faster compared to glucose deprivation alone (Fig. 3, diamond and circles). Working with 2-deoxyglucose as a transient chemostimulant Obeso et al. (1986) reported similar findings concerning substrate deprivation and pyruvate. We also noticed that both pH and sensory discharge to hypoxia were not abruptly abolished in glucose-free media or 2-deoxyglucose (Delpiano and Acker, 1985a). To elucidate if this delayed suppression, as shown in Fig. 2, was due to stored glucose supplied from glycogen, as assumed by Obeso et al., (1986) or simply due to rest glucose in bath (impaired glucose uptake), two inhibitors of glucose transport were used in presence of glucose: phlorizin and phloretin. As illustrated in Fig. 4, these inhibitors block both pH and nerve response to hypoxia. in Fig. 4A and B it can be seen, that phloretin is more effective than phlorizin to block these changes to hypoxia. Looking at the concentration for half-maximal inhibition phloretin K_{50} is about 10 mM and 30 mM for pH and nerve response respectively (Fig. 4A, B). This means that when pH is 50% inhibited nerve discharge to hypoxia is only 10%, i.e. glucose transport inhibition dissociates both responses. Phlorizin K_{50} was higher than 200 mM. These inhibitory effects occurred within few minutes when applied into the bath (not shown).

From these experiments it can be concluded that the delayed reduction of both pH and nerve response in glucose-free media is more probably due to impaired glucose uptake and not to stored glucose. Additionally, this uptake may be mainly carried out by a Na^+-independent mechanism since phloretin contrary to phlorizin does not bind to the Na^+/glucose cotransporter (Cheung and Hammerman, 1988; Toggenburger et al., 1982). If carotid body cells possess limited substrate storage, as concluded also by Spergel (this Symposium) and sensory response to hypoxia is blocked by phloretin, then glucose uptake, as substrate supply, must be essential for chemoreception. From these findings

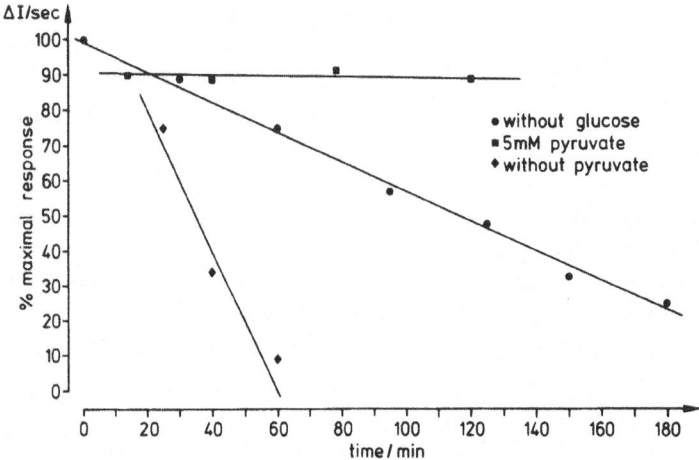

Fig. 3. % of maximal response of nerve discharge to hypoxia during superfusion with zero glucose (circles), after addition of pyruvate by 180 min (square) and after removal of pyruvate from superfusion (diamond).

and previously reported (Delpiano and Acker, 1985a, b; Delpiano, 1987) the main conclusion is that: (1) glycolysis seems to be a consequence rather than a prerequisite for the PO_2-sensing mechanism, (2) Both, glucose transport and utilization are necessary for expression of carotid body O_2-chemoreception, i.e. without energy supply chemoreception gradually vanishes (see Spengel, this Symposium), (3) Glucose may be transported into carotid body cells by a Na^+-independent mechanism, and (4) Tissue acidification related to carbohydrates metabolism seems to be a common metabolic response in the carotid body of other species.

Concerning glucose consumption, Obeso et al. (1989) also reported increased consumption stimulated by hypoxia in rabbit carotid bodies. The question remains to be elucidated, which cellular elements of the carotid body determine extracellular pH decrease, inasmuch as intracellular pH in type I cells does not change markedly during hypoxia (Rumsey et al., this Symposium).

Hayashida, Y., Koyano, H., and Eyzaguirre, C., 1980, An intracellular study of chemosensory fibers and endings, J. Neurophysiol., 44:1077.

Mills, E., and Jöbsis, F. F., 1972, Mitochondrial respiration chain of carotid body and chemoreceptor response to changes in oxygen tension, J. Neurophysiol., 35:405.

Obeso, A., Almaraz, L., and Gonzalez, C., 1986, Effects of 2-deoxy-D-glucose on in vitro cat carotid body, Brain Res., 371:25.

Obeso, A., Gonzalez, C., Dinger, B., and Fidone, S, 1989, Metabolic activation of carotid body glomus cells by hypoxia, J. Appl. Physiol., 67:484.

Rumsey, W. L., Iturriaga, R., Spergel, D., Lahiri, S., and Wilson, D. F., 1992, Intracellular pH and chemoreception in the isolated perfused and superfused cat carotid body (this Symposium).

Spergel, D., 1992, Glutamate as a metabolic substrate in O_2 chemoreception in the cat carotid body (this Symposium).

Toggenburger, G., Kessler, M., and Semenza, G., 1982, Phlorizin as a probe of the small-intestinal Na^+, D-glucose cotransporter, Biochim. Biophys. Acta, 688:557.

Webb, J. L., 1966, "Enzyme and Metabolic Inhibitors, Vol. III, Academic Press, New York.

chemoreceptor was made (Lahiri and DeLaney 1975). For close intra-arterial injections a catheter tip was placed close to the carotid sinus region through the thyroid artery. $CdCl_2$ (0.2 - 0.4 mmoles), $NiCl_2$ (0.2 - 0.5 µmoles) and $MnCl_2$ (1.1 - 3.0 µmoles) were given by close intra-arterial injections and the chemosensory response was tested. The pH of these solutions were measured and blank solutions were prepared matching the pH of the

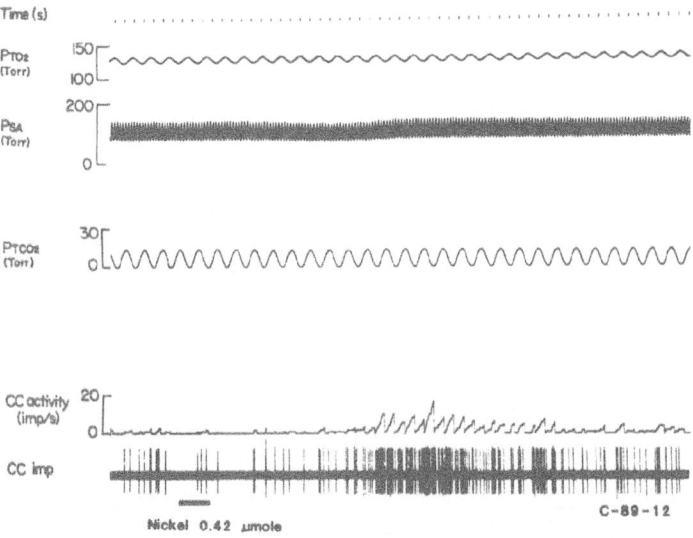

Fig. 1 Effect of Nickel on carotid chemoreceptor activity. Traces from the top are: time (s), tracheal PO_2 (PTO $_2$), arterial blood pressure (PSA), tracheal PCO_2 (PTCO $_2$) carotid chemoreceptor activity (impulse/sec) and impulses.

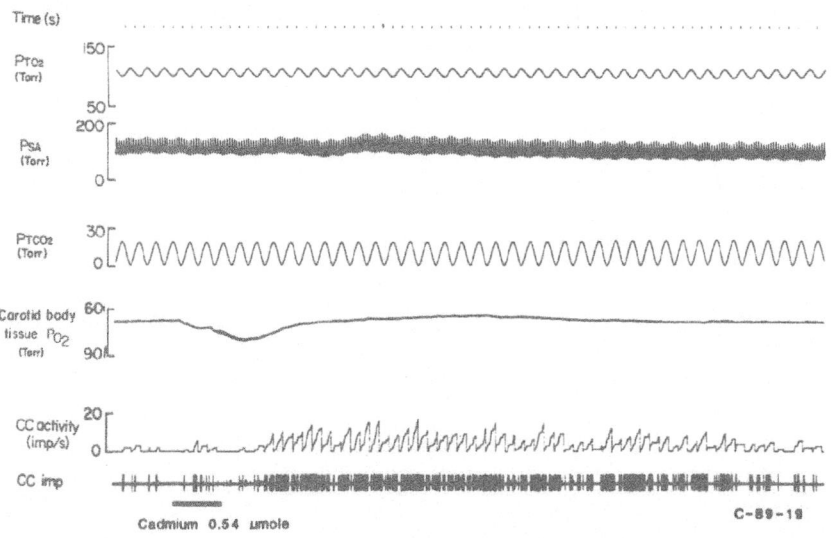

Fig. 2 Effect of Cadmium on carotid chemoreceptor activity. Traces from the top are: time (s), tracheal PO_2 (PTO $_2$), arterial blood pressure (PSA), tracheal PCO_2 (PTCO $_2$) carotid tissue PO_2, chemoreceptor activity (impulse/sec) and impulses. The increase in the carotid chemoreceptor activity was not due to any decrease in the tissue PO_2.

118

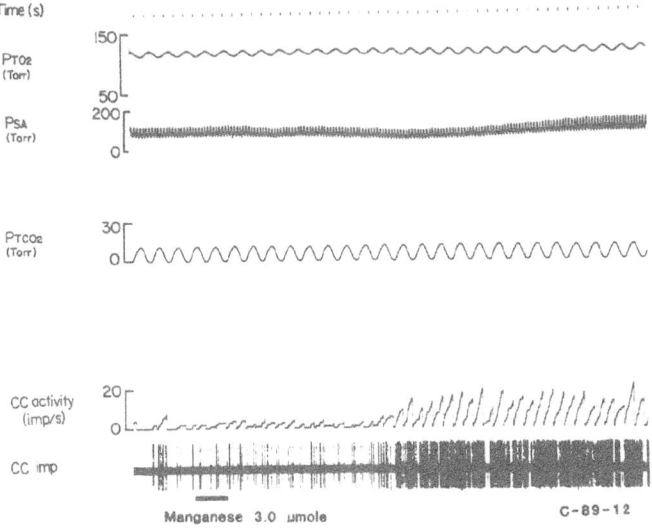

Fig.3 Effect of Manganese on carotid chemoreceptor activity. Traces from the top are the same as in Fig.1.

test solutions. Hypoxic response before and after the injections was also tested. To examine whether the effects mediated by the changes in the carotid body tissue PO_2, we also measured it in some experiments with an O_2 sensitive microelectrode (Di Giulio et al. 1990).

RESULTS

Fig.1 shows the chemosensory response to a close intra-arterial injection of Ni^{2+} during normoxia. The initial small decrease in the activity is followed by a significant stimulatory response lasting up to 20sec. The initial decrease in the chemosensory activity was also observed during the blank saline injection.

Fig.2 shows the effects of Cd^{2+} and chemosensory activity. Tissue PO_2 increased transiently following the close intra-arterial injection and the return to the control. Blank saline showed a similar effect on the PO_2 trace. This PO_2 increase presumably suppressed and delayed the development of the response. The increased activity lasted for many seconds. Similar results were found during hypoxia, indicating that the Cd^{2+} did not block hypoxic response .

The effect of Mn^{2+} is shown in fig.3. The increased chemosensory response lasted for a much longer period.

DISCUSSION

The observation that Ni^{2+}, Cd^{2+} and Mn^{2+} stimulated carotid chemosensory activity is consistent with those of Co^{2+} which belong to the same family of transitional metals . These divalent cations are also known to be Ca^{2+} channel (voltage-gated) blockers

(Hagiwara, S. and K.Takahashi 1967) and have been used to test the hypothesis that blockade of the channels would also block chemosensory activity in parallel. The results however did not support the hypothesis, and indicated that Ca^{2+} entry is not required for the chemosensory activity, at least for the duration of the cation administration. The stimulation is however not explained by the blockade of the Ca^{2+} channels. According to the neurotransmitter hypothesis a rise of intracellular Ca^{2+} is expected for the stimulation. This could happen if blockade of normal pathway of Ca^{2+} influx triggered a release of Ca^{2+} from the intracellular stores. There are instances in the literature that certain cell surface receptors are stimulated by the trace metal ions and show a transient increase in intracellular Ca^{2+} (Dwyer et al. 1991). The chemosensory responses resemble those calcium transients, and it is not impossible that the same mechanisms exist in the carotid body. The hypothesis here is that these divalent cations trigger phospholipase C activity which hydrolyses membrane phosphatidylinositol, increasing intracellular IP_3 and stimulating intracellular Ca^{2+} release (Berridge, M.J. and R.F.Irvine 1989). IP_3 has been found to increase concomitantly in these cells. The hypothesis can be tested for glomus cells also.

ACKNOWLEDGEMENT

We want to express our sincere thanks to Dr.D.G.Buerk, who participated in some of the experiments for the carotid body tissue PO_2 measurements.
Supported in part by the NIH grants HL-19737 and HL-43413.

REFERENCES

Berridge, M.J. and R.F.Irvine (1989). Inositol phosphates and cell signalling (A Review). 341:197-205.

Biscoe, T.J. and M.R.Duchen (1990). Responses of type I cells dissociated from the rabbit carotid body to hypoxia. J.Physiol. Lond. 428: 39-59.

Delpiano, M.A. and H.Acker (1989). Hypoxic and hypercapnic responses of Ca^{2+} and K^+ in the cat carotid body in vitro. Brain Res. 482: 235-246.

Di Giulio, C., W-X,Huang, S.Lahiri, A.Mokashi and D.G.Buerk (1990). Cobalt stimulates carotid body chemoreceptors. J.Appl.Physiol. 68:1844-1851.

Dwyer, S.D., Y.Zhuang and J.B.Smith (1991). Calcium mobilization by cadmium on decreasing Na^+ or pH in coronary endothelial cells. Expt. Cell Res. 192:22-31.

Eyzaguirre, C. and P.Zapata (1965). Pharmacology of pH effects on carotid body chemoreceptors in vitro. J.Physiol. Lond. 195: 557-588.

Hagiwara, S. and K.Takahashi (1967). Surface density of calcium ion and calcium spikes in the barnacle muscle fiber membrane. J.Gen.Physiol. 50:583-601.

Fidone, S.,C.Gonzalez and K.Yoshikazi (1982). Effects of low oxygen on the release of dopamine from the rabbit carotid body in vitro. J.Physiol.Lond. 333:93-110.

Lahiri,S. and R.G.DeLaney (1975). Stimulus interaction in the response of carotid body chemoreceptor afferent fibers. Respir. Physiol. 24:249-266.

Sato,M., K.Ikeda, K.Yoshikazi and H.Koyano (1991). Response of cytosolic calcium to anoxia and cyanide in cultured glomus cells of newborn rabbit carotid body. Brain Res. 55:327-330.

Shirahata,M. and R.S.Fitzgerald (1991). Dependency of hypoxic chemotransduction in cat carotid body on voltage-gated calcium channels. J. Appl. Physiol. 71:1062-1069.

18 THOSE STRANGE GLOMUS CELLS

C. Eyzaguirre

Department of Physiology, University of Utah School of Medicine, Salt Lake City, Utah 84108, U.S.A.

This article is based on a lecture given at the Symposium. It does not cover all the presented material, but contains some reflections on the subject of the talk.

There is little doubt that carotid body glomus (type I) cells are essential in chemotransduction. When the cells are eliminated, sensory transduction disappears (see [12]). However, multiple efforts by many authors have failed to clearly establish the mechanism by which these cells influence the carotid nerve terminals.

From the morphology of the receptor organ, a traditional electrophysiologist would assume that various stimuli (hypoxia, hypercapnia, acidity, etc.) primarily act on these cells. It seems a necessary step since this receptor is so unspecific. The cells respond to stimulation by releasing "transmitters" which in turn depolarize the nerve terminals, an essential condition to elicit the carotid nerve digcharge. The response of the nerve is predictable. All stimuli increase the sensory discharge in stimulus-dependent fashion. From this perspective, it could be reasoned that stimuli act directly on the nerve terminals and that the glomus cells are doing something else. However, this reasoning has some problems. For instance, it is known that hypoxia tends to depolarize nerve fibers whereas hypercapnia and acidity have membrane stabilizing effects. It is inconceivable, therefore, that membrane depolarizers and stabilizers would both increase the sensory discharge. Thus, we have to go back to the glomus cells and here is where the problems arise.

During stimulation, the glomus cells secrete [12]. But,they contain and release multiple chemicals such as catecholamines, ACh and neuropeptides. No wonder that pharmacological blockers have been so ineffective in suppressing transmission across the glomus cell-nerve ending synapses. If one agent is blocked, others take over. To complicate matters further, not all the released chemicals are excitatory or have the same function in different species. ACh excites the chemoreceptors of the cat but inhibits those in the rabbit. Dopamine inhibits the cat chemoreceptors but excites rabbit receptors[12]. This evidence suggests that the nerve terminals have a variety of receptor sites (excitatory and inhibitory) and that their proportion varies in different species. Because of technical difficulties, it has not been possible to study the nerve ending receptors in detail and this is an open field for future work.

What is less clear is how secretion occurs. Different investigators have used various electrophysiological techniques in an effort to solve this problem, namely, intracellular recordings with conventional and ion-selective microelectrodes, tight-seal whole cell recordings and cell patch clamping. Also, voltage- and ion-sensitive dyes have been employed to examine changes in membrane potential or intracellular ionic activities at rest and during stimulation. The preparations have been whole or sliced carotid bodies, and freshly isolated or cultured glomus cells. Clearly, these investigations cover only one

Neurobiology and Cell Physiology of Chemoreception
Edited by P.G. Data *et al.*, Plenum Press, New York, 1993

part of the complex function of glomus cells. Other studies have focussed on the effects of activators on the cell cytoplasm and intracellular organelles. The intracellular constituents, rather than the envelope, may play a more fundamental role in chemorecoption. We know practically nothing about the sustentacular (type II cells), which surround the glomus cells and the nerve terminals. They may play an important, if only modulatory, role in chemoreception.

With regard to glomus cell electrophysiology. When the chemoreceptors are stimulated and the glomus cell secrete, the cells are either depolarized or hyperpolarized [1,10,11,26]. Intuitively, one would expect the opening of channels, allowing the passage of ions across the cells membrane along their gradient. The resulting membrane potential (E_m) shift would depend on the equilibrium potential (E_{ion} or E_{rev}) of the ions involved. In other cells, one could expect that opening of Na^+ or Ca^{2+} channels would increase their intracellular levels, and induce cell depolarization since their E_{rev}s are likely to be very positive. On the other hand, movements of K^+ and Cl^- should induce less marked effects since their E_{rev}s are closer to tho cell's resting potential. Again, the shift will depend on where the E_{rev} is. Cells would depolarize if this level is more positive than the E_m and hyperpolarize if it is more negative. If cellular activation results in the simultaneous opening of various types of channels (as in post-synaptic membranes), the resultant E_m shift will be determined by the combined reversal potential of the ions involved. Also, if there is a resting current provided by a continous leak of one or more ions, channel blocking by a stimulus would also displace the E_m because of a shift in E_{rev}. In this context, glomus cells have turned out to be very complex as shown below.

<u>Sodium</u>. There is no information about the intracellular activity of Na^+ ions ($a_i Na^+$) in glomus cells. We know, however, that the cell's resting potential increases when Na_o is reduced or eliminated[2]. This suggets that, at rest, there is a continuous leak of Na^+ into the cell and that this ion may be responsible to some extent for the E_m. Experiments with $^{22}Na^+$ have shown sodium influx in resting glomus cell that increases considerably in the presence of veratridine, an effect blocked by tetrodotoxin (TTX) [31].

These experiments are in line with those of others, using tight-seal whole cell current and voltage clamping. Depolarizing pulses induce an inward Na^+ current which can lead to the generation of action potentials [3,8,20,34]. Nevertheless, during stimulation with hypoxia or cyanide this current does not change.

<u>Calcium</u>. In substitution experiments, this ion behaves very much like Na^+. Removing extracellular calcium, also induces cell hyperpolarization suggesting a passive inward leak under resting conditions[2]. Experiments using the fluorescent dye fura-2 have shown an $a_i Ca^+$ of about 50 nM[4,32]. Exposing the glomus cells to hypoxia or cyanide increases $a_i Ca^+$[4,29,32], although there is disagreement about the source of increased $a_i Ca^+$. Some have suggested that the increased $a_i Ca^+$ is due to entry from the extracellular fluid[29,32] whereas others have indicated that it comes from intracellular stores[4]. An intracellular source is more in line with whole cell voltage clamp experiments since the electrically evoked, and labile, inward calcium current[22] is not changed by hypoxia[8,18,20]. Quite recently, however, others have found a decrease in intracellular calcium activity during severe hypoxia induced by $Na_2S_2O_4$[19].

<u>Potassium</u>. The intracellular K^+ activity ($a_i K^+$) of glomus cells has been established as 33.0 mM for whole rabbit carotid bodies [24]. Higher values (73 mM) have been found in cultured rat glomus cells (L. Pang and C. Eyzaguirre, unpublished). The latter values

are closer to those obtained from other (e.g., heart muscle) cells. It is not clear what role K^+ plays in maintaining the cell E_m. In cat and rabbit carotid bodies, substitution experiments did not show clear effects of $a_o K^+$ on the E_m of glomus cells [2,24]. In the rat, however, it appears that K^+ plays a role in maintaining the E_m (L. Pang and C. Eyzaguirre, unpublished).

Whole-cell tight-seal voltage clamp has shown the importance of this ion in the response of glomus cells to hypoxia, cyanide or acidity. But, again, there is no consensus. Several groups of investigators have shown that outward K^+ currents (calcium- or voltage-dependent) are depressed by hypoxia and acidity [18,20,27,28]. This has been further refined by recording single K^+ ion channels in patch recordings [7,14]. These channels are not depressed, but the probability of their opening diminishes. However, another group has demonstrated an increase in the K^+ outward current during exposure to cyanide [3]. Unpublished data from this laboratory indicate that during severe hypoxia (PO_2, 5-10 Torr) induced by $Na_2S_2O_4$, there is a 40% loss of K^+ from the cell interior. These data would agree with the increase in the outward potassium current just mentioned.

Chloride. A drastic reduction of Cl^- ions from the extracellular fluid, markedly depolarized glomus cells and their resistance decreased [2]. Studies with chloride-sensitive microelectrodes have shown an $a_i Cl^-$ of 20.9 mM in rabbit glomus cells [25]. Higher values (45 mM) have been found in cultured rat glomus cells. There is a clear relationship between $a_i Cl^-$ and $a_o Cl^-$ and the cell's E_m. During severe hypoxia (PO_2, 5-10 Torr) induced by $Na_2S_2O_4$, extracellular Cl^- moves into the cells. $a_i Cl^-$ increases by about 26% (L. Pang and C. Eyzaguirre, unpublished). Voltage-independent chloride channels in the glomus cell membrane have been described [33].

Marked changes in intracellular K^+ (decrease) and Cl^- (increase) are not unusual when the central nervous system is exposed to ischemia and cyanide [35].

Hydrogen ions. Very different "normal" intracellular pH (pH_i) values for glomus cells have been reported by various investigators using different methods. In rat cultured glomus cells, we have obtained a pH_i of 6.3-7.2 with intracellular H^+ -selective microelectrodes. The extracellular pH (pH_o) was 7.32 to 7.54 in HEPES-NaOH buffer [16,17]. In freshly isolated glomus cells, Biscoe et al.[4] using BCECF and HEPES-buffered saline (pH_o 7-4) found a pH_i of about 6.9, which is close to our values. With the same method, Wilding et al.[36] and Cummins and Donnelly [6] have reported pH levels of 7.2 to 7.3. These values are close to those reported for the whole carotid body using DMO [15]. On the other hand, Buckler et al.[5] using carboxy-SNARF-1 and HEPES-buffered Tyrode (pH_o 7-4) on isolated glomus cells, report a pH_i of 7.77.

The pH_i of glomus cells follows pH_o changes [16,36]. However, H^+ ions are not passively distribute. In an acid environment, there seems to be a hydrogen pump expelling H^+ ions. This pump may be operating as a Na^+/H^+ exchanger. Toward alkalinity, such pumps usually do not work and it seems possible that we have an HCO_3/Cl^- exchanger [5,6,16,36]. The E_m of glomus cells seems to be influenced by the relationship between pH_i and pH_o. Cells depolarize if the equilibrium potential for H^+ ions [E_H= 60 x (pH_i - pH_o)] becomes more positive than E_m and hyperpolarize if this values goes in a negative direction [16]. Relative hypoxia (from 50% O_2 to air) decreased pH_i in about 60% of glomus cells, the rest showing an increased value. However, 5% CO_2 acidified about 90% of the cells studied [17]. This is not too surprising since glomus cells contain carbonic anhydrase [23,30]. During severe hypoxia or application of cyanide, some authors have found no changes in pH_i [3,15]. Others[19] report intracellular acidification when PO_2 falls to zero. We have found increased

3. *T.J. Biscoe and N.R. Duchen*, Electrophysiological responses of dissociated type I cells of the rabbit carotid body to cyanide, J. Physiol.(Lond.), 413:447 (1989).

4. *T.J. Biscoe, M.R. Duchen, D.A. Eisner, S.C. O'Neill and M. Valdeolmillos*, Measurements of intracellular Ca^{2+} in dissociated type I cells of the rabbit carotid body, J. Physiol.(Lond.), 416:421 (1989).

5. *K.J. Buckler, R.D. Vaughan-Jones, C. Peers and P.C.G. Nye*, Intracellular pH and its regulation in isolated type I carotid body cells of the neonatal rat, J.Physiol.-(Lond.), 436:107 (1991).

6. *T.R. Cummins and D.F. Donnelly*, Regulation of intracellular pH in rat glomus cells, Neurosci. Abstr., 17:102 (1991).

7. *M.A Delpiano and J. Hescheler*, Evidence for a PO_2 sensitive K^+ channel in the type-I cell of the rabbit carotid body, FEBS Lett., 249:195 (1989).

8. *M.R. Duchen, X.W. Caddy, G.C. Kirby, D.L. Patterson, J. Ponte and T.J. Biscoe*, Biophysical studies of the cellular elements of the rabbit carotid body, Neuroscience, 26:291 (1988).

9. *C. Eyzaguirre, Y. Hayashida and L. Monti-Bloch*, Effects of denervation on the glomus cell membrane, Brain Res., 524:164 (1990).

10. *C. Eyzaguirre, L. Monti-Bloch, N. Baron, Y. Hayashida and J.W. Woodbury*, Changes in glomus cell membrane properties in response to stimulants and depressants of carotid nerve discharge, Brain Res. 477:265 (1989).

11. *C. Eyzaguirre, L. Monti-Bloch and J.W. Woodbury*, Effects of putative neurotransmitters of the carotid body on its own glomus cells, Eur. J. Neurosci., 2:77 (1990).

12. *S.J. Fidone and C. Gonzalez*, Initiation and control of chemoreceptor activity in the carotid body, in: Handbook of Physiology. The Respiratory System, Vol.II, sect.3, Am. Physiol.Soc., Bethesda, MD (1986).

13. *R. Gallego, C. Eyzaguirre and L. Monti-Bloch*, Thermal and osmotic responses of arterial receptors, J.Neurophysiol., 42:665 (1979).

14. *N.D. Ganfornina and J. López-Barneo*, Single K^+ channels in membrane patches of arterial chemoreceptor cells are modulated by O_2 tension, Proc.Natl.Acad.Sci.USA, 88:2927 (1991).

15. *J. Garcìa-Sancho, F. Giráldez and C. Belmonte*, Absence of apparent cell pH variations during hypoxia in the carotid body chemoreceptors in vitro, Neurosci. Lett., 10:247 (1978).

16. *S-F. He, J-Y. Wei and C. Eyzaguirre*, Intracellular pH and some membrane characteristics of cultured carotid body cells, Brain Res., 547:258 (1991).

17. *S-F. He, J-Y. Wei and C. Eyzaguirre*, Effects of relative hypoxia and hypercapnia on intracellular pH and membrane potential of cultured carotid body cells, <u>Brain Res.</u>, 556:333 (1991).

18. *J. Hescheler, M.A. Delpiano, H. Acker, and F. Pietruschka*, Ionic currents on type-I cells of rabbit carotid body measured by voltage-clamp experiments and the effect of hypoxia, <u>Brain Res.</u>, 486:79 (1989).

19. *D. Xholwadwala and D.F. Donnelly*, Hypoxia decreases in intracellular calcium in rat carotid body glomus cells, <u>Neurosci. Abstr.</u>, 17:102 (1991).

20. *J. López-Barneo, J. López-López, J. Urena and C. Gonzàlez*, Chemotransduction in the carotid body: current modulated by PO2 in Type I chemoreceptor cells, <u>Science</u>, 241:580 (1988).

21. *L. Monti-Bloch and C. Eyzaguirre*, Effects of natural stimuli, chemical agents and transmitters on glomus cell membranes and intercellular communications, <u>in</u>: Arterial Chemoreception, C. Eyzaguirre, S.J. Fidone, R.S. Fitzgerald, S. Lahiri. and D.M. McDonald, eds., Springer-Verlag, New York (1990).

22. *L.J. Mullins*, Depolarization and calcium entry, <u>in</u>: The Biophysical Approach to Excitable Systems, W. J. Adelman, Jr., and D. E. Goldman, edg., Plenum Press, New York (1981).

23. *C.A. Nurse*, Carbonic anhydrase and neuronal enzymes in cultured glomus cells of the carotid body of the rat, <u>Cell Tissue Res.</u>, 261:65 (1990).

24. *Y. Oyama, J.L. Walker and C. Eyzaguirre*, Intracellular potassium activity, potassium equilibrium potential and membrane potential of carotid body glomus cells, <u>Brain Res.</u>, 381:40S (1986).

25. *Y. Oyama, J.L. Walker and C. Eyzaguirre*, The intracellular chloride activity of glomus cells in the isolated rabbit carotid body, <u>Brain Res.</u>, 368:167 (1986).

26. *L. Pang and C. Eyzaguirre*, Different effects of hypoxia on the membrane potential and input resistance of isolated and clustered glomus cells, <u>Brain Res.</u>, 575:167-173 (1992).

27. *C. Peers*, Hypoxic suppression of K^+ currents in type I carotid body cells: selective effect on the Ca^{2+} activated K^+ current, <u>Neurosci.Lett.</u>, 119:253 (1990).

28. *C. Peers and J. O'Donnell*, Potassium currents recorded in type I carotid body cells from the neonatal rat and their modulation by chemoexcitatory agents, <u>Brain Res.</u>, 522:259 (1990).

29. *F. Pietruschka*, Calcium influx in cultured carotid body cells is stimulated by acetylcholine and hypoxia, <u>Brain Res.</u>, 347:140 (1985).

30. *R. Rigual, C. Iniguez, J. Carreres and C. Gonzalez*, Carbonic anhydrase in the carotid body and carotid sinus nerve, Histochemistry, 82:577 (1985).

31. *M. Sato, K. Yoshizaki and H. Koyano*, Veratridine stimulation of sodium influx in carotid body cells from newborn rabbits in primary culture, Brain Res., 504:132 (1989).

32. *M. Sato, K. Ikeda, K. Yoshizaki and H. Koyano*, Response of cytosolic calcium to anoxia and cyanide in cultured glomus cells of newborn rabbit carotid body, Brain Res., 551:327 (1991).

33. *A. Stea and C.A. Nurse*, Chloride channels in cultured glomus cells of the rat carotid body, Am. J. Physiol. Cell Physiol., 26:C-174 (1989).

34. *A. Stea and C.A. Nurse*, Whole-cell and perforated-patch recordings from O_2-sensitive rat carotid body cells grown in short- and long-term culture, Pflügers Arch., 418:93 (1991).

35. *A. Van Harreveld*, Brain Tissue Electrolytes, Butterworths: Washington, DC (1966).

36. *T.J. Wilding, B. Cheng and A. Roos*, The relationship between extracellular pH (pH_o) and intracellular pH (pH_i) in adult rat carotid body glomus cells, Biophys. J., 59:184a (1991).

19   CAROTID BODY NEUROTRANSMISSION

R.S. Fitzgerald, and M. Shirahata

Departments of Environmental Health Sciences, Physiology
Medicine, and Anesthesiology/Critical Care Medicine
The Johns Hopkins Medical Institutions
Baltimore, Maryland 21205 USA

INTRODUCTION

The most profound dependence that a higher organism has is its dependence on oxygen. The system mediating the tissue needs of the higher organism for oxygen and the presence of oxygen in the environment is the cardiopulmonary system. The neural control of the anatomical structures of that system includes the carotid body as a receptor. When the organism is challenged by a decrease in the amount of oxygen available to it, which manifests itself in a decrease in the partial pressure of oxygen in the arterial blood, neural activity in the carotid body increases. Among the reflex effects of the carotid body's increase in neural activity is the increase in tidal volume, breathing frequency, cardiac output, cardiac contractility, and a less precipitous drop in the total peripheral resistance. The carotid body is the essential receptor in the cardiopulmonary system's response to hypoxia. The question is: "How does it transduce a decrease in low oxygen into increased neural activity? Current speculation postulates that hypoxia somehow depolarizes the neurotransmitter-containing Type I cell, calcium rises in the cytosol from external and perhaps internal sources, promoting the release of one or more excitatory neurotransmitters. The neurotransmitter proceeds to the apposed dendrite, depolarizes it, and the action potential proceeds central to the nucleus tractus solitarius. Subsequently, the cardiopulmonary reflex responses develop.

Much early attention focused on acetylcholine (ACh). Schweitzer and Wright, and Liljestrand, Zotterman, von Euler, and Landgren of the "Swedish School" (for review see refs. 3, 4, 5) presented data which was quite convincing in suggesting that ACh was an important excitatory neurotransmitter during hypoxia. On the other hand the "British School" was less convinced that ACh was involved in hypoxic chemotransduction. Neil and his coworkers, as well as Douglas[2] and Moe et al.,[8] seemed to be the most opposed to what has subsequently been termed the "Cholinergic Hypothesis"— ACh is the excitatory neurotransmitter during hypoxia. Eyzaguirre and his colleagues[10] in the early 1960's were a more recent group of investigators to espouse ACh as the excitatory neurotransmitter. They, too, presented data suggesting that ACh was involved. Again the explanation of the data implicating ACh as an excitatory neurotransmitter was not widely accepted. As a result of these developments a clear and universally acceptable excitatory neurotransmitter essentially involved in the carotid body's response to hypoxia has yet to be identified. Recently Substance P (SP) has been suggested as the excitatory neurotransmitter involved during physiological stimulations. Prahbakar and his colleagues[9] have worked extensively with this agent.

Neurobiology and Cell Physiology of Chemoreception
Edited by P.G. Data *et al.*, Plenum Press, New York, 1993

In view of the above history the present studies have once again explored the role for ACh and the possibility that SP might be involved with ACh in the hypoxic excitation of the carotid body. SP has been reported to be involved in cholinergic transmission in other systems.

METHODS

Cats were anesthetized with sodium pentobarbital (30 mg/Kg). A tracheal cannula was inserted, as were femoral arterial and venous catheters. A polyethylene loop was inserted into the left common carotid artery for periodic perfusions of the carotid body with warmed Krebs Ringer bicarbonate (KRB) solutions (Figure 1). Otherwise the cat's own arterial blood flowed through the catheter. The rest of the vessels were ligated except for the lingual artery which was cannulated for the collection of perfusate. The baroreceptors were mechanically or thermally destroyed and the whole carotid sinus nerve was prepared for recording.

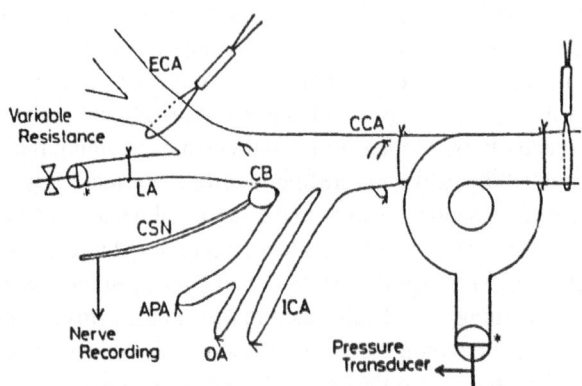

Figure 1. Perfusion Set-Up. APA = ascending pharyngeal artery; CB = carotid body; CCA = common carotid artery; CSN = carotid sinus nerve; ECA = external carotid artery; ICA = internal carotid artery; LA = lingual artery.

In the first set of experiments the protocol required that cats be exposed to 10% O_2 in N_2. Hypoxic blood perfused the carotid body for three minutes. Then the snares around the common carotid and external carotid arteries were drawn closed, the stopcock was turned, and the carotid body was perfused with warmed hypoxic KRB for two minutes. The stopcock was then adjusted to allow the hypoxic blood to perfuse the carotid body once again. The hypoxic KRB contained, or not, three cholinergic blockers (in µM): alpha-bungarotoxin (Bgt), 2; mecamylamine, 402; atropine, 942.

In the second set of experiments the carotid body of normoxic cats was perfused with warmed hypoxic KRB, warmed hypoxic KRB containing SP (10 µM), or warmed hypoxic KRB containing SP (10µM) and the above concentrations of the cholinergic blockers. Following the perfusions normoxic arterial blood was once again allowed to perfuse the carotid body.

RESULTS

First Set of Experiments

Table 1 presents the results of two perfusions. Neural activity increases in response to the arterial blood rendered hypoxic by the cat's breathing of the 10 O_2 in N_2 mixture. Switching to the hypoxic KRB did not significantly alter the neural activity, which remained the same when the carotid body was re-exposed to the hypoxic arterial blood.

Table 1. Neural Activity (µV)

EXPT. #1 (n=4) Perfusate	Control	After Breath- ing 3 min HPX	After 1 min of Perfu- sion	After 2 min of Perfu- sion	After 1 min Re-exp.	After 3 min Re-exp.
HPX KRB	1.8+0.3	5.2+0.6	5.0+0.6	5.2+0.6	5.5+0.5	5.7+0.6
HPX KRB plus Blockers	1.4+0.4	5.6+0.4	1.3+0.6	1.1+0.4	3.4+0.8	4.2+0.8

LEGEND: HPX = Hypoxic or Hypoxia; KRB = Krebs Ringer bicarbonate solution; Re-exp. = re-exposure to arterial blood; Blockers = 2 µM a Bgt, 402 µM mecamylamine, 942 µM atropine (X ± S.E.).

However, when the cholinergic blockers were included in the hypoxic KRB, the neural activity—elevated by the initial exposure to hypoxic arterial blood—decreased significantly, a reduction to 20% of the maximum response. Upon reexposure to hypoxic arterial blood the neural activity began to regain its original level.

Second Set of Experiments

Table 2 presents the results of three perfusions. In the normoxic cat the perfusion of hypoxic KRB significantly elevated the neural activity. Upon re-exposure to normoxic blood, the activity returned to control levels. To the second perfusion of hypoxic KRB which contained SP (10 µM) the response was essentially the same. But when the carotid body was re-exposed to the normoxic arterial blood after the perfusion, the neural activity remained elevated significantly. Finally, when the cholinergic blockers were included in the hypoxic KRB containing the SP, there was no significant change in the neural activity; nor did return to normoxic blood change the activity.

DISCUSSION

The neural activity clearly is decreased when the cholinergic blockers are included in the hypoxic KRB. The concentrations of blockers were somewhat in excess of concentrations used by investigators testing their efficacy by systemic administration. In this study, however, the blockers were administered only locally, and only for a very

Table 2. Neural Activity (μV)

EXPT. #2 (n=5) Perfusate	Control	After 30 sec Perfusion	After 90 sec Perfusion	After 1 min Re-exp.	After 2 min Re-exp.
HPX KRB	2.5+0.7	6.0+1.8	5.8+1.6	2.6+1.2	2.6+1.0
HPX KRB plus SP	2.6+0.7	5.8+1.5	6.2+1.6	4.0+1.4	3.9+1.3
HPX KRB plus SP plus Blockers	2.4+0.6	2.7+0.6	2.6+0.5	2.8+0.8	2.8+0.7

LEGEND: As in Table 1; SP - Substance P.

brief time. Hence, the concentrations of the blockers in the perfusing medium were somewhat elevated. It seems unlikely that the tissue concentrations ever reached equilibrium with the perfusate concentrations. However, the use of higher concentrations raises the possibility that the concentrations were such as to create a non-specific inhibition. This hypothesis was tested in several animals by leaving the baroreceptors intact and perfusing with the same two KRB solutions that were used in this study. Once again the chemoreceptor component of the whole nerve tracing disappeared while the baroreceptor component remained. On the contrary when 30μM lidocaine was included in the hypoxic KRB, both chemoreceptor and baroreceptor components of the trace decreased significantly within 25 seconds. Since the effect of the combination of blockers was so immediate and profound, it would seem incontestable that at least in this preparation cholinergic blockers generate a significant reduction in the carotid body's response to hypoxia.

It is an interesting note that in 1968 Joels and Neil concluded their brief overview[10] of the role for ACh with the statement: "It would seem, however, that acetylcholine cannot be discarded as a transmitter even though it may not be the sole transmitter of the glomus." This position seems to echo the earlier reservations of Douglas in 1954[2] who ended his rejection of an excitatory role for ACh during hypoxia with the statement "...the argument that acetylcholine is involved in chemosensory transmission appears to me to stand but poorly at present; but the theory is an attractive one and we have no convincing evidence against it." Somewhat paradoxically, therefore, is the statement of Heymans and Neil, chronologically sandwiched between the above two in their 1958 volume Reflexogenic Areas of the Cardiovascular System[6]: "It is the authors' opinion that ACh has nothing whatever to do with the normal transmission of the chemoreceptor impulses. The reasons for the conflict of evidence concerning the effects of anticholinesterases and ganglioplegic drugs on the activity and sensitivity of the glomus nerve endings to normal or pharmacological stimuli require further clarification." Efforts to clarify the chemotransductive mechanisms have proceeded. From the 1950's to the 1970's investigators have pursued and established for ACh most of the criteria which must be met in order to define a substance as a neurotransmitter. In conclusion those investigators who were least impressed with the data suggesting an excitatory role for ACh in hypoxic chemotransduction did seem to leave open the possibility that such might be the case.

Substance P has frequently been sssociated with cholinergic transmission. Livett and his colleagues[7] in their first study observed that SP inhibited the release of norepinephrine provoked by cholinergic agonists. But in a second study[1] they reported that SP protects against the desensitization of norepinephrine release produced by nicotinic agonists including ACh. These results applied to the excitation of the adrenal medulla by the splanchnic nerve suggest an interesting modulatory role for SP. Under stress large amounts of ACh would be released from the nerve and perhaps desensitize the response of the chromaffin cell to ACh. This would reduce the release of adrenal catecholamines. With the simultaneous release of SP from the nerve the initial release of adrenal catecholamines would be dampened. But SP would concomitantly protect against the nicotinic desensitization of the adrenal medulla. This would prolong the release of catecholamines during stress and prevent a large burst in catecholamines followed by little or nothing due to desensitization. Hence, it appears that in this system the effect of SP is that it modulates the postsynaptic effectiveness of ACh insuring a more even, long-term response to ACh. The data in the present study suggest that SP may well act through cholinergic mechanisms in the carotid body.

In the sympathetic ganglion peptides have been described as being responsible for a late slow excitatory postsynaptic potential. This depolarization lasts for minutes. Perhaps it is this feature which keeps the neural activity elevated in the carotid body after the inection of hypoxic KRB containing the SP.

In summary, then, the data presented herein suggest that the Cholinergic Hypothesis remains a viable hypothesis. In spite of the fact that the strongest critics of the Cholinergic Hypothesis failed to be convinced that ACh is the excitatory neurotransmitter in the carotid body, they did leave open the possibility that ACh might have a role in the excitation of the carotid body during hypoxia. And secondly, it appears that at least one mode of action for SP in the carotid body could be through a modulation of an excitatory cholinergic pathway.

ACKNOWLEDGEMENTS

This work was supported by HL 10342.

REFERENCES

1. *BOKSA, P., AND B.G. LIVETT.* Substance P protects against desensitization of the nicotinic response in isolated adrenal chromaffin cells. J. Neurochem. 42: 618-627, 1984.

2. *DOUGLAS, W.W.* Is there chemical transmission at chemoreceptors? Pharmacol. Rev. 6: 81-83, 1954.

3. *EYZAGUIRRE, C., R.S. FITZGERALD, S. LAHIRI, AND P. ZAPATA.* Arterial chemoreceptors. In: Handbook of Physiology Section 2: The Cardiovascular System. Vol III. Peripheral Circulation and Organ Blood Flow. Eds. J.T. Shepherd and F.M. Abboud. Bethesda, Md., American Physiological Society, 1983; pp. 557-622.

4. *FIDONE, S.J., AND C. GONZALEZ.* Initiation and control of chemoreceptor activity in the carotid body. In: Handbook of Physiology. Respiration. Vol II. Part 1. Control of Breathing. Eds. N.S. Cherniack and J.G. Widdicombe. Bethesda, Md., American Physiological Society, 1986; pp.247-312.

5. *HEYMANS, C., AND E. NEIL.* Reflexogenic Areas of the Cardiovascular System. Boston, Little, Brown and Co., 1958.

6. *HEYMANS, C., AND E. NEIL.* Op. cit., p.191.

7. *LIVETT, B.G., V. KOZOUSEK, F. NISOBE, AND D.M. DEAN.* Substance P inhibits nicotinic activation of chromaffin cells. Nature 278: 256-257, 1979.

8. *MOE, G.K., L.R. CAPO, AND B. PERALTA.* Action of tetraethylammonium on chemoreceptor and stretch receptor mechanisms. Amer. J. Physiol. 153: 601-605, 1948.

9. *PRABHAKAR, N.R., J.MITRA, AND N.S. CHERNIACK.* Role of substance P in hyypercapnic excitation of carotid chemoreceptors. J. Appl. Physiol. 63: 2418-2425, 1987.

10. *TORRANCE, R.W.* Arterial Chemoreceptors. Oxford and Edinburgh, Blackwell Scientific Publications, 1968.

20   CARBONIC ANHYDRASE AND THE CAROTID BODY

Robert E. Forster II

Department of Physiology, School of Medicine
University of Pennsylvania
Philadelphia, Pennsylvania, 19104-6085 U.S.A.

CARBONIC ANHYDRASE AND CO_2 REACTION KINETICS

In 1932[2] Meldrum and Roughton in Cambridge and Stadie and Obrien in Philadelphia knew that there must be a catalytic acceleration of the dehydration of HCO_3^- in the red cells in a lung capillary and were looking for an enzyme, which turned out to be carbonic anhydrase (CA). In contrast today we know there is CA in the carotid body, but we do not know its function. A fundamental assumption is that since nature carefully maintains CA in the carotid body, it is there for a definite purpose. Therefore I will briefly summarize our present knowledge about CA and list some possible ways in which it might be involved in chemosensing by the carotid body.

Carbonic anhydrase (CA) is an enzyme of about 30 KDa with one atom of zinc at the active site whose only known physiological function is to catalyze the reversible hydration-dehydration of CO_2-HCO_3^-, the first reaction in Equ. 1 below.

$$
\begin{array}{c}
k_u \qquad \qquad \infty rate \\
CO2+HOH \Leftrightarrow H2CO3 \Leftrightarrow [H+]+[HCO3-] \\
k_v
\end{array}
\qquad (1)
$$

where[5]: k_u = reaction velocity constant for the hydration of CO_2, 0.018 seconds^{-1} and
k_v = reaction velocity constant for the dehydration of H_2CO_3, 64.3 seconds^{-1}, both at 37 C and pH 7.4.

The hydration reaction is a second order process with the rate proportional to $[CO_2][HOH]$ but $[HOH]$ is constant at 55 M and ubiquitous in physiological conditions and in tissues, so that it is subsumed in k_u. H_2CO_3 ionizes to $[H+][HCO_3^-]$ with a half-time of approximately 10^{-8} seconds[5] which is instantaneous on a physiological time scale so the following equilibrium always holds.

$$
(2)
$$

$$
K_{HA} = \frac{[H^+][HCO_3^-]}{[H_2CO_3]}
$$

where $K_{HA} = 3.4 \times 10^{-4}$ M.

At these same physiological conditions the hydration-dehydration reaction will reach equilibrium in about 4 seconds at which point

$$\frac{k_u}{k_v} = \frac{[H_2CO_3]}{[CO_2]} \qquad (3)$$

$[H_2CO_3]$ is about 1/400 of $[CO_2]$ under physiological conditions.
    If we substitute $[H_2CO_3]$ from Equ. 2 in Equ. 3 we obtain

$$\frac{k_u}{k_v} = \frac{k'}{k_{HA}} = \frac{[H_2CO_3]}{[CO_2]} \qquad (4)$$

where K' is the classical Henderson-Hasselbalch constant. Carbonic anhydrase accelerates both k_u, and k_v by the same amount. It is convenient to define an analagous reaction velocity constant for the enzyme catalyzed fraction (of the hydration rate), called k_{enz} (sometimes k_{cat}), as described by the following equation

$$\frac{d[CO_2]}{dt} = -(k_u + k_{enz})[CO_2] \qquad (5)$$

where $d[CO_2]/dt$ is the initial rate of hydration of CO_2 assuming that $[HCO_3^-]$ is zero at time zero.
    It is also convenient and sometimes helpful to measure CA activity in terms of the acceleration, A, of the reaction produced by carbonic anhydrase, defined as follows:

$$A = \frac{k_{enz} + k_u}{k_u} \qquad (6)$$

    A can be misleading as an index of CA activity when the temperature changes because the uncatalyzed rate, k_u, which is the reference value, decreases more rapidly with a fall in temperature than the catalyzed velocity constant, k_{enz}, as shown in Table 1.
    The reversible reactions of CO_2 and H_2CO_3 are unusual in that their rate is significant in the complete absence of enzyme.Thus the minimal detectable CA activity in a sample is equivalent to the minimal error in measuring the uncatalyzed reaction velocity,which last is decreased 1 to 2 orders of magnitude by lowering the temperature to near 0 °C. Thus in many techniques for the measurement of CA activity, $d[CO_2]/dt$ is determined at O to 4 °C.
    CA can catalyze two other classes of chemical reactions[4].
    1.Hydration of an aldehyde to a diol.
    R-HCO + HOH $\Leftrightarrow$ R-HC(OH)$_2$

    This reaction might be of physiological importance, as for glyceraldehyde[15], but this has not be demonstrated experimentally.

138

2.Esterification

$$R_1\text{-COOH} + R_2\text{OH} \Leftrightarrow R_1\text{-COO-R}_1$$

The esterase effect of CA on nitrophenylacetates, which produces a color change, has been used as a convenient assay of its activity[4] but this action of CA has not been shown to be of physiological importance.

Table 1[6]. Effect of temperature on the uncatalyzed (k_u) and catalyzed (k_{enz}) reaction velocity constants for CO_2 hydration.

TEMP. ($^\circ$C)	k_u (sec^{-1})	k_{enz} normalized (sec^{-1})	k_u/k_{enz} normalized
0	0.0022	0.068	0.032
4	0.0037	0.097	0.038
10	0.014	0.26	0.054
25	0.051	0.46	0.110
37	0.19	1.0	0.19

The values of k_u are for pH 7.4 and ionic strength 0.150. The values of k_{enz} are for similar conditions using human CA II but have been normalized to the value at 37 $^\circ$C for simplicity.

k_u increases almost 100 fold from 0 to 37°C while K_{enz} increases only 14 times so there is a 6 fold increase in k_u/k_{enz} primarily because of the effect of temperature on k_u.

Carbonic Anhydrase Isozymes (Table 2)

In 1961 it was found[2] that there were two isozymes of CA in the human red cell which were given the names CA I and CA II.CA II is the archetypical high activity carbonic anhydrase critical to blood-gas CO_2 exchange, but it actually comprises only about 20% of the red cell CA, 80% of which is CA I. The latter isozyme has a similar molecular weight but 1/30th of the turnover number of CAII and in addition is inhibited by the normal [Cl$^-$] in the red cell, which CA II is not. Thus CA I contributes far less than half the CA activity of the human red blood cell. If it has a function other than in the transport of CO_2, it is not presently known.

CA III is found in very high concentration in red striated muscle, and in male rat liver, and constitutes that largest amount of CA in the body. It has an even lower turn over number than CA I and the distinguishing characteristic of being less sensitive to sulfonamides than any other CA. Its function is entirely unknown, although Gros and colleagues[7] suspect it is involved in excitation-contraction coupling and Dodgson and Lynch (personal communication) suspect it is important in fatty acid metabolism because of its presence in adipocytes.

139

CA IV is a membrane bound isozyme with a high turn over number approximating CA II, but with a larger molecular weight. The basic enzyme molecule appears to be much the same as CA II but has a carbohydrate chain added presumably to attach the CA IV to the phospholipid membrane structure. CA VI and VII also have additional carbohydrate chains attached increasing their molecular weight. While CA IV is presumably important in H+ secretion in the renal tubule, the functions of CA VI and VII are less well identified.

CA V is a high activity, basically CA II type isozyme found in liver and heart mitochondria, and in some striated muscle mitochondria. It appears to be involved in urea synthesis presumably supplying the HCO_3^- ion needed for the formation of carbamoyl phosphate. There is considerable evidence for its involvement in gluconeogensis and fatty acid synthesis.

Table 2[6]. Carbonic Anhydrase Isozymes.

Type	Location	Molecular Weight	Amino Acid Residues	TON (Relative)	Chloride Inhibition (Relative)	Sulfonamide Sensitivity (Relative)
I	Red cells & gut	30,000	260	0.03	High	High
II	Red cells	30,000	259	1.0	Low	High
III	Muscle: cytosol	30,000	259	0.003	Low	Very low
IV	Membrane: renal, lung, & brain	52-68,000	>260	0.5	Low	High
V	Mitochondria: liver, renal, heart, ± muscle	29,000	≈ 260	0.02?	Low	High
VI	Saliva	45,000	314	0.3-0.8	Low	High
VII	Salivary gland membrane	45,000	263	High	Low	High
VIII	Brain; Purkinje cells	>30,000	>260			

CA VIII was first identified as a CA from its DNA[9].

CA Genetics

The genes for the different CA isozymes are not all on the same chromosome. CA I, II and III are located on chromosome 8 at 8Q22, that for CA VI on the tip of chromosome 1, and that for CA VII on chromosome 16 at q21-23. Large segments of CA isozymes, minus the active site, are seen in some viruses, such as vaccinia, and some oncogenes. The reason for this is not clear. A phylogenetic tree for CA I,II and III can be constructed originating 400 to 300 million years ago since which time these isozymes have remained clustered in the same chromosome.

Mechanism of Carbonic Anhydrase Function

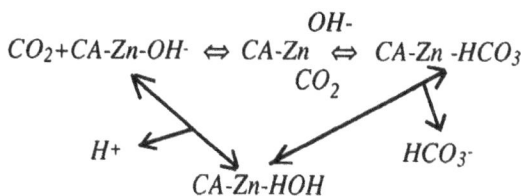

Carbonic anhydrase has the fastest turnover of any enzyme, approaching 1,000,000 and the actual catalytic step may be nearly a magnitude faster than that[10,18]. The presently accepted mechanism is diagrammed in the schema above. The cycle starts by the active zinc site taking up a HOH which is converted to OH^-, leaving an H^+ behind, next binding CO_2, then performing the catalytic step forming HCO_3^- which is released in the final step. The rate limiting step is not the catalysis but the removal, or accretion of the H^+ formed or taken up. The transport of H^+ appears to be divided into two phases[10]. The first is proton movement from the bulk solution to the bottom of the 15 Å well in the protein where the zinc is located. This process appears to be diffusive in nature but is not rate limiting unless the concentration of small buffer molecules, which presumably facilitate the flux of H^+, drops into the low μM range. The second phase is proton movement within the molecule itself, which involves shuttling among 3 imidazole residues in the neighborhood of the zinc.

The turnover number for CA pertains to the formation of H^+ and HCO_3^- from CO_2, or the reverse. The velocity of the actual catalytic step from zinc bound $OH^- + CO_2$ to HCO_3^- can be measured with nuclear magnetic resonance of ^{13}C labeled CO_2/HCO_3^- and is found to be some 8 fold faster than the turnover above[18]. Thus the internal movement of H^+, which apparently cannot be influenced by buffer, ultimately rate limits CA function.

CA inhibitors

In their early research Meldrum and Roughton investigated numerous inhibitors of CA and their general findings can be summarized as follows in terms of the dissociation constant of the inhibitor, K_i.

1. K_i from 10^{-2} to 10^{-1} M. This group consists of common anions, such as cyanates, halides, phosphate and borate.

2. K_i from 10^{-4} to 10^{-3} M. This group consists of ions that attack enzymes with a metal active center, such as azide, cyanide and sulfide.

3. K_i from 10^{-9} to 10^{-14} M. Sulfonamides, drugs with the active grouping $-SO_2NH_2$.

The discovery that sulfanilamide was a potent inhibitor by Keilin intiated an era in which a large series of sulfonamide inhibitors have been synthesized and found to have important actions, some not expected[13]. The properties of a sampling of these drugs is in Table 3.

Ca III is not significantly inhibited by any of these drugs except chlorzolamide. Benzolamide is useful as a drug of high affinity which remains outside cell membranes. Ethoxzolamide is about the opposite; it has a high affinity but is highly fat soluble and therefore enters cells rapidly. Acetazolamide, a widely used drug clinically, is not very permeable and at concentrations in plasma in the neighborhood of its K_i does not enter cells and inhibit the CA as much as expected.

The anions of many buffer systems accellerate the uncatalyzed CO_2/HCO_3^- reactions, that is they increase k_u. The amount they add to k_u is proportional to their concentration; the increment equals [anion] x constant. This constant is 30,000 for hypobromite, 10,000 for arsenate, 150 for borate and 8 for posphate. In other words 100 μM hypobromite will increease k_u by a factor of 3. These compounds may or may not affect CA activity as well. If the technique used to assay CA activity involves a comparison of catalyzed rate with the uncatalyzed rate, an increase in k_u will reduce the calculated activity of the sample.

Table 3. Sulfonamide inhibitors of human carbonic anhydrase II[17].

Inhibitor	K_i (microM)	Membrane Permeability (cm/sec)
Acetazolamide	0.01	4×10^{-7}
Benzolamide	0.0004	$\approx 10^{-7}$
Chlorzolamide	0.0004	0.6×10^{-4}
Ethoxyzolamide	0.002	0.6×10^{-4}

Methods of measuring CA activity [5]

The hydration of CO_2 (Equation 1) increases $[H^+]$, lowers $[CO_2]$ increases the ions in solution thereby increasing electrical conductivity and liberates heat. All these properties can be used to measure the progress of the reaction, but the change in pH is the most widely employed, with a glass pH electrode or pH sensitive dyes.

At 37 °C the uncatalyzed reaction has a half-time in the range of several seconds, too fast to be measured with any accuracy by simple mixing and stirring techniques; more complicated rapid mixing apparatus is required. The common solution is to work at low

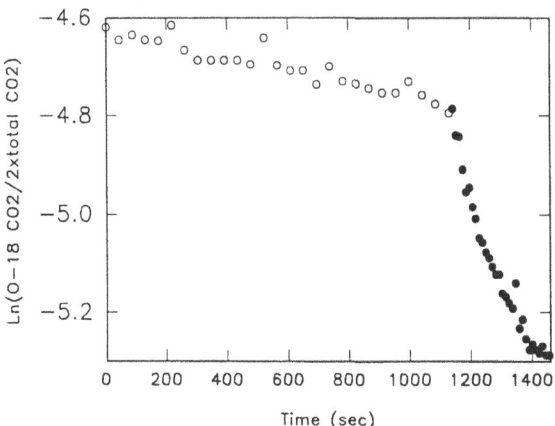

Figure 1. Semi-logarithmic graph of the decrease in α, which = ([C^{18}O^{16}O]/2 + [C^{18}O_2]/total CO_2,versus time in 3 ml of Krebs-Ringer including 25 mM NaHCO_3 enriched 2% with ^{18}O at 4 °C and pH 7.4 before,the open circles,and after,the solid circles,the sample of pooled pulverized carotid bodies was added.(Data of Lin, Ono and Lahiri).

temperatures (see Table 1), slowing the uncatalyzed rate by almost two orders of magnitude but lowering the uncatalyzed rate much less, so that measurements can be made of pH in a mechanically stirred vessel with relative impunity. Of course this is not physiological. Stop-flow rapid mixing apparatus, or less commonly continuous-flow rapid mixing apparatus, can be used at physiological temperatures with pH sensitive dyes or a rapidly responding glass electrode[5]. We have developed a method which measures the rate of ^{18}O exchange between the CO_2 and HCO_3- in a solution and water by following the disappearance of $C^{18}O^{16}O$, which requires a mass spectrometer. The method can be carried out at physiological temperatures, pH is constant during the reaction, the activity of CA inside a membrane can be measured (this is not possible by other techniques because H^+ accumulates and stops the reaction) as well as a measure of HCO_3^- permeability of the membrane. Using a variation of this method in which the sample is exposed to a gas phase containing $C^{18}O^{16}O$, as little as 2.4 x 10^{-15} moles of CA can be detected, the most sensitive technique applied so far. Figure 1 shows the disappeance of mass 46 CO_2,$C^{18}O^{16}O$, from labelled CO_2 in the presence of several milligrams of cat carotid body.

THE FUNCTION OF CARBONIC ANHYDRASE IN THE CAROTID BODY

The known physiological functions of CA all involve acceleration of the reversible reaction, $CO_2 + HOH \Leftrightarrow H_2CO_3$, and include at least the followin
1.Facilitation of CO_2 exchange between blood and alveolar gas or tissue.
2.Transport of H^+ across phospholipid membranes.
3.Buffering changes in P_{CO2} in alveolar gas and in tissue.
4.Facilitated transport of CO_2 within cells and tissues.
5.Provision of HCO_3 for synthesis,such as of urea, glucose and fatty acid.

1. Facilitation of CO₂ transport between blood and alveolar gas or tissue

Seventy percent of the total $[CO_2]+[HCO_3^-]$ in blood is carried as HCO_3^- in the plasma. In the pulmonary capillaries most of the HCO_3^- eliminated as CO_2 has to enter the red cell, exchanging for chloride ion in the process, combine with H^+ to form H_2CO_3, which then dehydrates under the influence of CA to give CO_2 which diffuses freely into the alveolar gas. Metabolic processes form CO_2, not HCO_3^-, in the cells of the carotid body which has to be transported into the capillary blood as in any other tissue, but the CA involved is inside the red cells. It is hard to see why CA in the carotid body itself can be necessary for the elimination of this CO_2.

In the lungs and some tissues there is a CA available to the capillary plasma, presumably a CA IV type enzyme attached to the luminal surface of the endothelium[14], which brings plasma $[H+][HCO_3^-]$ into equilibrium with $[CO_2]$ within the plasma transit time, or shortly therafter. Without this CA, the re equilibration after elimination of HCO_3^- would have to be brought about by the uncatalyzed reactions which would require a number of seconds and might not be complete by the time the blood reached peripheral tissues. Mirror images of this mechanism take place in tissues, and would be expected in the carotid body. However, the magnitude of any change in plasma $[HCO_3^-]$ necessary to produce equilibrium in the plasma is extremely small and its importance in total CO_2 exchange, of a tissue is questionable. Some of the CA found in the carotid body may be on the capillary endothelium, but it is an unlikely explanation for the large CA activity found.

2. Transport of H^+ across phospholipid membranes

An increase in arterial P_{CO2} stimulates nerve traffic out of the carotid body and increases its sensitivity to a drop in P_{O2}[11]. While an increased $[H^+]$ has many effects on chemical reactions in the cell, an increased P_{CO2} per se has no obvious significant reactions, except possibly a combination with uncharged amine groups to form carbamino compounds [8]. Therefore the most attractive hypothesis for the stimulatory action of arterial CO_2 on the carotid body is that it speeds the increase in intracellular $[H^+]$ but not affect the steady state stimulus, which is borne out by experiment[11]. An increase in extra cellular $[H^+]$ independent of a change in P_{CO2} cannot of itself cross a phospholipid cell membrane but can produce an increased extra cellular P_{CO2} by reacting with the ubiquitous HCO_3^- ion to form first H_2CO_3 and then CO_2 which could then enter the cell and form H^+ and HCO_3^-. CA in the blood or extracellular fluid would accelerate the first step in the process. This H^+ transport will cease when the intracellular $[H^+][HCO_3^-]$ reaches equilibrium with P_{CO2}. If the HCO_3^- produced can exchange for an extracellular anion, such as Cl^-, the total amount of H^+ moved into the cell will be far greater. CA is present in number and/or cytosol of cells which secrete H^+ or OH^-, including those that regulate internal $[H^+]$. The mechanism of H^+ transport though the membrane presumably involves these interactions with CO_2 movement, except where H^+ is exchanged for another cation and is a promising function for carotid body CA.

3. Buffering changes in P_{CO2} in alveolar gas and in tissue

The presence of CA in lung parenchymal tissue maintains chemical equilibrium between $[H+][HCO_3^-]$ in the tissue and alveolar PCO2 . Since there is at least twice as much $CO_2 + HCO_3^-$ in lung parenchyma as CO_2 in alveolar gas, the lung tissue pool acts as a buffer to dampen swings in alveolar PCO_2 caused by alternating inspiration and expiration[3]. However, this action of CA would appear to have no pertinence to the carotid body.

4. Facilitated transport of CO_2 within cells and tissues

CO_2 transport through a layer of cell or tissue with a CO_2 permeable but HCO_3^- and H^+ impermeable boundary is magnified or facilitated by the formation of a high concentration of HCO_3^-/CO_2 at one boundary, diffusion of this $HCO_3^- + CO_2$ through the layer and then dehydration of the $HCO_3^- + H^+$ at the second boundary to reform CO_2[7,12]. The total transport of CO_2 through the layer equals that of CO_2 that of HCO_3^-. If the layer is sufficiently thin, the overall CO_2 transport is rapid and the facilitation becomes rate limited by the rate of hydration/dehydration of CO_2/HCO_3^-. CA speeds up the reaction and enhances the facilitation[16]. It is not obvious how this facilitation of CO_2 flux would be important for the carotid body. While it has its own metabolic CO_2 to dispose of, and facilitation of CO_2 diffusion would reduce any resulting P_{CO2} gradients within the carotid body, they should be small and unrelated to changes in arterial P_{CO2}.

5. Provision of HCO_3^- for syntheses such as those of urea, glucose and fatty acid

There is no evidence and no reason to suspect that the carotid body synthesizes urea or glucose. Presumably like many tissues it synthesizes some fatty acid de novo for membrane production or repair. However it appears unlikely that this function of CA is important in the carotid body.

6. $Ca^{++}/H+$ co transport

Inhibitors of CA block neuromuscular contraction in frog nerve-muscle preparations[11] and Geers et al.[8] propose that CA is involved in the uptake of H^+ that takes place during the release of Ca^{++} by the sarcoplasmic reticulum in striated muscle excitation-contraction coupling. This presumably involves membrane transport of Ca^{++}, either from the extracellular fluid or from the sarcoplasmic reticulum, which may involve CA. Thus although the mechanism which involves CA in H^+ or Ca^{++} transport is not entirely clear, it is a fruitful avenue to explore in the carotid body.

REFERENCES

1. Buckler, K.J., R.D.Vaughan-Jones, C.Peers and P.C.G.Nye,Intracellular pH and its regulation in isolated type I carotid body cells of the neonatal rat, J.Physiol. (Lond). 436:107,1991.

2. Dodgson, S.J.,The carbonic anhydrases: overview of their importance in cellular physiology and in molecular genetics, in: "The carbonic anhydrases: cellular physiology and molocular genetics", S.J.Dodgson, R.E.Tashian,G.Gros and N.D.Carter.eds., Plenum, New York, 1991.

3. *DuBois, A.B., W.O.Fenn and A.G.Britt*. CO2 dissociation curve of lung tissue.<u>J. App. Physiol.</u> 5:13 (1952).

4. *Edsall, J.T*, The carbonic anhydrases of erythrocytes, <u>Harvey Lect.</u>, 62:191 (1968).

5. *Forster, R.E.II*,Methods for the measurement of carbonic anhydrase activity,<u>in</u>: "The carbonic anhydrases:cellular physiology and molocular genetics", S.J.Dodgson, R.E.Tashian,G.Gros and N.D.Carter.eds., Plenum, New York, 1991.

6. *Forster,R.E.*, Buffering in blood with emphasis on Kinetics.<u>In</u>: "The kidney:physiology and pathophysiology",2nd edition, D.W.Seldin and G.Giebisch,eds.,Raven,New York,to be published in 1992.

7. *Geers,C., and G.Gros*. Muscle carbonic anhydrascs:function in muscle contraction and in the homeostasis of muscle pH and pCO2, <u>in</u>: "The carbonic anhydrases:cellular physiology and molecular genetics", S.J.Dodgson,R.E.Tashian,G.Gros and N.D.Carter,eds., Plenum, New York, 1991.

8. *Gros,G., H.S.Rollema and R.E.Forster*,The carbamate equilibrium of a-and e-amino groups of human hemoglobin at 37 °C.<u>J.Biol.Chem.</u> 256:5471 (1981).

9. *Hewett-Emmett,D., and R.E.Tashian*, Structure and evolutionary origins of the carbonic anhydrase multigene family, <u>in</u>: "The carbonic anhydrases:cellular physiology and molecular genetics", S.J.Dodgson,R.E.Tashian,G.Gros and N.D.Carter.eds., Plenum, New York, 1991.

10. *Khalifah,K.G., and D.N.Silverman*. Carbonic anhydrase kinetics and molecular function, <u>in</u>: "The carbonic anhydrases:cellular physiology and molecular genetics", S.J.Dodgson, R.E.Tashian,G.Gros and N.D.Carter.eds., Plenum, New York, 1991.

11. *Lahiri,S.*, Carbonic anhydrase and chemoreception in carotid and aortic bodies, <u>in</u>: "The carbonic anhydrases : cellular physiology and molecular genetics", S.J.Dodgson, R.E. Tashian, G. Gros and N.D.Carter.eds., Plenum, New York, 1991.

12. *Longmuir,I.S., R.E.Forster and C.Woo*. Diffusion of carbon dioxide through thin layers of solution. <u>Nature</u> 209:393 (1966).

13. *Maren,T.H.,Carbonic anhydrase*: chemistry, physiology and inhibition, <u>Physiol.Rev.</u> 47:595,1967.

14. *Nioka,S., R.P.Henry and R.E. Forster*. Total CA activity in isolated perfused guinea pig lung by [18] O-exchange method. <u>J.Appl.Physiol.</u> 65:2236 (1988).

15. *Pocker,Y., and S.*Sarkanen.Carbonic anhydrase;structure, catalytic versatility and inhibition, <u>Adv.Enzymol.</u> 47:149 (1978).

21   Ca^{2+} DYNAMICS IN CHEMORECEPTOR CELLS: AN OVERVIEW

C. González, J.R. López-López, A. Obeso, A. Rocher, and J. García -Sancho

Dpto. Bioquímica y Biología Molecular y Fisiología, Facultad de Medicina, Universidad de Valladolid, 47005 Valladolid, Spain

The carotid body (CB) was defined as a sensory organ by De Castro in 1928. Two years later, Heymanns and coworkers demostrated that the organ was sensitive to alterations in blood gases and pH, in such a way that a decrease in blood PO$_2$ or pH or an increase in blood PCO$_2$ produced activation of the CB and, reflexely, hyperventilation. De Castro postulated that glomus cells were the sensor structures and that they should release some substance to transmit the stimulus to the sensory nerve endings (De Castro, 1928). De Castro's point of view, was widely accepted, and therefore the CB was considered a secondary sensory receptor. As a consequence, the principal aims of many workers in the chemoreception field have been to define the nature of the sensing mechanims (sensory transduction process) and to identify the substances released by chemoreceptor cells.

The initial experimental approaches to these questions were based on the use of "in vivo" preparations. The physiological role of CB in systemic reflexes was established and the "in vivo" preparation led also to some hypothesis about the sensory transduction process. The metabolic and the acidic hypothesis were developed (Fidone and Gonzalez 1986). However, the truly intimate functional aspects of the chemoreception process remained undefined. It became necessary to look for alternative experimental approaches, and an elegant "in vitro" preparation was developed by Eyzaguirre and Lewin in 1961. However Eyzaguirre's preparation was not well accepted; the results obtained in "in vitro" preparations were criticized, arguing that the CB was not functional or that it was dead (see discussion in Torrance, pp. 248 and 298 and in Acker *et al.*, 1977 pp. 76-77). General acceptance of the preparation had to await until the seventies, when Whalen and Nair (1976) and Starlinger and Lübbers (1976) demostrated that the CB "in vitro" exhibited PO$_2$ levels and O$_2$ consumption rates very similar those found "in vivo". Ultrastructural analysis of the CB after "in vitro" superfusion showed excellent preservation of the tissue (Verna *et al.*, 1981). The "in vitro" preparation offered a well controlled system to study basic mechanisms, including neurotransmitter synthesis and release.

In the later sixties, a considerable background existed in the field of neurotransmission and secretion in general. Katz and Miledi (1965) had demostrated the role of Ca^{2+} in the release of acetylcholine in the neuromuscular junction. Studies performed in the adrenal medulla led to similar results and to the proposal of the concept of stimulus-secretion coupling by Douglas in 1968. In this scheme, acetylcholine would produce an increase in membrane permeability, an influx of Ca^{2+} and the secretion of catecholamines. Douglas viewed stimulus-secretion coupling as a general concept, that should be common to many different cells; the differences would be the nature of the stimulus and of the released substances. Secretion studies carried out in other preparations as neurohypophysis submaxillary gland or mast cells confirmed Douglas's theory. Sensory synapses should display similar properties, and therefore, chemotransduction in the CB should be considered as a particular case of stimulus-secretion coupling.

The first studies about the role of Ca^{2+} in chemotransduction process appeared in the Symposium on Arterial Chemoreceptors held in Oxford in 1966 (Eyzaguirre and Zapata, 1968). At that time, the idea that chemoreceptor nerve endings were stimulated by some substance released by chemoreceptor cells was prevalent. There was not agreement about the identity of the neurotransmitter, but acetylcholine and catecholamines were the two candidates at hand. In that meeting, Eyzaguirre and Zapata showed, using the "in vitro" preparation that the CB output (carotid sinus nerve discharges) was indeed Ca^{2+} dependent. Bathing the CB in Ca^{2+}-free solutions prevented the CB to respond to hypoxia, acidity or interruption of flow. These results should indicate, according to the stimulus-secretion coupling theory, that Ca^{2+} was necessary to release some substance from chemoreceptor cells during stimulation. In 1973, in the Bristol meeting, Eyzaguirre (Eyzaguirre, Fidone and Nishi, 1975) confirmed the role of Ca^{2+} studying, also in the "in vitro" preparation, the generation of mass receptor potentials. He showed that acetylcholine was unable of reversing the receptor blockade produced by Ca^{2+}-free solutions, a finding that should be considered contrary to the notion that acetylcholine was the transmitter between chemoreceptor cells and nerve endings .

This research program, which tacitly settled CB function in the context of stimulus-secretion coupling theory, suffered an unfortunate drawback in the early seventies. Biscoe and coworkers repeated De Castro's degeneration experiments and found that nerve endings in synaptic contact with glomus cells were efferent. Therefore glomus cells should be glandular cells activated by their secretomotor innervation. The chemoreceptor structures should be a very fine sensory fibers present between sustentacular cells (Biscoe, 1971). However, although this theory prevailed during almost a decade, the chemoreceptor nature of chemoreceptor cells was re-established with different approaches by many laboratories (for a detailed discussion see Fidone and Gonzalez, 1986). In the early eigthies Fidone and coworkers demostrated, using an "in vitro" preparation of rabbit CBs, that hypoxic stimulation activated synthesis and release of dopamine in proportion to the stimulus intensity, and that the release was highly dependent on extracellular Ca^{2+} (Fidone et al., 1982a, 1982b). These results supported clearly the original point of view considering chemotransduction as a particular case of stimulus-secretion coupling.

Keeping this idea in mind, Gonzalez's group has performed a large series of neurochemical experiments directed to check the role of Ca^{2+} in the chemotransduction process. The experiments were designed with a double aim: first, it should be demonstrated that dopamine was released in proportion to the intensity of every stimuli; second, it should be demonstrated that the stimulus-induced release was Ca^{2+}-dependent. These two premises imply that the release response was a valid index of chemoreception and also a valid index of $[Ca^{2+}]_j$ (Baker and Knight, 1984). Rigual et al. (1984, 1986 and 1991) demostrated that the cat CB "in vitro" was able to synthetize dopamine and to release it in response to acidic and hypoxic stimulation, and that the amount of dopamine released was proportional to the intensity of stimulus and paralleled by the frequency of sensory discharges. They also showed that hypoxic-and acid-induced release of dopamine was strongly dependent of extracellular Ca^{2+}. A typical experiment of Ca^{2+}-dependence of hypoxic stimulus-induced release is shown in figure 1A. Simultaneously, Almaraz et al. (1986) found that extracellular high K^+ was a very powerful stimulus for the cat CB promoting a great release of dopamine and that the effect of K^+ was also mediated by extracellular Ca^{2+}.

A typical experiment showing the Ca^{2+} dependence of high K^+-induced release of dopamine is showed in figure 1B. It was known (Kirpekar and Prat, 1979; Kikodoro and Richie, 1980) that K^+ induced neurotranmitter release by depolarizing the secretory cells and promoting Ca^{2+} entry through voltage-dependent channels. Therefore, Almaraz's results demostrated two important aspects of chemoreceptor cell function. Firstly, they established that membrane potential of chemoreceptor cells should be dependent of K^+ a notion questioned up to the middle eighties, and secondly, they strongly supported the existence of voltage dependent Ca^{2+} channels in these cells.

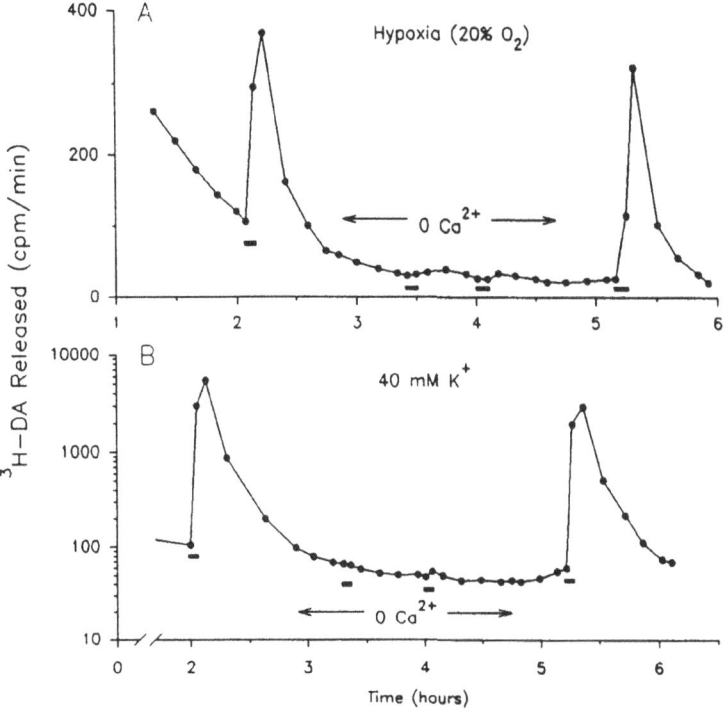

Fig. 1. Ca^{2+} dependence of the ^3H-dopamine induced release. The time course of the release of ^3H-dopamine from two different cat carotid prelabelled with ^3H-tyrosine are shown. In (A) hypoxia (20 % O_2 control 100% O_2) and in (B) High K^+ (60 mM) were applied at horizontal bars. Ca^{2+} free media was used during the marked time. Taken from Rigual's Doctoral Thesis, 1984 (A) and from Almaraz's Doctoral Thesis, 1983 (B).

As the stimulus-secretion coupling theory predicted, all the known stimuli of the CB were able to produce release of dopamine in a Ca^{2+}-dependent manner (Obeso et al., 1986, 1989). The release was always proportional to the intensity of stimulus, and the frequency discharges of carotid sinus nerve increased in parallel to the increase release of dopamine. This parallelism disappeared in Ca^{2+}-free solutions. In this conditions,

151

different stimuli augmented sensory discharges by about half of that seen in control Ca^{2+}-containning solutions, but the release of dopamine was reduced by 80-95%. Fidone et al. (1982b) explainned this lack of parallelism using the hazy concept of "a great synaptic safety factor" in the aminergic synapses. In fact, it was known that the affinity of dopaminergic antagonists diminished in Ca^{2+}-free solutions (Van Buskirk and Dowling, 1982) and that the effects of neurotransmitters were also stronger in Ca^{2+}-free media (Kanno et al., 1976; Kato and Narashami, 1982).

It was clear that extracellular Ca^{2+} was necessary for dopamine release. Therefore, the understanding of the transduction process itself could go a step further if we were able of understand the cellular mechanisms involved in the influx of Ca^{2+} during stimulation. The existence of voltage-dependent Ca^{2+} channels was confirmed by Obeso in 1984 in her Doctoral Thesis (see Obeso et al., 1987). She tested the effects of Nitrendipine 0.5 mM (a specific blocker of L-Type voltage dependent Ca^{2+} channels) on dopamine release induced by hypoxia, high K^+ and hypercapnic acidosis, using an "in vitro" preparation of cat CB. The results are shown in figure 2.

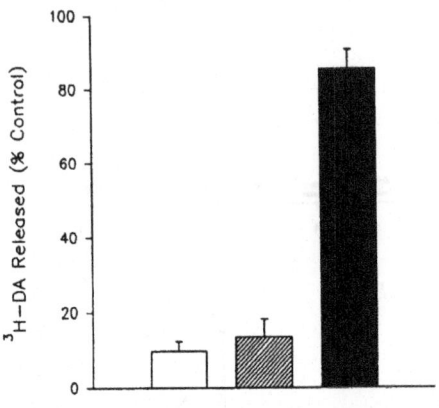

Fig. 2. Effects of Nitredipine (0.5 mM) on ^3H-dopamine release from cat carotid body induced by 60 mM K^+ (open bar), 20% O_2 (crossed bar) and 20% CO_2, pH 6.8 (filled bar). Control solution was 95 % O_2, 5% CO_2. Data are mean±SEM, n=4. Taken from Obeso's Doctoral Thesis,1984.

Hypoxic and high-K^+-induced release were inhibited by nitrendipine by almost a 90% suggesting that hypoxia, as high K^+, depolarize chemoreceptor cells in order to activate the voltage dependent Ca^{2+} channels. Similar results for hypoxic stimulation were obtained by Shaw et al. (1989) measuring ^{45}Ca fluxes. On the contrary, acidic stimulus induced release was unaffected by nitrendipine. These findings indicated that, although as already mentioned acidic stimulus induced release was Ca^{2+} dependent, Ca^{2+} did not enter the cells through Ca^{2+} channels. Therefore, the mechanisms involved in the transduction of hypoxic and acidic stimuli seem to be different and other pathways for Ca^{2+} entry, different than voltage dependent Ca^{2+} channels, had to be investigated (see below).

Rocher in 1989 in her Doctoral Thesis (see Obeso et al., 1992) expanded Obeso's results in a "in vitro" preparation of rabbit CBs, testing the effects of different dihidropyridines (BayK 8644 or Nisoldipine) and inorganic blockers of Ca^{2+} channels (Cd^{2+} or Co^{2+}) on dopamine induced release. They demostrated also the presence of Na^+ channels in chemoreceptor cells studying the effects of veratridine and tetrodotoxin on release of dopamine (Rocher et al., 1988).

Direct verification of the existence of voltage dependent channels has been obtained with the patch clamp tecnique in different laboratories using isolated chemoreceptor cells from different animal species. Ca^{2+} currents with properties of L-type channels have been described in isolated cells from adult rabbit (Duchen et al., 1988;

Ureña et al., 1989), neonatal rabbit (Hescheler et al., 1989) and neonatal rats (Peers, 1990). The hypothesis of hypoxic induced depolarization gained also support from patch clamp experiments. López-Barneo et al. (1988) identified in chemoreceptor cells a K^+ current selectively inhibited on lowering PO_2 in the bathing solution. It was also observed an increase in action potentials firing frequency of chemoreceptor cells during low PO_2 stimulation and it was proposed a cause-effect relationship between low PO_2, K^+ current inhibition and the increase in firing frequency (López-López et al., 1989). Hypoxic K^+ current inhibition has been confirmed by other authors, although there are species and age-related differences in the nature of the K^+ channels inhibited (Hescheler et al., 1989; Peers, 1990; Stea and Nurse, 1991).

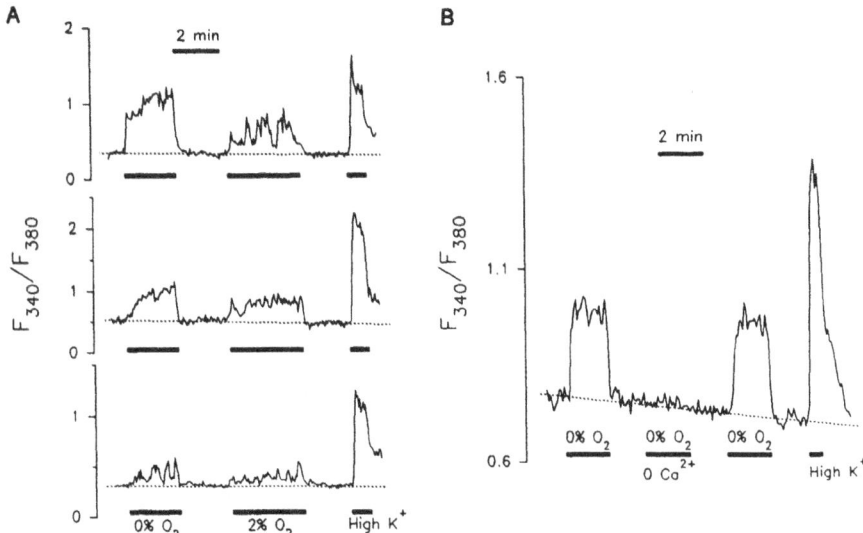

Fig. 3. Intracellular Ca^{2+} variations under hypoxic and high K^+ stimulation of chemoreceptor cells isolated from adult rabbit carotid body. Ca^{2+} levels are represented as the ratio of the fluorescences obtained at 340 and 380 nm in single cells loaded with fura-2. (A) shows the time course of the intracellular Ca^{2+} level in three different cells stimulated with anoxia, 2% O_2 and 60 mM K^+. (B) shows the extracellular Ca^{2+} dependence of Ca^{2+} increase induced by anoxic stimulation in a different chemoreceptor cell.

A new experimental approach to the study of changes in cytosolic Ca^{2+} during stimulation in isolated chemoreceptor cells is now available. Computer assisted image analysis allows to quantify modifications in intracellular Ca^{2+} in single cells using fluorescent dyes as fura-2. Figure 3A shows the relative intracellular Ca^{2+} levels in three different cells during normoxia and two intensities of hypoxic stimulation, anoxia and 2%O_2 -equilibrated media. It can be seen that hypoxia increased the intracellular Ca^{2+} in a dosis-dependent manner. In some cells, during hypoxia the intracellular calcium levels showed an oscillatory behavior. The increase in calcium induced by high extracellular potassium is shown at right. The cell response was fast in onset and was mantained during the 30 seconds of high K+ superfusion. Figure 3B shows the modifications in intracellular Ca^{2+} during hypoxia both in the presence and in the absence of Ca^{2+} in the bathing media.

In Ca^{2+}-free media hypoxia was not able to change intracellular Ca^{2+}. These data, although very preliminar, confirmed the neurochemical observations described above. Using similar experimental approaches (i.e. patch clamp and intracellular Ca^{2+} imaging), however, Biscoe and coworkers obtained different results, postulating different hypoxic transduction mechanisms. Briefly, they postulated that mitochondria should be both, the sensor structure and the source of Ca^{2+} to increase cytosolic Ca^{2+} levels. A full discussion comparing both, the plasma membrane and the mitochondrial models, could be find in Biscoe and Duchen (1990) and in Gonzalez et al. (1992).

As it has been mentioned above, acidic transduction seems to involve different mechanisms. Rigual et al. (1991) had shown that the effective stimulus during both metabolic or respiratory acidosis were the changes in intracellular pH. Rocher et al. (1991), using the secretory response as a measure of intracellular Ca^{2+} levels, investigated the pathways for Ca^{2+} entry into the cells under acidic stimulation. Dopamine release induced by acidic stimuli was dependent on extracellular Na^+ and Ca^{2+}, and it was also inhibited to a great extent by Na^+/H^+ exchanger blockers, as amiloride or EIPA, and by bicarbonate reduction and Cl^- removal from the bathing solution. These results led to the proposal of a model for acidic transduction in which the coupled activation of different antiporters would produce the required increase in intracellular Ca^{2+} to elicit neurotransmitter release. The existence of these antiporters has been confirmed with direct measurements of intracellular Ca^{2+} and pH by other groups (Biscoe et al., 1989; Buckler et al., 1991). Results obtained with these tecniques (Sato et al., Buckler et al. and Vaughan-Jones et al., present meeting), or in a "in vitro" preparation of cat CB, measuring nerve discharges (Iturriaga and Lahiri, present meeting) are also consistent with the fundamental aspects of this model.

Departing from the last sixties, these have been explained many aspects of the role of Ca^{2+} in the special stimulus-secretion coupling that represents the chemotransduction process. However, there is not yet a general agreement about Ca^{2+} routes and dynamics in chemoreceptor cells and a lot of questions have to be solved experimentally in the next years. Among others, aspects such as the role of Ca^{2+} present in intracellular stores and the nature of the stimulus that induce its release, the role of second messengers modulating the Ca^{2+} entry pathways, the antiporters or the release of Ca^{2+} from intracellular stores should be studied.

ACKNOWLEDGEMENTS

Thanks to María de los Llanos Bravo for technical assistance. Work supported by grants DGICYT 89/0358 and Junta de Castilla y León 1102/89.

REFERENCES

Acker, H., Fidone, S., Pallot, D., Eyzaguirre, C., Lübbers, D.W., and Torrance, R.W., 1977, eds., "Chemoreception in the Carotid Body".

Almaraz, L., Gonzalez, C., and Obeso, A., 1986, Effects of high potassium on the release of ^3H-dopamine from the cat carotid body in vitro, J. Physiol., 379:293.

Baker, P.F., and Knight, D.E., 1984, Calcium control of exocytosis in bovine adrenal medullary cells, TINS, 4:120.

Biscoe, T.J., 1971, Carotid body: structure and function. Physiol. Rev., 51:427.

Biscoe, T.J., and Duchen, M.R. 1990, Cellular basis of transduction in carotid chemoreceptors, Am. J. Physiol., 258:L271.

Biscoe, T.J., Duchen, M.R., Eisner, D.A., O'Neill, S.C., and Valdeolmillos, M., 1989, Measurements of intracellular Ca^{2+} in dissociated type I cells of the rabbit carotid body, J. Physiol. 416:421.

Buckler, K.J., Vaughan-Jones, C., Peers, C., and Nye, P.C.G., 1991, Intracellular pH and its regulation in isolated type I carotid body cells of the neonatal rat, J.Physiol., 436:197.

De Castro, F., 1928, Sur la structure et l'innervation du sinus carotidien de l'homme et des mammifères. Nouveaux faits sur l'innervation et la fonction du glomus caroticum. Etudes anatomiques et physiologiques, Trab. Lab. Invest. Biol. Univ. Madrid, 25:331.

Douglas, W.W., 1968, Stimulus-secretion coupling: The concept and clues from chromaffin and other cells, Br.J.Pharmacol., 34:451.

Duchen, M.R., Caddy, K.W., Kirby, G.C., Patterson, D.L., Ponte, J., and Biscoe, T.J., 1988, Biophysical studies of the cellular elements of the rabbit carotid body, Neuroscience 26:291.

Eyzaguirre, C., and Lewin, J., 1961, Chemoreceptor activity of the carotid body of the cat, J. Physiol. 159:222.

Eyzaguirre, C., and Zapata, P., 1968, A discussion of possible transmitter or generator substances in carotid body chemoreceptors, in: "Arterial chemoreceptors," R.W., Torrance, ed., Blackwell, Oxford.

Eyzaguirre, C., Fidone, S., and Nishi, K., 1975, Recent studies on the generation of chemoreceptor impulses, in: "The Peripheral Arterial Chemoreceptors," M.J., Purves, ed., Cambridge Univ. Press., London.

Fidone, S.J., and Gonzalez, C., 1986, Initiation and control of chemoreceptor activity in the carotid body, in: "Handbook of Physiology. The respiratory system II," A.P., Fishman, ed., Am. Physiological Society, Bethesda.

Rocher, A., Obeso, A., González, C., and Herreros, B., 1991, Ionic mechanisms for the transduction of acidic stimuli in rabbit carotid body glomus cells, J. Physiol., 433:533.

Rocher, A., Obeso, A., Herreros, B., and Gonzalez, C., 1988, Activation of the release of dopamine in the carotid body by veratridine. Evidence for the presence of voltage-dependent Na^+ channels in type I cells. Neurosci. Lett. 94:274.

Shaw, K., Montague, W. & Pallot, D.J., 1989, Biochemical studies on the release of catecholamines from the rat carotid body in vitro, <u>Biochim. Biophys. Acta.</u>, 1013:42.

Starlinger, H., and Lübbers, D.W., 1976, Oxygen consumption of the isolated carotid body tissue (cat), <u>Pflügers Arch.</u>, 366:61.

Stea, A., and Nurse, C.A, 1991, Whole-cell and perforated-patch recordings from O_2-sensitive rat carotid body cells grown in short- and long-term culture, <u>Pflügers Arch.</u>, 418:93.

Torrance, R.W., 1968, ed., "Arterial Chemoreceptors".

Ureña, J., López-López, J.R., Gonzalez, C., and López Barneo, J., 1989, Ionic currents in dispersed chemoreceptor cells of the mammalian carotid body, <u>J. Gen. Physiol.</u>, 93:979.

Van Buskirk, R., and Dowling, J.E., 1982, Calcium alters the sensitivity of intact horizontal cells to dopamine antagonists., <u>Proc. Natl. Acad. Sci. USA.</u>, 79:3350.

Verna, A., Roumy, M., and Leitner, L.M., 1981, Ultrastructural features of the carotid body after in vitro experiments: correlation with physiological results, <u>J.Neurocyol.</u>, 10:659.

Whalen, W.J., and Nair, P., 1976, PO_2 in the carotid body perfused and/or superfused with cell-free media, <u>J. Appl. Physiol.</u>, 41:180.

22 SPECTROPHOTOMETRIC ANALYSIS OF HEME PROTEINS IN OXYGEN SENSING CELL SYSTEMS

A. Görlach, B. Bölling, E. Dufau, G. Holtermann and H. Acker

Max-Planck-Institut für Systemphysiologie, Rheinlanddamm 201, 4600 Dortmund 1, Germany

INTRODUCTION

The mechanism by which specialized cells in the body are able to sense the microenvironmental PO_2 and to form a corresponding signal for avoiding hypoxic cell damages is not elucidated. There are, however, many indications that part of the oxygen sensing mechanism is brought about by heme proteins. We have, therefore, carried out spectral analysis of carotid body cells as well as hepatoma cells (HepG2) to characterize their inherent heme proteins. Both cell systems are very well known for their capability to transduce changes of the oxygen pressure into a biological signal like the chemoreceptor discharge stimulating respiration or circulation or the hormone erythropoietin stimulating proliferation and maturation of the erythrocytic progenitors in the bone marrow.

The idea of heme proteins actin as PO_2 sensor in the carotid body has been published by several groups [1,2,3]. In their classical paper Mills and Jöbsis[3] analyzed photometrically a cytochrome aa_3 in the carotid body with a low as well as with a high O_2-affinity component. In a more recent paper Biscoe and Duchen,[4] supported the idea of a specialized cytochrome aa_3 by a model which located the O_2 sensor into carotid body mitochondria responding to oxygen changes due to a low O_2 affinity far above the critical mitochondrial PO_2 with a mitochondrial membrane depolarization and a subsequent calcium release. Intensive studies of the different cytochromes in the respiratory chain, however, have revealed affinity values for oxygen with a $PO2_{50}$ below 1 Torr only[5,6], which questioned $PO2_{50}$ values of about 90 Torr as given by Mills and Jöbsis,[3] for the specialized cytochrome aa_3 component in the carotid body. Cross et al.[7] carried out a detailed photometric analysis of the rat carotid body to give more information about heme protein characteristics in this tissue. They detected a measurable heme signal with absorbance maxima at 560 nm, 518 nm and 425 nm suggesting the presence of a b-type cytochrome. This was confirmed by pyridine haemochrome and CO spectra. This heme protein is capable of H_2O_2 formation and seems to possess, therefore, similarities with the cytochrome b of the NAD(P)H oxidase in neutrophiles[8].

Erythropoietin production (Epo) is triggered in kidney and liver by hypoxia with the cortical peritubular cells in the kidney as the main source[9,10]. The oxygen sensing process of this hypoxia-mediated hormone production was intensively investigated on the molecular level in the hepatoma cell lines Hep3B and HepG2[11]. Hypoxia and cobalt chloride both markedly increase expression of Epo RNA as well as biological active Epo in these cells. Since the increased Epo production under hypoxia could be inhibited by carbon monoxide the involvement of a heme protein in the oxygen-sensing mechanism was suggested[11]. A cytochrome b type was proposed as a possible candidate for the oxygen sensor since phenobarbital as an inductor of cytochrome P_{450} stimulated Epo production similar to hypoxia[12].

This paper will give indications that the oxygen-sensing mechanism in both cell systems might be similar with a cytochrome b as a major component, which does not participate in the energy production of the respiratory chain and is capable of H_2O_2 formation.

MATERIALS AND METHODS

Carotid Body Preparation

After prolonged flushing of the common carotid arteries with macrodex 6% (Knoll, Glandorf, FRG) to eliminate red blood cells from the tissue, carotid bodies with the intact sinus nerve and the immediate vessels were excised from rats anaesthetised with a mixture of 2% chloralose, 25% urethan (0.5 ml/100g body weight) and heparinized with 1300 units/rat. The carotid bodies were denuded of all other structures to be clearly visible for the photometric studies and placed in a small lucite superfusion chamber mounted on the stage of an upright microscope (Olympus, Hamburg, FRG) for spectral analysis.

Hepatoma Cell (HepG2) Tissue Culture

HepG2 cells (ATCC HB 8065) were grown in multicellular spheroid tissue culture using the spinnerflask technique (see for review Acker et al.[13]). Spheroids were grown at 37° C with a spin rate of 40 rpm in Dulbecco Eagle's basal medium and 10 % fetal bovine serum. The flagks were cylindrical (11 cm diameter * 24 cm high) and contained 250 ml of medium which was replenished three times a week. For photometry HepG2 multicellular spheroids were taken with a diameter between 500 and 800 µm containing 17,000 to 66,000 cells per spheroid.

Photometry

Carotid bodies and spheroids were located in a superfusion chamber on a small bench with little holes having the diameter of these small organs. The superfusion chamber was described in detail elsewhere[7], therefore, a brief description only is given below. Physiological salt solution (Locke's solution) containing 5 mM glucose and equilibrated with different O_2/CO_2 mixtures for adusting PO_2 at various levels and pH at 7.42 flowed at a rate of 10 ml per minute through the chamber. The organs were supplied symmetrically with nutrients by this procedure. Temperature was maintained at 36°C. Oxygen tension, pH and temperature in the Locke's solution were continuously controlled. The superfusion chamber was mounted on the stage of a light microscope (Olympus, Hamburg, FRG) for light absorption measurements. Light from a halogen lamp (12 V, 100 W) transilluminated the spheroid or carotid body only, since the bench was made intransparent by sputtering with gold, which has the advantage of avoiding uncharacteristic light scattering. Light crossing the spheroid or the carotid body was recorded by a photodiodearrayspectrophotometer (MCS 210, Zeiss, Köln, FRG) which was connected to the third ocular of the microscope trinocular head via a light guide.

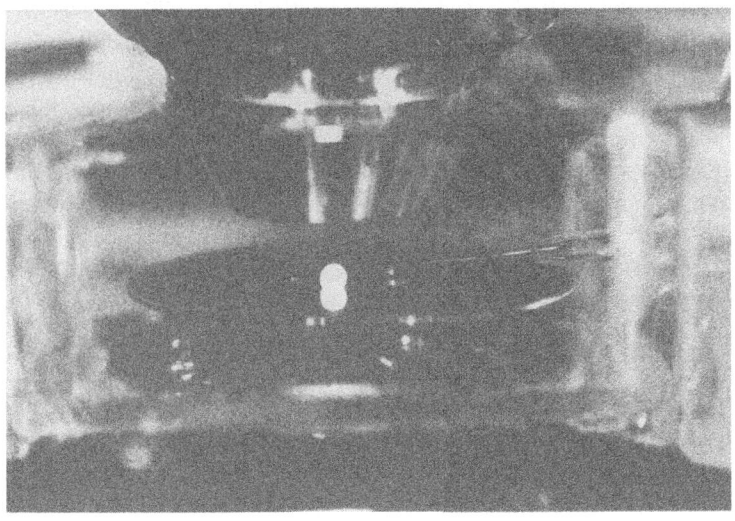

Fig. 1. Experimental set up for measuring light absorption changes in a HepG2 spheroid

Measurement of H₂O₂ Production

According to Cross et al.,[7] dihydrorhodamine 123 (Molecular Probes, Eugene, OR, USA) was dissolved in dimethylsulphoxide to give stock solutions of 50 mM. The dihydrorhodamine was stored under N_2. The non-fluorescent dihydrorhodamine is converted to rhodamine in the reaction with H_2O_2. This fluorescent positively charged dye is taken up by cells. Images of the fluorescent cells were obtained using a Bio Rad MRC 500 confocal scanning optical microscope (Bio Rad, München, FRG) mounted on a Zeiss IM405 inverted microscope (Zeiss, Koln, FRG). For measurements carotid bodies or HepG2 spheroids stained with dihydrorhodamine were placed as described above in the superfusion chamber which was mounted on the stage of the inverted microscope.

RESULTS

The first figure shows the location of a HepG2 spheroid in the superfusion chamber for light absorption measurements. The illuminated spheroid resting in a hole of an otherwise intransparent round size shaped bench is clearly visible. An oxygen sensitive microelectrode is pointing with its tip to the spheroid for controlling the PO_2 in the superfusion bath very close to the tissue. Light from a halogen lamp comes fom the bottom through a condensor and is transmitted by the spheroid tissue acting as a light guide to the obective and continues to the photodiodearrayspectrophotometer. This experimental procedure reults in a very specific and sensitive photometrical analysis of the tissue since there is no light travelling beside the tissue biasing as a separated light compartment the signal of the heme proteins being analyzed.

Figure 2 shows a typical example of light absorption characteristics of different heme proteins in a carotid body (right side) and a HepG2 spheroid (left side). A wavelength range between 400 and 650 nm is shown. From top to bottom are to be seen: the difference spectrum between oxidized and hypoxia-reduced tissue, the difference spectrum between oxidized and cyanide-reduced tissue and the difference spectrum between cyanide and hypoxia-reduced tissue. Significantly distinguishable peaks could be identified according to the literature[5], as cytochrome b at 430 and 559 nm, as cytochrome c at 550 nm and as cytochrome aa_3 at 445 and 603 nm. Cytochrome c and cytochrome aa_3 can be characterized as components of the respiratory chain since they are getting reduced under hypoxia as well as cyanide poisoning. In contrast to hypoxia cyanide induces at 559 nm a small increase of the light absorption only. The major part of cytochrome b responses to hypoxia at 430 nm and 559 nm in spite of the complete reduction of the respiratory chain by cyanide.

Figure 3 gives indications that carotid body cells (left side) and HepG2 cells (right side) are producing H_2O_2. The intracellular detection of fluorescent rhodamine 123 generated by the reaction between dihydrorhodamine and H_2O_2 is accomplished by confocal laser microscopy. The confocal microscope scans and collects emitted light from only within the plane of focus of the obective lens. Therefore the image obtained is the fluorescence from a discrete slice of about 10 μm within the whole tissue. Non fluorescent cells are not visible[7]. The highly fluorescent cells are in the plane of the focus (white colours). The fluorescent cytosol round the nucleus is clearly visible.

DISCUSSION

Cyanide insensitivity and the capability to produce O_2 radicals with a subsequent dismutation to H_2O_2 is a characteristic feature of a cytochrome b[8,14] which is described to be located in the outer cell membrane, the plasma membrane of the endoplasmatic reticulum as well as the Golgi apparatus, the outer mitochondrial membrane and the nucleus membrane. Even the cytochrome b as a member of the respiratory chain is desribed to produce oxygen radicals amounting up to 5% of the total oxygen consumption. Spectral charateristics as published in the literature are in very good accordance with the presented results underlining the specificity of our spectral analysis of the cytochrome b in carotid body - and HepG2 cells (see for review[14]).

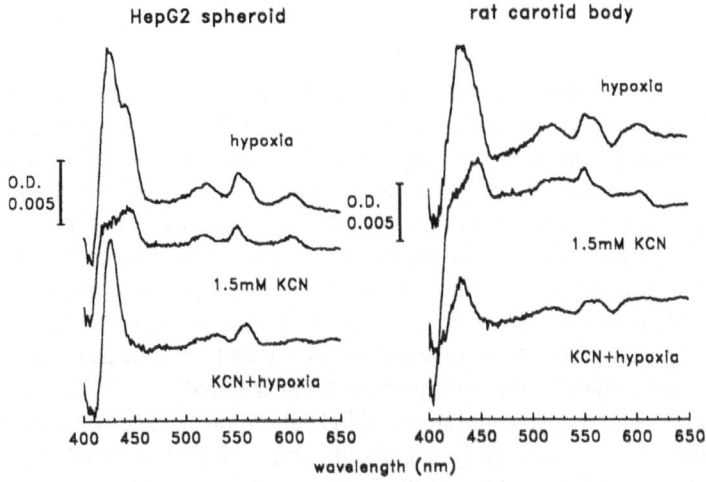

Fig. 2. Characteristic light absorption changes of carotid body and HepG2 cells under different reducing conditions

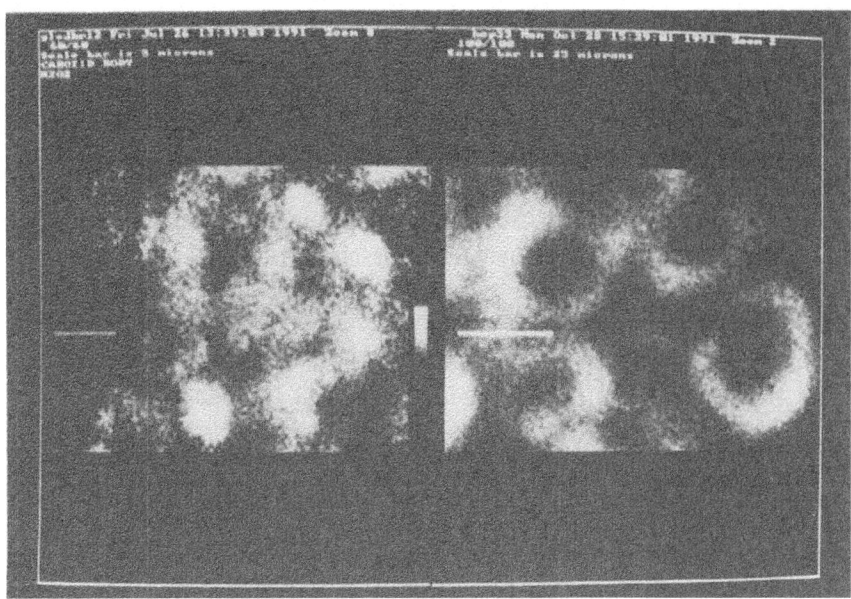

Fig. 3. After incubation with dihydrorhodamine cells in the carotid body (left side) and HepG2 spheroids (right side) were stained with rhodamine 123. Fluorescence intensity is shown with white as highest and black as lowest intensity

Several indications in the literature might support the importance of a H_2O_2 producing cytochrome b for the oxygen sensing process. Ueno et al.[15] could show that exogenously applied H_2O_2 or glucose oxidase as a direct H_2O_2 generator caused a significant enhancement of erythropoietin production in a renal carcinoma cell line. Also experiments as described by Fandrey et al.[12] that inducing cytochrome P_{450}, a member of the cytochrome b family, in HepG2 cells by phenobarbital application enhanced erythropoietin production, hint to the possibility that H_2O_2 generated by cytochrome P_{450} triggers this process. It might be interesting to mention that the participation of H_2O_2 in oxygen sensing is discussed also in the hypoxic vasoconstriction of the lung vessels[16]. In lung vessels H_2O_2 induces a catalase-dependent guanylate cyclase activation by interaction of the heme groups of the two enzymes probably leading to the cGMP mediated opening of K^+ channels with a subsequent relaxation of pulmonary arteries. This process is reversed under hypoxia when H_2O_2 production arrests leading to a vasoconstriction[16]. The involvement of a cytochrome b as a PO_2 sensor can be speculated for the carotid body with the model that the H_2O_2 production, which is scavenged by glutathione, decreases under hypoxia triggering herewith a change in the GSH/GSSG couple. This change might lead to an altered ion channel conductivity[17] since glutathione-gated K^+ channels are described which close under reduced conditions[18] like the PO_2-dependent K^+ channels of type-I cells[19,20]. The requirement of an increased H_2O_2 level in HepG2 cells under hypoxia for an increased erythropoietin production[15] could mean that hypoxic HepG2 cells have unknown mechanisms to stimulate H_2O_2 production by cytochrome b or to decrease the activity of H_2O_2 scavenging systems like catalase probably by interaction of its heme group with lower levels of oxygen.

REFERENCES

1. *B.B.Lloyd, D.J.C.Cunningham, and R.C.Goode*, Depresion of hypoxic hyperventilation in man by sudden inspiration of carbon monoxide, in: "Arterial Chemoreceptors," R. W. Torrance, ed., J Blackwell, Oxford, (1968).

2. *S. Lahiri and R. G. DeLaney*, Stimulus interaction in the response of carotid body chemoreceptor single afferent fibres, Respir. Physiol. 4:229 (1975).

3. *E. Mills and F. F. Jobsis*, Mitochondrial respiratory chain of carotid body and chemoreceptor response to changes in oxygen tension, J. Neurophysiol. 35:405 (1972).

4. *T. J. Biscoe and M.R. Duchen*, Monitoring PO_2 by the carotid chemoreceptor, NIPS 5:229 (1990).

5. *M. Oshino, T. Sugano, R. Oshino, and B. Chance*, Mitochondrial function under hypoxic conditions: the steady states of cytochrome aa_3 and their relation to mitochondrial energy states, Biochem. Biophys. Acta 368:298 (1974).

6. *D.F.Uilson, U.L.Rumsey, Th.J.Green, and J.M.Vanderkooi*, The oxygen dependence of mitochondrial oxidative phosphorylation measured by a new optical method for measuring oxygen concentration, J. Biol. Chem. 236:2712 (1988).

7. *A.R. Cross, L. Henderson, 0. T. G. Jones, M. A. Delpiano, J. Hentschel, and H. Acker*, Involvement of an NAD(P)H oxidase as a PO_2 sensor protein in the rat carotid body, Biochem. J. 272:743 (1990).

8. *O.T.G. Jones, A. R. Cross, J. T. Hancock, L. M. Henderson, and V. B. O'Donnel*, Inhibitors of NAD(P)H oxitase as guides to its mechanism, Biochem. Soc. Trans. 19:70 (1991).

9. *H. Scholz, H. J. Schurek, K. U. Eckhardt, A. Kurtz, and C. Bauer*, Oxygen dependent erythropoietin production by the isolated perfused rat kidney, Pflugers Arch. 418:228 (1991).

10. *H.Pagel, W.Jelkmann, and C. Weiss*, Isolated serum-free perfused rat kidneys release immunoreactive erythropoietin in response to hypoxia, Endocrin 128:2633 (1991).

11. *M.A. Goldberg, S. P. Dunning, and H.F. Bund*, Regulation of the erythropoietin gene: Evidence that the oxygen sensor is a heme protein, Science 242:1412 (1988).

12. *J. Fandrey, F. P. Seydel, C. P. Siegers, and W. Jelkmann*, Role of cytochrome P_{450} in the control of the production of erythropoietin, Life Science 47:127 (1990).

13. *R. M. Sutherland ant R. E. Durand*, Growth and cellular characteristics of multicell sphoroids, in: "Spheroids in Cancer Research," H. Acker, J. Carlsson, R. Durand, R. M. Sutherlant, eds., Rec. Res. Canc. Res., Springer Verlag, (1984).

14. *T. Galeotti, S. Borello, and L. Masotti*, Oxygen radical sources, scavenge systems ant membrane damage in cancer cells, in : "Oxygen Radicals: Systemic Events and Disease Processes," D. K. Das, W. B. Essman, ets., S. Karger AG, Basel (1990).

15. *M. Ueno, J. Brookins, B. S. Beckmann, ant J. U. Fischer*, Effects of reactive oxygen metabolites on erythropoietin production in renal carcinoma cells, Biochem. Biophys. Res. Comm. 1S:773 (1988).

16. *P. D. Cherry, H. A. Omar, K. A. Farrell, J. S. Stuart, ant M. S. Wolin*, Superoxide anion inhibits cGMP associatet bovine pulmonary arterial relaxation, . J. Physiol. 259:H1056 (1990).

17. *H. Sies*, Peroxisomal enzymes ant oxygen metabolis in liver, in: "Tissue Hypoxia and Ischemia," M. Ruvich, R. Coburn, S. Lahiri, B. Chance, eds., Plenum Publishing Corp., New York, (1988).

18. *J. P. Ruppersberg, M. Stocker, O. Pongs, St. H. Heinemann, R. Frank, and M. Koenen*, Regulation of fast inactivation of cloned ammalian IK(A) channels by cysteine oxidation, Nature 352:711 (1991).

19. *J. Hescheler, M. A. Delpiano, H. Acker, ant F. Pietruschka*, Ionic currents on type-I cells of the rabbit carotid body mesured by voltage clamp experiments ant the effect of hypoxia, Brain Res. 486:79 (1989).

20. *J. Lopez-Lopez, C. Gonzalez, J. Urena, and J. Lopez-Barneo*, Low PO_2 selectively inhibits K^+ channel avtivity in chemoreceptor cells of the mammalian carotid body, J. Gen. Physiol. 93:1001 (1989).

23 NEUROCHEMICAL AND MOLECULAR BIOLOGICAL ASPECTS ON THE RESETTING OF THE ARTERIAL CHEMORECEPTORS IN THE NEWBORN RAT

H. Holgert[1,2],T. Hertzberg[1],A. Dagerlind[2],T. Hökfelt[2] and H. Lagercrantz[1]

1)Pediatric Clinic and 2) Department of Histology and Neurobiology, Karolinska Institute, S-104 01 Stockholm, Sweden

INTRODUCTION

The role of the carotid body chemoreceptors in the fetus and immediately newborn has been a matter of debate over the past decades. Studies on fetal sheep suggested that the chemoreceptors are active in utero (Barcroft and Karvonen, 1948; Itskovitz and Rudolph, 1982, 1987), and this was confirmed by direct recordings from the sinus nerve (Blanco et al., 1984). However, the chemoreceptor response to hypoxia was lower in the fetus than in the newborn lamb (Blanco et al., 1984). This is consistent with the observation that the strength of the chemoreflex response is increased in babies a few days after birth as compared to the immediately newborn (Hertzberg and Lagercrantz, 1987). Thus, the arterial chemoreceptors seem to adapt from the low PaO_2 prevalent in utero and reset their sensitivity after birth, when arterial oxygen tension is rising. As a consequence, the chemoreceptors are quiescent during the first day or so after birth. Accordingly, onset of breathing after birth appear to be dependent on other drive mechanisms, e.g. thermal, visual and tactile stimuli. The postnatal resetting of the chemoreceptors has been hypothesized to be due to changes in neurotransmitter expression and turnover in the carotid body. Substance P and the catecholamines are currently among the most studied messenger molecules. Substance P-like immunoreactivity was found in nerve fibers prenatally, while immunoreactive cells were not seen until after birth in kittens (Scheibner et al., 1988). In the adult cat the stimulation of carotid body chemoreceptors by hypoxia, but not by hypercapnia, was blocked by a substance P-antagonist (Prabhakar et al., 1984, 1987). Dopamine, the most abundant catecholamine in the carotid body of most species, has been shown to modulate chemoreceptor activity (Cardenas and Zapata, 1981; see also Hellström et al., 1984). In the following we would like to discuss some experiments aiming to define the role of these two neurotransmitters in the carotid body.

DOPAMINE AND RESETTING OF CHEMORECEPTOR SENSITIVITY

A series of experiments were undertaken to see whether or not dopamine levels in the carotid body are involved in the changes in chemoreceptor sensitivity around birth. In the newborn rat, the increase in the strength of the chemoreflex occurring after birth is preceded by a marked drop in dopamine levels (Hertzberg et al., 1990). Chemoreceptor sensitivity in utero seems to be determined by the prevalent PaO_2 (Blanco et al., 1988). To mimic the low PaO_2 levels in utero we let rat pups to be born in a hypoxic environment (FiO_2 0.11-0.13). We found that rat pups kept in hypoxia had weaker chemoreflexes than normoxic ones and that the weak drive from the arterial chemoreceptor was paralleled by a sustained turnover rate of carotid body dopamine. Moreover, when rat pups born and reared in hypoxia were allowed to breathe air, the strength of the chemoreflex increased, and the turnover rate of carotid body dopamine dropped. This further supports the notion that the chemoreceptor sensitivity is governed by the PaO_2 in the neonate and that this is mediated by the rate of carotid body dopamine turnover (Hertzberg et al., 1992). Thus, the drop in dopamine utilization reflects the lifting of an inhibitory dopaminergic mechanism present at the low PaO_2 in the fetus.

IN SITU HYBRIDIZATION

The technique for hybridization of antisense oligonucleotide probes with mRNAs in situ (see Young, 1990) offers the possibility to visualize presence and levels of mRNAs for various gene products at the cellular level. We here present some preliminary in situ hybridization results from the carotid body. Spragley-Dawley rat pups from ages of one day before birth (fetal), 1-10 h, 1-2 d and 4-7 d (tyrosine hydroxylase mRNA) and 4 d and 21 d (preprotachykinin mRNA) were used. They were killed by a sharp blow against the head and the carotid bodies were rapidly dissected out and frozen on dry ice. The fetal group was obtained by caesarean section after sacrificing the mother by cervical dislocation. Pups were immediately put on ice and the carotid bodies dissected out. The tissue was cut in a cryostat at 14mm and thaw mounted onto microscopic slides. Probes complementary to nucleotides 1441-1488 of rat tyrosine hydroxylase mRNA (Grima et al., 1985) and nucleotides 145-192 of rat preprotachykinin A (SP) mRNA (Krause et al., 1987) were labelled with a ^{35}s-dATP at the 3'-end . In situ hybridization was performed essentially as described by Dagerlind and coworkers (1990; see also Young, 1990). Slides were incubated at 42°C for 16 h with 10^6 cpm of the labelled probe per 100 ml hybridization solution. After hybridization sections were rinsed, air-dried and dipped in liquid emulsion for autoradiography. Slides were exposed for 6 weeks, developed and densiometric analysis was done by using a DAGE-MTI 72 CCD series camera (DAGE-MTI, USA) coupled to a Nikon Microphot-FX microscope (Nikon, Japan). The image processing was performed with Image 1.16 software (courtesy of Dr. Wayne Rasband, NIMH, Washington, D.C., USA). Areas subjected for measurements were first identified in the microscope and the borders were interactively defined and a grain density value was computed. Rat brain stems and ganglion nodosum were used as control tissue for the preprotachykinin A probe, and these tissue sections were mounted on the same slides as the carotid bodies, thus hybridized together with the carotid bodies.

We found mRNA coding for tyrosine hydroxylase in all carotid bodies examined, where a hybridization signal was demonstrated over parenchymal cells (Fig. 1), and the preliminary results show a developmental decrease in tyrosine hydroxylase mRNA levels in the carotid bodies (Fig. 2). Although hybridization for the preprotachykinin A mRNA was seen in the control tissue, i.e. nucleus raphe magnus in the brain stem and over single cells in ganglion nodosum, no labelling was observed in the examined carotid bodies (Fig. 1).

The present study demonstrates that mRNA coding for tyrosine hydroxylase is present in the carotid body of the newborn rat, thus providing evidence for catecholamine synthesis within the glomus tissue of the organ.

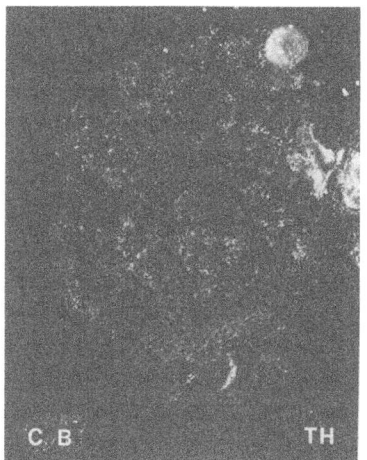
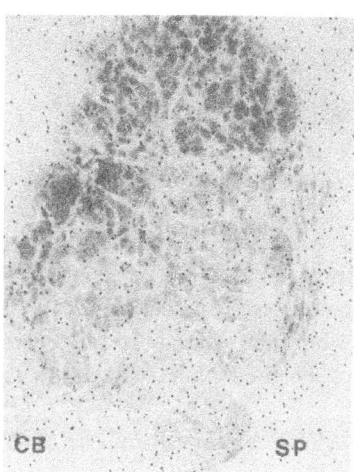
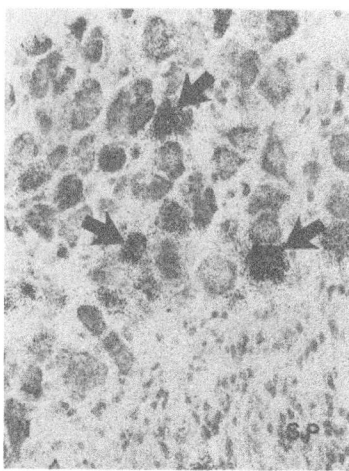

Fig. 1. Autoradiographs of emulsion dipped tissue sections of rat carotid bodies (CB) and ganglion nodosum (N) hybridized with probes complementary to rat tyrosine hydroxylase (TH) or preprotachykinin A (SP) mRNA. Slides are photographed in darkfield (TH) and in brightfield after toulidine blue staining (SP). Labelling is seen as light (TH) and dark (SP) grains over the tissue. Note the absence of labelling over the 21 d carotid body using the preprotachykinin A probe. Approximately 150x enlargement.

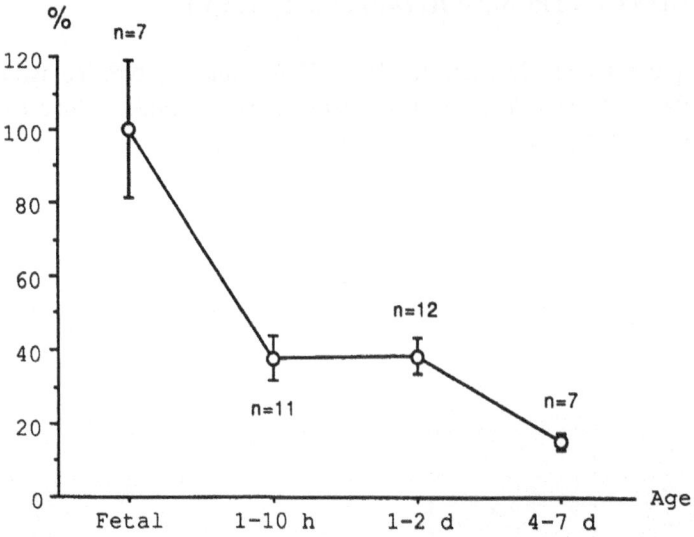

Fig. 2 Postnatal levels of tyrosine hydroxylase mRNA relative to fetal (E21) levels (100%). Mean±SE.

Although earlier results from our group suggest that dopamine is involved in the resetting of chemoreceptor sensitivity occuring after birth, it is far from clear how this is accomplished (Hertzberg et al., 1990, 1992). The present findings of decreased levels of tyrosine hydroxylase mRNA as the rat pups became older (Fig. 2) is consistent with our previous observations of a drop in dopamine turnover.

NO EVIDENCE FOR EXPRESSION SUBSTANCE P mRNA

In the present study, there was no evidence for the presence of mRNA coding for the substance P precursor preprotachykinin A in carotid bodies from 4 d and 21 d old rats. Substance P has been found in type-1 cells and nerve fibers in carotid bodies from cat by using immunohistochemistry (Lundberg et al., 1979; Cuello and McQueen, 1980; Scheibner et al., 1988; Prabhakar et al., 1989; Heym and Kummer, 1989), and substance P has been measured in cat by radioimmunoassay (Prabhakar et al., 1989). Substance P was found in nerve fibers also in rat, however no type-1 cells were labelled (Jacobowitz and Helke, 1980). Taken together these results suggest that substance P may not be synthesized in glomus cells in rats but rather in cells located in the petrosal ganglion or in the superior cervical ganglion which innervate the carotid body. Thus, substance P could be transported via nerve fibers. Another possibility is that substance P is synthesized at very low levels in the rat carotid body. There seems to be a clear species differences. Therefore results from studies on substance P in the carotid body of cats may not be valid for the rat

CONCLUSION

The present results, although preliminary, show that tyrosine hydroxylase mRNA is expressed in the carotid body of the newborn rat. They also indicate that a developmental decrease in tyrosine hydroxylase mRNA levels takes place during the first week of life. This supports the hypothesis of an involvment by dopamine in the postnatal resetting of the arterial chemoreceptors (Hertzberg et al., 1990, 1992). The absence of labelling of mRNA coding for preprotachykinin A suggests primarly that substance P mRNA is present below levels of detection using our in situ hybridization procedure, alternatively that substance P is not synthesized in the rat carotid body at these stages of development. Future studies are needed to elucidate whether substance P is synthesized or not in the carotid body of the newborn rat. It is also of interest to evaluate to what extent the synthesis rate of dopamine and substance P are affected by environmental factors such as hypoxia.

REFERENCES

Barcroft J. and Karvonen M.J., 1948, The action of carbon dioxide and cyanide on foetal respiratory movements; The development of chemoreflex function in sheep, J. Physiol., 107: 153-161.

Blanco C.E., Dawes G.S., Hanson M. and McCooke H.B., 1984, The response to hypoxia of arterial chemoreceptors in the fetal sheep and newborn lambs, J Physiol., 351: 25-37.

Blanco C.E., Hanson M.A. and McCooke H.B., 1988, Effects on carotid chemoreceptor resetting of pulmonary ventilation in the fetal lamb in utero, J. Dev. Physiol., 10, 167-174.

Cardenas H. and Zapata P., 1981, Dopamine induced ventilatory depression in the rat, mediated by carotid nerve afferents, Neurosci. Letts., 24: 29-33.

Cuello A. C. and McQueen D.S., 1980, Substance P: a carotid body peptide, Neurosci. Letts., 17: 215-229.

Dagerlind Å., Schalling M., Eneroth P., Goldstein M. and Hökfelt T., 1990, Effects of reserpine on phenylethanolamine-Nmethyltransferase mRNA levels in rat adrenal gland: role of steroids, Neurochem. Int., 17: 343-356.

Grima B., Lamouroux A., Blanot F., Faucon-Biguet N., and Mallet J., 1985, Complete coding sequence of rat tyrosine hydroxylase mRNA, Proc. Nat. Acad. Sci. USA, 82: 617-622.

Hellström S., Hanbauer I., Commissiong J., Karoum F. and Koslow S., 1984, Role and regulation of catecholamines in carotid body, in: Dynamics of Neurotransmitter Function. Editor; Hanin I. pp. 31-38. Raven Press. N.Y., USA.

Hertzberg T. and Lagercrantz H., 1987, Postnatal sensitivity of the perpheral chemoreceptors in newborn infants, Arch. Dis. Childh., 62: 1238-1241.

Hertzberg T., Hellström S., Lagercrantz H. and Pequignot J-M., 1990, Development of the arterial chemorefelx and turnover of carotid body catecholamines in the newborn rat, J. Physiol., 425: 211-225.

Hertzberg T., Hellström S., Holgert H., Lagercrantz H. and Pequignot JM., 1992, Ventilatory response to hyperoxia in newborn rats born in hypoxia - possible relationship to carotid body dopamine, J. Physiol. 456: 645-654.

Heym C. and Kummer W., 1989, Immunohistochemical distribution and colocalisation of regulatory peptides in the carotid body, J. Elect. Micr. Technol., 12: 331-342.

Itskovitz J. and Rudolph A. M., 1982, Denervation of arterial chemoreceptors and baroreceptors in fetal lambs in utero, Am. J. Physiol., 242, H916-922.

Itskovitz J. and Rudolph A. M., 1987, Cardiorespiratory response to cyanide of arterial chemoreceptors in fetal lambs. Am. J. Physiol., 252, H916-922.

Jacobowitz D. W. and Helke J. C., Localization of substance P immunoreactive nerves in the carotid body. Brain Res Bull., 5, 195-197 .

Krause J.E., Chirgwin J.M., Carter M.S., Zu Z.S. and Hershey A.D., 1987, Three rat preprotachykinin mRNAs encode the neuropeptides substance P and neurokinin A. Proc. Nat. Acad Sci. USA, 84: 881-885.

Lundberg J.M., Hokfelt T., Fahrenkrug J., Nilsson G. and Terenius L., 1979, Peptides in the cat carotid body (glomus caroticum): VIP-, enkephalin-, and substance P-like immunoreactivity, Acta Phys. Scand., 107 : 279-281 .

Prabhakar N.R. Runold M., Yamamoto Y., Lagercrantz H. and von Euler C., 1984, Effects of substance P antagonist on the hypoxia-induced carotid chemoreceptor activity, Acta Phys . Scand., 121 : 301-303 .

Prabhakar N.R., Mitra J. and Cherniack N.S., 1987, Role of substance P in hypercapnic excitation of carotid body chemoreceptors, J. Appl. Physiol ., 63 : 2418-2425 .

Prabhakar N.R., Landis S.C., Kumar G.K., Mullikin-Kilpatrick D., Cherniack N.S. and Leeman S., 1989, Substance P and neurokinin A in the cat carotid body: localization, exogenous effects and changes in content in response to arterial PO_2, Brain. Res , 481: 205-214.

Scheibner T., Read D. and Sullivan C.E., 1988, Distribution of substance P-immunoreactive structures in the developing cat carotid body, Brain . Res ., 453 : 72-78 .

Young W.S. III., 1990, In situ hybridization histochemistry. In: Handbook of Chemical Neuroanatomy, Volume 8, Ch. 19, Editors, Björklund A., Hokfelt T., Wouterlood F.G., Van Der Pol A.N., Analysis of Neuronal Microcircuits and Synaptic Interactions. Elsevier Science Publisher B.V.

24 CARBONIC ANHYDRASE AND CAROTID BODY CHEMORECEPTION IN THE PRESENCE AND ABSENCE OF CO_2-HCO_3^-

R. Iturriaga and S. Lahiri

Department of Physiology, University of Pennsylvania, Philadelphia,
PA 19104-6085, USA

INTRODUCTION

Carbonic anhydrase (CA) is present in the glomus cells [1] of the carotid body (CB). The enzyme catalyzes $CO_2 + H_2O \Leftrightarrow H_2CO_3$ which dissociates into H^+ and HCO_3^- [2]. Since CA is involved in the cellular production of H^+ and HCO_3^- and since these ions participate in the regulation of the intracellular pH (pH_i) [3], the function of CA in the CB may extend beyond its simple catalytic function. The ionic species, H^+ and HCO_3^- are rapidly supplied by the action of CA, as the membrane exchangers transport these ions away in the dynamic ionic balance. In this sense, the function of cellular CA is expected to be dependent on the function of the ion exchangers, particularly on the Cl^-/HCO_3^- exchangers.

Carbonic anhydrase inhibitor has been used in vivo to measure the physiological function of CA [4,5,6]. All the results showed that CA inhibitors diminished the response to both hypercapnic and hypoxic stimuli, as measured in the sampled arterial blood. However, disequilibrium in the CO_2 hydration reaction in the circulating blood after CA inhibition in vivo [6] makes it impossible to accurately measure the stimulus-response relationship. The reaction of the sampled blood goes to completion slowly, raising PCO_2 and [H^+], and hence the measured values are falsely higher which result in underestimation of the responses. To overcome this problem we used cat CB perfused and superfused in vitro with physiological solution pre-equilibrated with respiratory gases, and studied the effects of methazolamide, a permeable CA inhibitor [2], on the hypoxic and hypercapnic chemosensory responses. Using Tyrode with and without CO_2-HCO_3^- at constant pH_o and hypercapnia, with and without acidosis, we attempted to understand the role of extra and intracellular CO_2 hydration on carotid body chemoreception.

METHODS

Isolated CBs with the carotid sinus nerve (CSN) were excised from cats anesthetized with sodium pentobarbitone (35 mg/kg ip). The CBs were perfused and superfused at $36.5 \pm 0.5\,°C$ with a modified Tyrode and chemosensory discharges were recorded as previously described [7,8]. The composition of the Tyrode was (in mM) Na^+ 154; K^+ 4.7; Ca^{2+} 2.2; Mg^{2+} 1.1; Cl^- 123; glutamate 42.0; and glucose 5.0, and dextran 5 g/l. The medium was buffered with HEPES (5 mM) to pH 7.40. To study the effects of CO_2-HCO_3^- (PCO_2=35 Torr and pH 7.40) the Tyrode was modified by adding 21.4 mM $NaHCO_3^-$, replacing the same amount of NaCl. Chemosensory responses were studied before and then during perfusion with methazolamide (10-30 mg/l, pH 7.40) for 30-60 min. The protocol consisted of perfusion with acidic hypercapnia ($PCO_2 = 55$ Torr, PO_2 = 120 and pH 7.15 - 7.20) for 2 min, isohydric hypercapnia ($PCO_2 = 55$ Torr, $PO_2 = 120$ Torr and pH 7.40, $HCO_3^- = 40$ mM) for 2 min, and hypoxia ($PO_2 = 25$-30 Torr and pH 7.40) for 2-8 min. PO_2, PCO_2 and pH of the perfusate and superfusate were measured with a gas and pH analyzer at 37.0 °C.

RESULTS

Effects of methazolamide on baseline chemosensory discarge

Methazolamide promptly reduced the baseline chemosensory activity in presence of CO_2-HCO_3^- in the perfused-superfused media (from 55.4 ± 6.2 imp/s to 10.0 ± 2.4 imp/s n = 13, p < 0.05) but not in the absence of CO_2-HCO_3^- (12.1 ± 2.1 imp/s vs. 12.1 $\pm$ 1.7 imp/s, n = 13).

Effects of methazolamide on chemosensory responses to acidic and isohydric hypercapnia

Figure 1 shows the effects of methazolamide (30 mg/l) on chemosensory response to acidic (PCO_2=55 Torr, HCO_3^-=21.4 mM and pH 7.25) and to isohydric hypercapnia (PCO_2=55 Torr, HCO_3^-=40 mM and pH=7.40). Both acidic and isohydric hypercapnia produced peak responses followed by adaptation. The magnitude of the response to isohydric hypercapnia was considerably smaller than that of acidic hypercapnia (Fig. 1A). Methazolamide eliminated the initial peak responses and reduced the late responses to acidic and isohydric hypercapnia (Fig 1B).

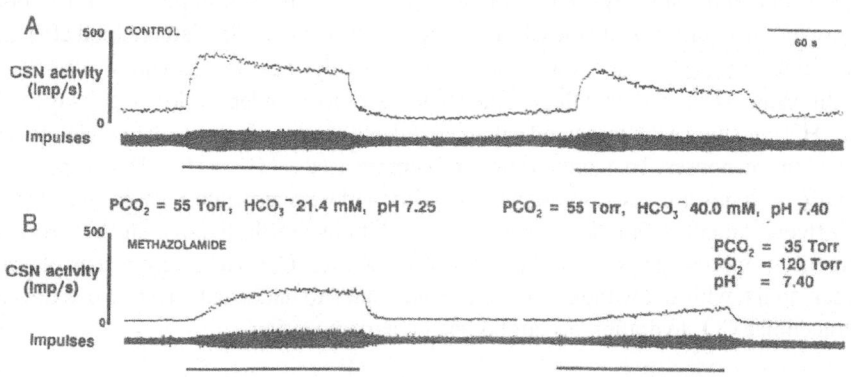

Fig. 1. Effects of methazolamide on CB chemosensory responses to acidic (left hand, PCO_2=55 Torr, HCO_3^- =21.4 mM, pH 7.25) and isohydric hypercapnia (right hand, PCO_2=55 Torr, HCO_3^-=40.0 mM, pH 7.40). Before (A) and during (B) methazolamide. Bars indicate duration of hypercapnia.

Effects of methazolamide on chemosensory responses to hypoxia

Figure 2 shows the effects of methazolamide (30 mg/l) on carotid chemosensory activity during transition from PO_2 of 120 Torr to 30 Torr (PCO_2 of 35 Torr and pH 7.41

were maintained). In the presence of CO_2-HCO_3^- hypoxia promptly increased the chemosensory discharges (Fig. 2A). Methazolamide reduced the baseline chemosensory activity and the speed of the response (Fig. 2B). The net magnitude of the response (maximal-baseline) was not significantly reduced. In 6 CBs, the activities increased in response to hypoxia before and during methazolamide, respectively to 343 ± 36 imp/s and to 284 ± 29 imp/s (p < 0.05). The net hypoxic responses were 285 imp/s and 274 imp/s respectively with and without methazolamide (p > 0.05). Methazolamide increased the halftime of the hypoxic response from 10.2 ± 1.9 s to 31.6 ± 5.5 s (p < 0.05).

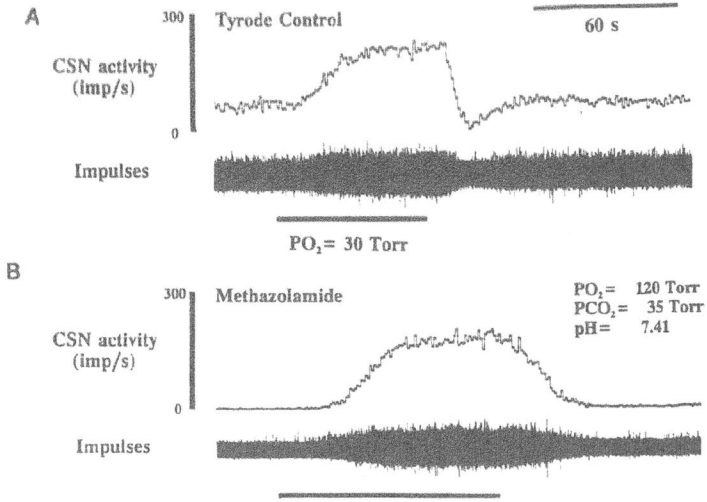

Fig. 2. *Effects of methazolamide on CB chemosensory responses to hypoxia ($PO_2 = 30\,Torr$, $PCO_2 = 30$ Torr, pH 7.41). Before (A) and during (B) methazolamide. Bars mark duration of hypoxia.*

Figure 3 shows the effects of methazolamide (10 mg/l) on carotid chemosensory response to hypoxia (PO_2=30Torr and pH 7.40) in the nominal absence of CO_2-HCO_3^-. Hypoxia slowly increased chemosensory discharges (Fig. 3A). Perfusion with methazolamide for 30 min increased the latency but the maximal chemosensory response was not reduced (Fig. 3B). In 6 CBs, hypoxia (PO_2= 20-30 Torr) produced similar increases of chemosensory activity before and during methazolamide (respectively, to 167 ± 46 imp/s and to 187 ± 34 imp/s, p > 0.05). However, the half-time of the response increased significantly from 204.6 ± 45.5 s to 340.8 ± 88.0 s (p < 0.05). Methazolamide eliminated the response to bolus injection (0.2 ml) of Tyrode equilibrated at PCO_2 of 38 Torr and pH 7.40 (Fig. 3C vs 3D).

DISCUSSION

The discussion of the results is based on the model that the glomus cells are the chemoreceptor cells which receive sensory innervation, and the sensory discharges reflect the events occurring in the glomus cells as a result of imposed changes in the microvasculature of the CB[9].

Methazolamide decreased the baseline chemosensory activity in the presence of CO_2-HCO_3^- but not in its absence, eliminated the initial peak and reduced the late response to acidic and isohydric hypercapnia and delayed the response to hypoxia in the absence or presence of CO_2-HCO_3^-. A significant decrease in the baseline activity with methazolamide in the presence of CO_2-HCO_3^- indicates that CA is necessary to maintain an acid pH_i in the chemoreceptor cells and hence the baseline activity. Indeed CA inhibition in the glomus cells increased the pH_i[10]. In the absence of CO_2-HCO_3^-, when the HCO_3^--dependent exchangers are not adequately functional, the role of CA is diminished.

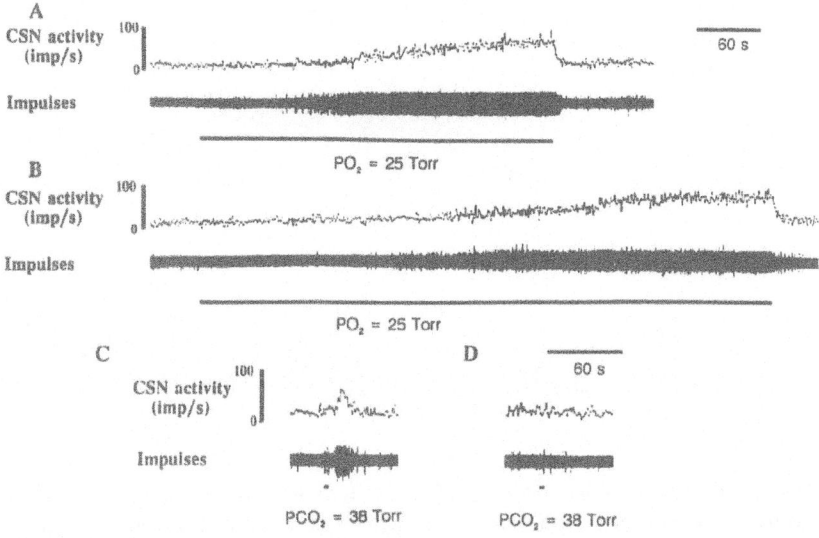

Fig. 3. Effects of methazolamide on carotid chemosensory responses to hypoxia (upper two panels, PO_2 = 25 Torr) and to transient CO_2 stimulus (lower two panels, PCO_2 = 38 Torr) in the absence of CO_2-HCO_3^-, before (A and C) and during methazolamide (B and D). Bars mark duration of hypoxia.

If the hydration of CO_2 mediated by CA occurs only in the chemoreceptor cells, hypercapnia should be followed by a rapid and enhanced intracellular acidosis which would finally be dissipated and regulated, depending on the level of extracellular acidosis. The hypothesis is that at lower $[H^+]_o$ the initial peak $[H^+]_i$ will be rapidly dissipated followed by a sustained level during hypercapnia. The sensory activity would correspond to the intracellular acidity. The chemosensory responses to isohydric and acidic hypercapnia supported this hypothesis, and followed the pattern of the expected $[H^+]_i$ change which was also influenced by $[H^+]_o$. The observation that methazolamide prevented the rapid initial but not the late response support the hypothesis that it is the intracellular compartment containing CA and not the extracellular compartment that determined the initial response. If CO_2 hydration were equal in the cell and in the blood vessels, a rapid dissipation of H^+ from the cell would be opposed by pH_o and the sensory activity would be better sustained. The observation that the chemosensory response was better maintained during acidic relative to isohydric hypercapnia indicates that CO_2 was not catalytically hydrated in the blood vessels of the CB.

It is known that hypoxic chemosensory response is augmented by $CO_2\text{-}HCO_3^-$. Consistent with this finding is the fact that in the nominal absence of $CO_2\text{-}HCO_3^-$ the hypoxic response developed slowly (compare figs 2 and 3) because of a likely high pH in the chemoreceptor cells [10]. However, the maximal response to hypoxia in the absence of $CO_2\text{-}HCO_3^-$ and the net response (maximal - baseline) in the presence of $CO_2\text{-}HCO_3^-$, were not reduced by methazolamide. Thus O_2 chemoreception was not directly dependent on the CA activity. We have previously shown that the presence of $CO_2\text{-}HCO_3^-$ at constant pH_o improves the response to hypoxia [7]. This facilitation is likely due to the function of the Cl^-/HCO_3^- exchanger which acts as an acid loader [3,10]. Accordingly, the presence of $CO_2\text{-}HCO_3^-$ makes the pH_1 acid in addition to that produced by CO_2 hydration. This net acid effect is reflected in the increased baseline activity in the presence of $CO_2\text{-}HCO_3^-$ at an external constant pH_o. This acidification is diminished by CA inhibition because of the reduced rate of production of H^+ and HCO_3^- and the consequent Cl^-/HCO_3^- swap. To what H^+ exerts its effect through membrane K^+ channel [11] and anion exchangers [12] and Ca^{2+} mobilization is not clear.

In summary, the results of this study predict a parallelism between the effects of CA inhibition on chemoreceptor cell pH and chemosensory activity, importance of CA in the chemoreceptor cell allowing pH_1 effect to precede that of pH_o. Absence of $CO_2\text{-}HCO_3^-$ delayed the responses to hypoxia, limiting the normal function of CA. Accordingly an interdependence of the functions of CA and Cl^-/HCO_3^- exchanger in the chemoreception is suggested. The hypoxic chemoreception is modulated but not mediated by the function of CA and the ion exchangers.

ACKNOWLEDGMENTS

Supported by NIH grants HL-43413 and HL-19737.

REFERENCES

1. *R. Rigual, C. Iñigez, J. Carreres, and C. Gonzalez*, Carbonic anhydrase in the carotid body and carotid sinus, Histochemistry 82: 557-580 (1985).

2. *T. H. Maren*, The general physiology of reactions catalyzed by carbonic anhydrase and their inhibition by sulfonamide, Ann. New York Acad. Sci. 429: 246-258 (1984).

3. *A. Roos, and W. F. Boron*, Intracellular pH, Physiol Rev. 61: 296-434 (1981).

4. *A.. M. S. Black, D.I. McCloskey, and R.W. Torrance*, The response of carotid body chemoreceptors in the cat to sudden changes in hypercapnic and hypoxic stimuli, Respir. Physiol. 13: 36-49 (1971).

5. *M. A. Hanson, P. C. G. Nye, and R .W.Torrance*, The exodus of an extracellular bicarbonate theory of chemoreception and the genesis of an intercellular one in:

Arterial Chemoreceptors, C. Belmonte, D. Pallot, H. Acker, and S. Fidone, eds, University Press, Leicester, p 403-416, 1981.

6.*S. Lahiri*, Carbonic anhydrase and chemoreception in carotid and aortic bodies in: The Carbonic Anhydrases Cellular Physiology and Molecular Genetics, S.D. Dogdson, R.E. Tashian, R.F., Gros R.F., and N.D. Carter, eds, Plenum Press, New York, p 341-344, 1991.

7.*R. Iturriaga and S. Lahiri,* Carotid body chemoreception in the absence and presence of CO_2-HCO_3^-, Brain Res. 568: 253-260 (1991).

8.*R. Iturriaga, W.L. Rumsey, A. Mokashi, D. Spergel, D.F. Wilson, and S. Lahiri,* In vitro perfused-superfused cat carotid body for physiological and pharmacological studies, J. Appl. Physiol. 70:1393-1400 (1991).

9.*C. Eyzaguirre, R. S. Fitzgerald, S. Lahiri, and P. Zapata*, Arterial Chemoreceptors. in: Handbook of Physiology. The Cardiovascular System. Peripheral Circulation and Organ Blood Flow. Sect. 2, Am. Physiol. Soc. , Williams and Wilkins, Baltimore, p 557-621, 1983.

10.*K. J. Buckler, R. D. Vaughan-Jones, C. Peers, and P. C. G. Nye*, Intracellular pH and its regulation in isolated type I carotid body cells of neonatal rat, J. Physiol. London. 436: 107-129 (1991) .

11.*C. Peers, and F. K. Green,* Inhibition of Ca^{2+}-activated K^+ current by intracellular acidosis in isolated type I cells of the neonatal rat carotid body, J. Physiol. London. 437: 589-602 (1991).

12.*A. Rocher, A. Obeso, C. Gonzalez, and B. Herreros,* Ionic mechanisms for the transduction of acidic stimuli in rabbit carotid body glomus cells, J. Physiol. London. 433: 533-548 (1991).

25 ROLE OF ION-EXCHANGERS IN THE CAT CAROTID BODY CHEMOTRANSDUCTION

S. Lahiri and R. Iturriaga

Department of Physiology, University of Pennsylvania, Philadelphia, PA 19104-6085, USA

INTRODUCTION

CO_2-O_2 stimulus interaction in the responses of the carotid body (CB) chemosensory discharges is well known [1]. Consistent with this is the observation that the response to hypoxia can be eliminated by diminishing the strength of CO_2 stimulus [2]. On the other hand, PCO_2 and H^+ stimuli can elicit a response without tissue hypoxia. Accordingly it may be hypothesized that the hypoxic response is based on and mediated by the CO_2-dependent H^+ stimulus. The hypothesis, however, cannot be fully tested in vivo because CO_2 cannot be eliminated to study the effects of hypoxia alone. In vitro experiments offers the advantage that CO_2-HCO_3^- can be eliminated in the external milieu, and the extracellular $[H^+]$ can be manipulated to test the idea that hypoxic response is dependent on CO_2-HCO_3^- and is mediated by cellular $[H^+]$. The current proposal is that decreases in glomus cell pH or PO_2 initiate intracellular Ca^{2+} changes, perhaps preceded by its membrane responses and followed by the release of neurotransmitters which, acting on the respective receptors, lead to neural chemosensory discharges [1]. If the chemosensory responses are determined by events in the glomus cells, several predictions can be made and tested regarding chemoreception in these cells.

In this presentation we propose to focus particularly on the anion exchanger in pH regulation and O_2 chemoreception. Cl^-/HCO_3^- exchanger is an acid loader [3] and it is present in the glomus cells [4,5]. This exchanger cannot function adequately in the absence of CO_2-HCO_3^- which would make the pH_i alkaline, at constant pH_o of 7.4, and could attenuate but not eliminate the response to hypoxia. Low $[Cl^-]$ in the external medium could likewise raise pH_i by the loss of Cl^- and gain of HCO_3^-, and attenuate chemosensory discharge without impeding O_2 chemoreception.

METHODS

Carotid bodies were excised from cats anesthetized with sodium pentobarbitone (35 mg/kg, ip) and were perfused and superfused at $36.5 \pm 0.5\,°C$ with a modified Tyrode solution as described in details previously [6]. The composition of the Tyrode was (in mM) Na^+ 154; K^+ 4.7; Ca^{2+} 2.2; Mg^{2+} 1.1; Cl^- 110; glutamate 42.0; and glucose 5.0, and dextran 5 g/l. The medium was buffered with HEPES (5 mM) to pH 7.40. One series of experiments was performed with and without CO_2-HCO_3^-, and another with normal and low substituted $[Cl^-]_o$. To study the effect of CO_2-HCO_3^-, we replaced 21.4 mM of NaCl with the same amount of $NaHCO_3$. To reduce $[Cl^-]_o$, NaCl was replaced by sodium gluconate. The tests consisted of responses to hypoxia, hypercapnia and bolus injection of nicotine. The latter was used to test the effects at the neurotransmitter receptor level, downstream from the site of O_2 and CO_2-H^+ chemoreception.

RESULTS

Effects of CO_2-HCO_3^- on the chemosensory response to hypoxia

Figure 1 shows the chemosensory response to hypoxia with and without CO_2-HCO_3^- in the perfusate and superfusate at constant pH_o. In contrast to the result in nominal absence of CO_2-HCO_3^-, the baseline chemosensory activity was higher with CO_2-HCO_3^- and the responses to hypoxia were augmented in three aspects; speed, rate of rise and absolute maximal activity. The magnitude of the latter was mostly due to the increase in the baseline activity. The summarized results are shown in Fig. 2. Clearly, presence of CO_2-HCO_3^- raised the baseline activity and the responses to hypoxia. The same contrast was present in the response to perfusate flow interruption (not shown). The speed and magnitude of the response to nicotine (1-4 μg) was the same with or without CO_2-HCO_3^-.

Effects of low external chloride on the hypoxic response

Figure 3 shows rapresentative results of low $[Cl^-]_o$ (10 mM) on the chemosensory response to hypoxia. Low $[Cl^-]_o$ decreased the baseline activity, and the speed, rate and the magnitude of the responses to hypoxia. The initial peak response to hypercapnic acidosis ($PCO_2 = 55$-60 Torr at pH 7.20) was also attenuated.

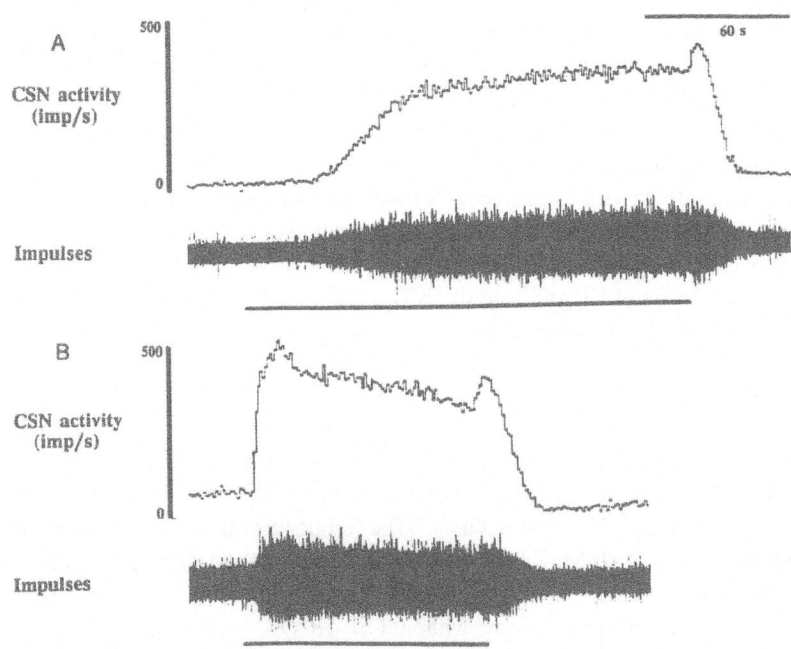

Fig. 1 Effects of CO_2-HCO_3^- on carotid chemosensory responses to hypoxia ($PO_2 = 25$ Torr antd pH 7.38).
A, without CO_2-HCO_3^-, and B, with CO_2-HCO_3^- ($PCO_2 = 35$ Torr and pH 7.38). PO_2 was about 120 Torr before and after the test. Bars indicate duration of hypoxia.

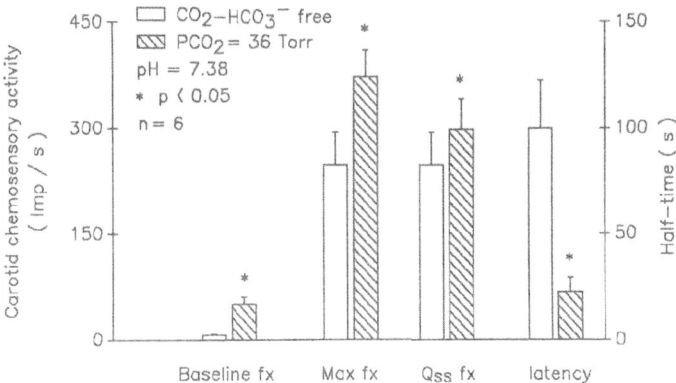

Fig. 2. *Summary of effects of CO_2-HCO_3^- on carotid chemosensory nerve responses to hypoxia ($PO_2=25$-30 Torr). Values are mean + SEM of 6 CBs. Max fx corresponds to the maximal response. Qss fx is the quasi steady-state response, measured at the end of the hypoxic perfusion (*, p < 0.05, Wilcoxon test).*

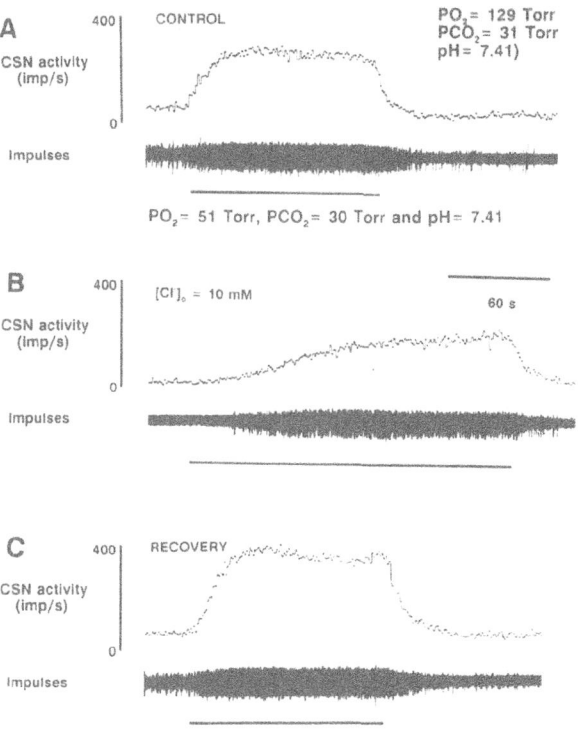

Fig. 3. *Effects low $[Cl^-]_o$ on carotid chemosensory responses to hypoxia ($PO_2 = 50$ Torr). A, control hypoxic perfusion. B, with $[Cl^-]_o = 10$ mM and C, after recovery from low $[Cl^-]_o$. Bars mark duration of hypoxia.*

DISCUSSION

These results demonstrate that functional blockade of the Cl^-/HCO_3^- exchanger (absence of CO_2-HCO_3^- or low $[Cl^-]_0$) diminished the baseline chemosensory activity, attenuated the responses to hypoxia but did not prevent O_2 chemoreception. Accordingly, these results differ from those of Shirahata and Fitzgerald[7] who claimed that CO_2-HCO_3^- is essential for O_2 chemoreception. The difference however, may be more apparent than real because we believe that cellular alkalosis in the absence of CO_2-HCO_3^- increased the threshold for hypoxic response. Since Cl^-/HCO_3^- exchanger is an acid loader [3], its functional blockade is expected to increase pH_i of the glomus cells. Indeed, recently Buckler et al. [4] reported that glomus cell pH (neonatal rat) turned alkaline upon removal of CO_2-HCO_3^- or Cl^-. Thus, the prediction from the chemosensory activity is consistent in general with the experimental findings in the glomus cells. However, the pH_i responses to CO_2-HCO_3^- found by Buckler et al. [2] are not in full agreement with those of Wilding et al. [5] or of Cummins and Donnelly [8].

Acid has been reported to decrease Ca^{2+}-dependent K^+ conductance of the glomus cell [9]. Conversely, alkaline pH may increase K^+ conductance. How these membrane current changes are linked to chemoreception is still unclear and disputed [10,11]. But chemosensory responses are the consistent expression of chemoreception and can be used as a guide to predict and test the possible link between chemoreception and membrane K^+ current of the glomus cells. A possible correlation between K^+ conductance change and chemosensory discharge with several unrelated stimuli would mean that K^+ conductance is not specific to H^+ and O_2 , and K^+ channel may not be the site of the chemoreception. This interpretation of K^+ conductance is contrary to that proposed by López-Barneo et al. [10] and is congruent with the view of Biscoe and Duchen [11]. However, more definitive experimental results are needed for better understanding.

The neurotransmitter hypothesis requires that acidosis releases neurotransmitter through initial rise in $[Ca^{2+}]_i$. Accordingly, the presence of CO_2-HCO_3^- would influence changes in $[Ca^{2+}]_i$ and neurotransmitter release. These predictions are testable. The response to hypoxia was not blocked by low $[Cl^-]_0$ or absence of CO_2-HCO_3^-. Accordingly, steady-state hypoxia would diminish K^+ conductance, increase $[Ca^{2+}]_i$ and release neurotransmitter in the presence or nominal absence of CO_2-HCO_3^- or in the presence of low external $[Cl^-]$. Although experiments designed to test this prediction have not been performed, there are relevant observations. Biscoe and Duchen [11] briefly reported that CO_2-HCO_3^- or HEPES buffer made little or no difference to $[Ca^{2+}]_i$ rise in response to hypoxia in the glomus cell. Also, release of CB dopamine by hypoxia has been reported both in the presence and absence of CO_2-HCO_3^- [1]. But there may have been a quantitative difference. Taken together, it is reasonable to suggest that these manifestation of O_2 chemoreception and therefore the initiation of O_2 chemoreception was not blocked during inadequate function of the Cl^-/HCO_3^- exchanger. However, quantitative observation regarding the effects of hypoxia with and without blockade of the Cl^-/HCO_3^- exchanger on K^+ conductance and $[Ca^{2+}]_i$ are not available. But from the chemosensory responses it is expected that presence of CO_2-HCO_3^- at pH_0 of 7.40 would increase $[Ca^{2+}]_i$.

Recently we measured intracellular pH dependent fluorescence in the CB during hypoxia and hypercapnia using BCECF/AM [12]. Hypoxia did not change the fluorescence unlike hypercapnia. Also, application of acetate in the nominal absence of CO_2-HCO_3^- stimulated the chemosensory discharge, as fluorescence signal indicated cellular acidosis. Thus, O_2 chemoreception did not seem to be mediated by intracellular acidosis. With this information at hand, we present the following explanation for the O_2-CO_2 stimulus interaction and for interdependent O_2-CO_2 stimulus threshold. Clearly, the chemosensory response to CO_2-H^+ cannot be eliminated by raising PO_2 alone. But the response to hypoxia could be diminished by lowering $[H^+]_i$. Accordingly the two stimuli do not work through a common mechanism. If K^+ conductance change is the expression of chemoreception of both O_2 and CO_2-H^+ by two different mechanism, the two responses should add, and not show stimulus interaction.

The mechanism of O_2 chemoreception is still unclear. At the chemosensory level[13] there is considerable evidence in favor of the metabolic hypothesis coming from the use of metabolic inhibitors. Biscoe and Duchen produced parallel evidence at the glomus cell level [11]. Their hypothesis that the $[Ca^{2+}]_i$ is derived from the mitochondria is not however shared by us, partly because oligomycin which leaves the calcium transport across the mitochondrial membrane intact, blocked the effect of hypoxia on the chemosensory discharges[13].

In summary, the results showed that O_2 chemoreception occurred in the nominal absence of CO_2-HCO_3^- in the extracellular medium but CO_2-HCO_3^- raised the baseline activity and improved the responses. The effect of CO_2-HCO_3^- could be attributed to intracellular acidification through the Cl^-/HCO_3^- exchanger mechanism. The basic mechanism of hypoxic chemotransduction seem not to be mediated by cellular acidification, although the dynamic sofits responses are influenced by pH_i through CO_2-HCO_3^-. Accordingly, CO_2-HCO_3^- in vivo makes important contribution to cellular physiology of chemotransduction in the carotid body.

ACKNOWLEDGMENT

We would like to thank Mr. A. Mokashi for his assistance. Supported in part by NIH grants HL-19737 and HL-43413

REFERENCES

1. C. Eyzaguirre, R.S. Fitzgerald, S. Lahiri, and P. Zapata, Arterial Chemoreceptors. in: Handbook of Physiology.The Cardiovascular System.Peripheral Circulation and Organ Blood Flow. Sect. 2, Am. Physiol. Soc., Williams and Wilkins, Baltimore, p 557-621, 1983

2. S. Lahiri, and R. G. DeLaney, Stimulus interaction in the responses of carotid body chemoreceptors single fibers, Respir. Physiol. 24: 249-266 (1975)

3. A. Roos, and W. F. Boron, Intracellular pH, Physiol. Rev. 61: 296-434 (1981)

4. K. J. Buckler, R. D. Vaughan-Jones, C. Peers, and P.C.G. Nye, Intracellular pH and its regulation in isolated type I carotid body cells of neonatal rat, J. Physiol. London 436: 107-129 (1991)

5. T. J. Wilding, B. Cheng, and A. Roos, The relationship between extracellular pH (pH_o) and intracellular pH (pH_i) in adult rat carotid body glomus cells, Biophys. J. 59: 184a (1991)

6. *R.Iturriaga, W.L.Rumsey, A.Mokashi, D.Spergel, D.F.Wilson, and S.Lahiri,* In vitro perfused-superfused cat carotid body for physiological and pharmacological studies J. Appl.Physiol. 70: 1393-1400 (1991)

7. *M. Shirahata, and R. S. Fitzgerald,* The presence of CO_2-HCO_3^- is essential for hypoxic chemotransduction in the in vivo perfused carotid body, Brain Res. 545:297-300 (1991)

8. *T.R.Cummins, and D.F. Donnelly,* Regulation of intracellular pH in the rat glomus cells, Soc. Neurosci. Abstr. 17 102 (1991)

9. *C.Peers, and F. K. Green,* Inhibition of Ca^{2+}-activated K^+ current by intracellular acidosis in isolated type I cells of the neonatal rat carotid body, J. Physiol. London 437: 589-602 (1991)

10. *J. R. López-López, C,.Gonzalez, J. Ureña, and J. López-Barneo,* Low PO_2 selectively inhibits K channel in chemoreceptor cells of the mammalian carotid body, J. Gen. Physiol. 93: 1001-1015 (1989)

11. *T.J.Biscoe, and M.R.Duchen,* Responses of type I cells dissociated from the rabbit carotid body to hypoxia, J. Physiol. London 428: 39-50 (1990)

12. *R.Iturriaga, W.L. Rumusey, S, Lahiri, D. Spergel, and D.F. Wilson,* Intracellular pH and oxygen chemoreception in the cat carotit body in vitro, J. Appl. Physiol. 72: 2259-2266 (1992)

13. *E. Mulligan, S. Lahiri, and B. T. Storey,* Carotid body O_2 chemoreception and mitochondrial oxidative phosphorylation, J. Appl. Physiol. 51: 438-446 (1981)

26 DOPAMINE METABOLISM IN THE RABBIT CAROTID BODY IN VITRO: EFFECT OF HYPOXIA AND HYPERCAPNIA

L-M. Leitner

Univ. P.-Sabatier, Fac. Méd., Lab. Physiol., UA CNRS 649, 133 Rte de Narbonne, 31062 Toulouse Cedex and Ecole Nat. Vét. Toulouse, Lab. Physiol. Pharmacol. 23, Chemin des Capelles, 1076 Toulouse, France

SUMMARY

Dopamine (DA) and noradrenaline (NA) were measured in the rabbit carotid body (CB) in vitro bv HPLC-ED under the following experimental conditions: 1h superfusion in normoxic, hypoxic ($10\%\ O_2$ in N_2) or hypercapnic ($8\%\ CO_2$, $20\%\ O_2$, $72\%\ N_2$) medium, 5h superfusion in normoxia or hypoxia. The contents of DA and NA were decreased by hypoxia and hypercapnia after 1 h and 5h indicating a possible DA and NA secretion. Under the same experimental conditions synthesis of DA and NA and catabolism of DA were studied with enzymatic inhibition of tyrosine hydroxylase and monoamine oxidase (MAO) respectively. In hypoxia (1 h and 5h) the rate constant of DA synthesis was the same as in normoxia; however NA synthesis was decreased after 1h hypoxia. On the contrary, hypercapnia, appeared to be a very effective stimulus of DA and NA synthesis.

INTRODUCTION

In the rabbit carotid body, DA is probably the most important amine although its role is not very well defined. On the one hand a possible involvement in excitatory or inhibitory neurotransmission or neuromodulation has heen attributed to DA, but the data are contradictory depending on the species (cat, rabbit, rat) and the techniques (in vivo, in vitro) used: for a review see Fidone and Gonzalez (1986). On the other hand, depletion of DA by reserpine and by α-methyl-para-tyrosine decreased very severely the carotid sinus nerve activity and the chemorecoptor response to hypoxia and hypercapnia (Leitner and Roumy, 1986) in the rabbit and in the cat in vitro. If DA is released by the CB during hypoxia or hypercapnia and if its action should modify chemoreception, a short term regulation of DA synthesis would occur through an end-product inhibition of tyrosine hydroxylas, the rate-limiting enzyme in DA and NA synthesis (Nagatsu et al. 1964). DA and sometimes NA synthesis have been studied in the carotid body of different species (rat, rabbit, cat) in vivo and in vitro (Hanbauer and Hellström, 1978; Fidone et al, 1982a,b; Brokaw et al., 1985; Rigual et al., 1986, Péquignot et al., 1987). The techniques for measuring catecholamine metabolism varied: enzymatic inhibition, radioactive tracer and all have their own drawbacks (Freeman and Gibson, 1986). The results obtained showed, most of the time, that hypoxia, especially of long duration, increased DA but not NA synthesis and that hypercapnia was ineffective (Fitzgerald et al., 1983; Hellström et al., 1989). Are these results specific of the CB ? Apparently not, since measurements of DA metabolism in vivo in the nervous system, central or peripheral, during hypoxia, essentially made in the rat although yielding contradictory results (Kuno et al., 1981 ; Miwa et al., 1986 ; Feinsiver, 1987) established that a long duration hypoxia always increased DA synthesis (Dalmaz et al., 1988, Finsilver et al., 1987).

In an experiment on the fate of catecholamines in vitro (Roumy et Leitner, 1988) the amount of DA and 3.4-dihydroxyphenylacetic acid (DOPAC) in the rabbit CB was found to decrease by about half in the first hour of superfusion. This decline was followed by a stable period of at least 4 hours and it is this period which has been chosen to study the metabolism of DA and NA under normoxic, hypoxic and hypercapnic conditions. To compare the results obtained in vitro with previous in vivo metabolic data (Leitner et al. 1986), the inhibition of monoamine oxidase by pargyline was used. Sine during the superfusion period as in previous electrophysiological experiments, no precursor (tyrosine or L-dioxyphenylalanine) was added to the medium, it was necessary to evaluate, also the catecholamine synthesis and this was done by inhibiting tyrosine hydroxylation.

Results showed that during the steady state period of 4 hours which followed CB dissection, DA synthesis rate did not change both in normoxia and in hypoxia but was increased by hypercapnia. NA synthesis was decrased by short (lh) hypoxia but not by long hypoxic stimulation.

METHODS

Experiments were done on New Zealand white rabbits anaesthetized (Nembutal IV, 36 ml/Kg) paralysed and artificialy ventilated. The carotid bifurcations on both sides were rapidly removed and the CB was cleaned from surrounding tissues in an ice-cold solution of the same composition as that used for superfusion. The CB were then placed in small (6µl) chambers (Roumy et al., 1988) and superfused at 38°C during 1, 2 or 5 h at 70 - 80 µl/min.) Hypoxic medium was equilibrated with 10 % O_2 in N_2 and contained (in mM) NaCl: 110; KCl: 5; $CaCl_2$: 2.2 , $MgCl_2$: 0.5 . gluose: 5.5, sucrose: 54; Hepes 5 (pH was adjusted to 7.4 at room temperatur with 1N NaOH). During hypercapnic stimulation (8 % CO_2, 21 % O_2. 71 % N_2 the following medium was used : NaCl :75.7; KCl: 5, $CaCl_2$,2.2; $MgCl_2$: 0.5; glucose: 5,5; sucrose: 54; Hepes: 5; $NaHCO_3$: 34,3, pH was adjusted to 7.4 at 38-C with 1N NaOH.

Catabolism of DA was studied by adding pargyline (20µM), a MAO inhibitor, to the superfusing medium for 15, 30 and 45 min. DOPAC was measured at time 0 and at each of these periods with high performance chromatoraphy and electrochemical detection (HPLC-ED) as described earlier (Leitner et al., 1986). The slope of the exponential decay of DOPAC measured the rate constant of DA catabolism.

Synthesis of DA was blocked by α-methyl-p-tyrosine (AMPT), 100µM. According to Brodie et al., (1966) DA and NA concentrations in the central nervous system follow a single exponential decay after treatment by AMPT. The CB were incubated for 2.4 and 8 min in the AMPT containing medium to which ascorbic acid (1mM) was added to prevent AMPT oxidation. The exponential decrease in DA content was followed with HPLC-ED and the regression line was calculated. Each point of the regression lines L_nDA, $L_n NA$, L_nDOPAC versus time were the mean of 6 to 12 measurements made on 6 to 12 different CB except for O time where the measurements varied from 16 to 31 CB.

The data obtained after each superfusion periods were compared with non parametric Mann-Whitney U test.

RESULTS

Prior to metabolism measurements it was necessary to know the amounts of DA, DOPAC and NA contained in CBs after 1 h superfusion in normoxia, hypoxia and hypercapnia and after 5h superfusion in normoxia and hypoxia. The figures obtained will be used as the values at O time in the enzymatic inhibitions studied later. As shown in fig. 1 the DA and NA contents were decreased by hypoxia or by hypercapnia after 1h incubation whereas the DOPAC content increased only in hypercapnia. This decrease in amine content could mean that the two amines have been secreted by this CB in response to the hypoxic and hypercapnic stimuli. However after 1 h hypercapnia, DOPAC was significantly increased, indicating an augmentation of DA utilization and this could explain the crease in DA content observed in spite of an increases synthesis (see further). After 5h hypoxia, DA and NA were also decreased and the DOPAC level was not modified.

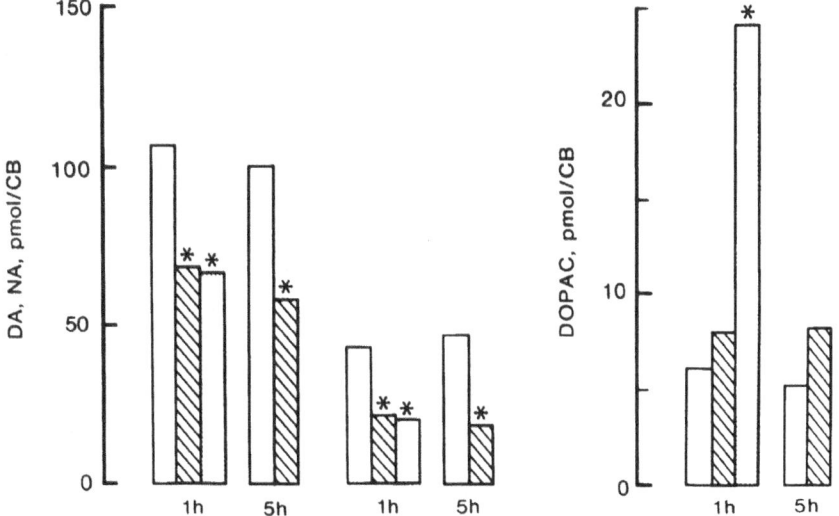

Figure 1 . DA, DOPAC and NA content (median)of rabbit CB under normoxia (white bars), hypoxia crossed bars and hypercapnia (dotted bars), after 1h and 5h in vitro. * different from normoxia Mann-Whitney U-test.

DA catabolism

After 1h and 5h of normoxic superfusion, the rate constants of DOPAC exponential decay were similar (-4.1/h and -4.41/h respectively) and identical to that found in vivo (-4.1/h) (Roumy et al., 1986). This seems to indicate that in vitro normoxic conditions have not perturbed catecholamine metabolism, in keeping with previous results obtained by Verna et al. (1981) on the variability of the in vitro preparation. After 1h. superfusion the rate constant was not changed by hypoxia (-4.6/h) but was decreased by hypercapnia (-1.8/h) and so was the catabolism in this last situation (fig. 3). After 5h incubation in the hypoxic medium, the rate constants (-4.41h) was not different from the value found in normoxia (-4.6h). This result could indicate that the metabolic capacity of type I cells has adapted to such a relatively long lasting stimulation. The turnover time did not vary under the different experimental conditions in vivo and in vitro except that it was increase by hypercapnia. The catabolism of NA could not be calculated since under these experimental conditions the measurement of 3-methoxy-4-hydroxy-phenylglycol (MHPG) was not reliable.

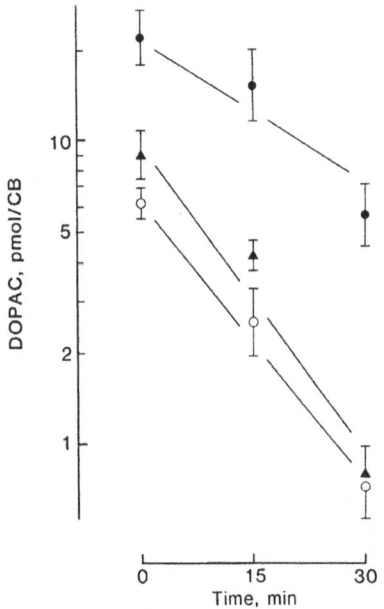

Figure 2. DOPAC exponential decay in vitro after pargyline (20μM) superfusion. O normoxia, Δ hypoxia,
   • hypercapnia.

DA and NA synthesis

The rate constant ot DA disapperance after 1h superfusion was unchanged by hypoxia but was increased by hypercapnia although the DA content was decreased significantly in both hypoxia and hypercapnia as it is indicated by the increase in DOPAC/DA ratio from 0.06 to 0.17 (Table 1). Compared to normoxia, DA synthesis was a little decreased in hypoxia but stayed constant in hypercapnia and the turnover time was very much decreased in this last situation: 0.07/h., compared to 0.12/h and 0.11/h for normoxia and hypoxia respectively.

After 5h superfusion, the rate constants of DA decrease in hypoxia and normoxia were not different from each other and the synthesis were the same despite a lower value for the median in hypoxia.

The kinetics of NA disappearance exhibited simlar characteristics as that of DA concerning the effect of 1 h superfusion with hypercapnic medium (Table 2) but since the NA content at O time was notably decreased in hypercapnia, NA synthesis was not increased. On the contrary the rate constant was significantly decreasd by hypoxia and so was the synthesis. However, after 5h hypoxia the rate constant has increased to reach a value similar to that obtained in normoxia but the median value of NA has decreased and the synthesis was consequently decreased: it thus appeared that a long lasting hypoxia was more harmful to NA synthesis than to DA synthesis.

Table 1. DA synthesis in vitro.

	NORMOXIA 1h	HYPOXIA 1h	CO_2 1h	NORMOXIA 5h	HYPOXIA 5h
k/h	- 8.3	- 8.7	* - 1.4	- 6.6	- 10.5
Synthesis, (pmol/CB/h)	887.0	589.0	906.0	657.0	614.0
mDAo	106.3 n=26	* 67.7 n=20	* 64.5 n=31	99.5 n=25	58.5 n=16

k = rate constant; mDAo, median values at time 0.* different from normoxia, $p < 0.001$; [†] hypoxia 5h different from hypoxia 1 h, $p < 0.001$. Mann-Whitney U-test for the comparisons between mDAo. Comparisons of the slopes of this regression lines indicated the existence or absence of significant difference between rate constants.

Table 2. NA synthesis in vitro.

	NORMOXIA 1h	HYPOXIA 1h	CO_2 1h	NORMOXIA 5h	HYPOXIA 5h
k/h	- 8.0	* - 1.6	* -13.9	- 6.9	† - 6.2
Synthesis, (pmol/CB/h)	350.0	34.2	292.0	312.6	111.6
mNAo	43.7 n=26	* 21.4 n=20	* 21.0 n=31	45.3 n=25	* 18.0 n=16

k = rate constant, m = median, * different from normoxia, $p < 0.005$; [†] 5h hypoxia different from 1 h hypoxia, $p < 0.005$, Mann-Whitney U-test.

DISCUSSION

It has been interesting to show that the CB preparation in vitro was still capable of synthesizing catecholamines without an external substrate and, surprisingly, the time course of tyrosine hydroxylase inhibition was very much shorter (< 8 min) than previously found in the rat CB in vivo (at least 3h) by Hertzberg et al. (1990). In the rabbit CB in vivo, APT (100 mg/kg IP) decreased DA content by bout half after a 18 hours delay (Leitner and Roumy, 1986) but even when this delay was reduced to 6h no intermediate values significantly different from control and half depleted CB have been found (unpublished data). No hypothesis can be put forward to explain such a difference in the AMPT effect on DA content between in vivo and in vitro situations.

The present results are in greement with others in that short term hypoxia did not increase DA synthesis, however longer lasting stimulation (5h) was not more effective and this is in contradiction with Fidone et al. (1982 b), possibly because the experimental set up was different and in particular the flow rate which was very important, about 70 µl/ min for a volume chamber of about 6 µl in our own experiments. Actually the decrease in DA content due to hypoxia was the same after 1h and 5h superfusion and that seems to indicate that a mechanism is at play which maintains a level of DA synthesis compatible with a good functioning of the CB as it has been constantly observed in our electrophysiological experiments. The absence of increase in DA synthesis might explain the decrease in DA content observed at 1h and 5h hypoxia if DA were secreted by the CB in response to the stimulation, secretion already observed by Fidone et al. (1982 b).

The response of the CB to hypoxia does not seem different from what is found in the nervous system. Prolonged but not acute hypoxia in vitro increased tyrosine hydroxylase activity in phaechromocytoma cells in culture (Feinsilver et al., 1987). In addition, a long term hypoxia increased DA turnover in the rat superior cervical ganglion, in vivo (Dalmaz et al., 1988). Indeed tyrosine hydroxylase did not show a simple relationship to the oxygen partial pressure but appeared to be influenced by pH changed (Crisson et al., 1977). This relationship seems to be in keeping with the present results showing that hypercapnia increased DA and NA synthesis in the CB but in contradiction with the results of Hellstrom et al. (1989) concerning the rat CB. No reasonable explanation can be given for such a discrepancy except that the very small number of measurements done during this last experiment does not seem to take into account the great spontaneous variability of DA and NA content in the CB of all species encountered. (Stone, 1983, Brown et al., 1974) . Other experiments have shown on rat hypothalamus, medulla and pons (Stone, 1983), or on the whole brain (Brown et al., 1974) that hypercapnia (30 min to 2.5 h) stimulated NA synthesis.

Does the metabolism of catecholamines in the carotid body exhibit specific features which could be a link between this metabolism and the chemoreceptors response to natural stimuli ? There is a good reason to think that hypoxia and hypercapnia induced a catecholamine secretion as already shown in the rabbit (Fidone et al.,1982 b) and in the cat (Rigual et al., 1984,1986) but also that this secretion is only partially compensated for by a synthesis which is not increased in hypoxia compared to normoxia. When the hypoxic stimulus is prolonged (5 h) an adaptation seems to adjust DA synthesis to the secretion but this does not seen to be the case for NA which was severely decreased. Finally, if the turnover time of DA and NA synthesis is accelerated by a short lasting (1 h) hypercapnia, the adaptation to a long-lasting (at least 5h) hypoxic stmulation is a probable link between catecholamine metabolism and carotid body chemoreception.

ACKNOWLEDGMENT

I thank Ms G. Costes for her competent technical assistance and MsC. Fortun for typing the manuscript and making the illustrations.

REFERENCES

Brodie B.B., Costa E., Dlabac A., Neff N.N., Smookler H.H., 1966, Application of steady state kinetics to the estimation of synthesis rate and turnover time of tissue catecholamines. J. Pharmacol Exp. Ther., 154: 493.

Brokaw J J. . Hansen J.T., Christie D.S., 1985, The effects of hypoxia on catecholomine dynamics in the rat carotid body. J. Auton. Nerv. Syst., 13: 35.

Brown R.M., Snider S.R., Carlsson A., 1974, Changes in biogenic amine synthesis and turnover induced by hypoxia and/or foot shock stress. 11. The central nervous system. J. Neural Trans., 35:293.

Carlsson A., Holmin T., Lindqvist M.J Siesjö B.K., 1977, Effect of hypercapnia and hypercarbia on tryptophan and tyrosine hydroxylation in rat brain. Acta Physiol. Scand., 99: 503.

Dalmaz Y. , Péquignot J.-M., Tavitian E., Cottet- Emard .M., Peyrin L., 1988, Long-term hypoxia increase the turnover of dopamine but not norepinephrine in rat sympathetic ganglia. J. Auton. Nerv. Syst., 24: 57.

Davis J.N., Carlsson A., 1973, The effect of hypoxia on monoamine synthesis, levels and metabolism in rat brain. J. Neurochem., 21 783.

Feinsilver S.H., Wong R., Raybin D.M., 1987, Adaptations of neurotransmitter synthesis to chronic hypoxia in cell culture Biochim. Biophys. Acta, 928: 56.

Fidone S., Gonzalez C., 1982a, Catecholamines synthesis in rabbit carotid body in vitro. J. Physiol. (London), 333: 69.

Fidone S., Gonzalez C., Yoshizaki K., 1982b, Effects of hypoxia on catecholamine synthesis in rabbit carotid body in vitro. J. Physiol. (London), 333 81.

Fidone S.J., Gonzalez C., 1986, Initiation and control of chemoreceptor activity in the carotid body; In, Handbook of Physiology, Section 3: The Respiratory system, Vol. 11. Control of Breathing, Part I, pp 247, AP. Fishman ed. Am. Physiol. Soc. Bethesda. Md.

Freeman G.B., Gibson G.E., 1986, Effect of decreased oxygen on in vitro release of endogenous 3,4dihydroxyphenylethylamine from mouse striatum. J. Neurochem., 47: 1924.

Hanbauer I., Hellström S., 1978, The regulation of dopamine and noradrenaline in the rat carotid body and its modification by denervation and by hypoxia. J. Physiol. (London), 282: 21.

Hellström S., Péquignot J.M., Dahlqvist A., 1989, Catecholamines in the carotid body are unaffected by hypercapnia. Neurosci. Lett., 97: 280.

Kuno T., Marukawa A., Fujiwara H., Tanaka C. ,1981 . Dopamine accumulation in the mouse brain during hypoxia. Japan. J. Pharmacol.. 31: 503.

Leitner L.-M., Roumy M. , 1986, Chemoreceptor response to hypoxia and hypercapnia in catecholamine depleted rabbit and cat carotid bodies in vitro. Pflügers Arch., 406: 419.

Leitner L.-M., Roumy M., Ruckebusch M., Sutra J.F., 1986, Monoamines and their catabolites in the rabbit carotid body. Effects of reserpine, sympathectomy and carotid sinus nerve section. Pflügers Arch., 406 : 552.

Miwa S., Fujiwara M., Inoue M., Fujiwara M., 1986, Effects of hypoxia on the activities of noradrenergic and dopaminergic neurons in the rat brain. J. Neurochem., 47: 63.

Nagatsu T., Levitt M., Udenfriend S. ,1964, Tyrosine hydroxylase. The initial step in norepinephrine biosynthesis. J. Biol. Chem.. 239: 2910.

Rigual R., Gonzalez E., Fidone S., Gonzalez C., 1984, Effects of low pH on synthesis and release of catecholamines in the cat carotid body. Brain Res., 309 :178.

Rigual R., Gonzalez E., Gonzalez C., Fidone S., 1986, Synthesis and release of catecholamines by the cat carotid body in vitro. Brain Res., 374 : 101.

Roumy M., Armengaud C., Ruckebusch M., Sutra J.F., Leitner L.-M.,1988, Fate of the catecholamine stores in the rabbit carotid body superfused in vitro. Pflügers Arch., 411- 436.

Roumy M., Ruckebusch M., Sutra J.F., Leitner L.-M., 1986, Rate of dopamine metabolism in the rabbit carotid body in vivo. Pflügers Arch., 407 : 575.

Stone E.A.,1983, Rapid adaptation of the stimulatory effect of CO2 on brain norepinephrine metabolism. Naunyn-Schmiedeberg's Arch. Pharmacol., 324: 313.

In Vitro

The methodological rationale behind the biochemical processing used was this. The CB tissue was the source of the enzyme studied. The substrate for the assay was either $[^3H]PI$ or $[^3H]PIP_2$ (Amersham, UK). The PLC activity was assessed from the measured formation of a radioactive inositol phosphate.

The assay procedures for the determination of PI- and PIP_2-specific PLC were in general similar. The major steps were the following. The CB tissue was homogenized. A stock solution was prepared, consisting of the unlabeled and labeled PI or PIP_2. The organic phase was dried under nitrogen and dissolved in a 100 mM TRIS-HCl buffer of pH 7.8 for PI and of pH 6.6 for PIP_2. Next, the substrate, 10 mM LiCl, and 0.1% sodium deoxycholate were added and adjusted to a final volume of 200 µl. The mixture was vortexed and incubated at 37°C for 30 min for PI and for 15 min for PIP_2. The reaction was stopped with 1 ml chloroform/methanol/concentrated HCl; then 0.3 ml of water were added to separate the phase. After centrifugation, 0.4 ml aliquots of the upper aqueous phase were taken for the measurement of radioactivity in the Bray scintillation fluid.

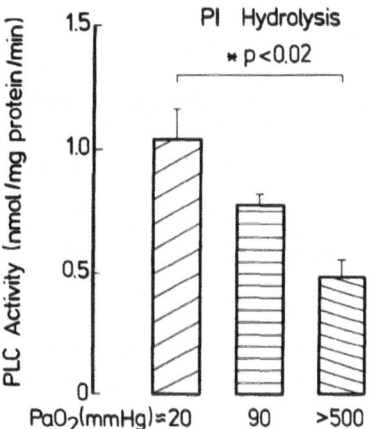

Fig. 1. Phosphatidylinositol-specific phospholipase C activity in the carotid body tissue at various PaO_2 levels.
* Significant difference between hypoxia and hyperoxia (F-test followed by Scheffe's test).

All experiments were carried out in duplicate. PLC activity is expressed in nmoles per milligram protein per minute. Data are presented as means ±SE. The significance of differences among the means across the three PaO_2 conditions studied: hypoxia, normoxia, and hyperoxia, was tested with a one-way analysis of variance, followed by Scheffe's test for multiple comparisons. P < 0.05 was deemed significant.

RESULTS

The effects on a phosphoinositide-specific phospholipase C of the studied levels of PaO_2 in the CB tissue are displayed in Fig. 1 and Fig. 2.

Phosphatidylinositol-specific PLC

The basal normoxic level of PI-specific PLC was 0.78 ± 0.04 nmol/mg protein/min. In comparison to this control level, PLC activity exhibited a 33% increase in hypoxia and a 37% decrease in hyperoxia (Fig. 1). The difference between the hypoxic and hyperoxic levels was significant at $P < 0.02$.

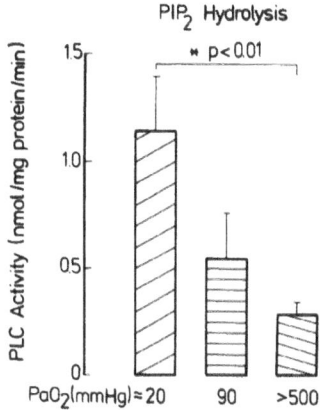

Fig. 2. *Phosphatidylinositol-4,5-bisphosphate-specific phospholipase C activity in the carotid body tissue at various PaO_2 levels. * Significant difference between hypoxia and hyperoxia (F-test followed by Scheffe's test). PLC activity was markedly increased in hypoxia and decreased in hyperoxia.*

Phosphatidylinositol-4,5-bisphosphate-specific PLC

The basal normoxic level of PIP_2-specific PLC was 0.55 ± 0.21 nmol/mg protein/min. The PLC's changes were qualitatively similar to, but more pronounced than, those of PI-specific PLC. It increased 2.1-fold in hypoxia and decreased by 48% in hyperoxia. The difference between the hypoxic and hyperoxic levels was significant at $P < 0.01$ (Fig. 2).

DISCUSSION

This study demonstrates that phosphoinositide-specific phospholipase C activity in the carotid body tissue was linked to the level of arterial PO_2. This activity was markedly increased in hypoxia and decreased in hyperoxia. The opposite effects of hypoxia and hyperoxia clearly argue for more than a casual association between the activity of PLC and the level of oxygenation.

These results are some of the first on PLC activity in the CB. The data on the effects on PLC activity of hypoxia in other tissues are sparse and not firm. Hypoxia increases the formation of IP_3 in the in vitro, and plausibly also in vivo, myocardium, which ought to be bound to enhanced PLC activity (Heathers et al., 1989). In contrast, the phosphatidylinositol transduction system which controls the contraction force of smooth muscle seems blunted in hypoxia (Coburn et al., 1988). In the neural tissue, Huang and Gibson (1989) showed that an in vitro histotoxic hypoxia did not alter the basal turnover of phosphatidylinositol.

On the basis of the present results we posit the following scheme of events in the CB chemoreceptor tissue in response to low PaO_2. Hypoxia increases cytosolic-free calcium, which stimulates the release of DA (Gibson et al., 1989; Shaw et al., 1989). Interaction between DA and its receptors with the possible involvement of the GTP-binding proteins activates the PLC. The PLC, degrading the phosphoinositides, leads to the formation of IP_3 and DAG messenger molecules which are essential for initiating the signal cascade leading to cellular responses. The IP_3 causes mobilization of calcium from intracellular storage sites. The DAG activates protein kinase C with subsequent regulatory proteins phosphorylation (Majerus et al., 1986). The calcium mobilizing action of messenger molecules formed promotes further DA release. Dopamine, in turn, may enhance PLC activation and consequent IP_3 and DAG formation (Rodrigues and Dowling, 1990), amplifying and perpetuating the whole process. This scheme assumes that dopamine receptors in the carotid body are linked to phospholipase C, which remains yet to be demonstrated.

In summary, while we believe we have shown that the phosphoinositides participate in the transduction of the hypoxic signal in the carotid body tissue, the exact intracellular determinants of chemotransduction and their interrelations remain to be further elucidated.

REFERENCES

Coburn, R. F., Baron, C., and Papadopoulos, M. T., 1988, Phosphoinositide metabolism and metabolism-contraction coupling in rabbit aorta, Am. J. Physiol.. 255:H1476.

Gibson, G. E., Manger, T., Toral-Barza, L., and Freeman, G., 1989, Cytosolic-free calcium and neurotransmitter release with decreased availability of glucose or oxygen, Neurochem. Res.. 14:437.

Gomez-Nino, A., Dinger, B., Gonzalez, C., and Fidone, S. J., 1990, Differential stimulus coupling to dopamine and norepinephrine stores in rabbit carotid body type I cells, Brain Res.. 525:160.

Harrison, D. C., Lemasters, J. J., and Herman, B., 1991, A pH-dependent phospholipase A_2 contributes to loss of plasma membrane integrity during chemical hypoxia in rat hepatocytes, Biochem. Biophys. Res. Comm.. 174:654.

Heathers, G. P., Evers, A. S., and Corr, P. B., 1989, Enhanced inositol trisphosphate response to alpha-adrenergic stimulation in cardiac myocytes exposed to hypoxia, J Clin. Invest.. 83:1409.

Huang, H-M. and Gibson, G. E., 1989, Phosphatidylinositol metabolism during in vitro hypoxia, J. Neurochem.. 52:830.

Majerus, P. W., Conolly, T. M., Deckmyn, H., Ross, H., Bross, T. S., Ishii, T. E., Bansal, V. S., and Wilson, D. B., 1986, The metabolism of phosphoinositide-derived messenger molecules, Science. 234:1519.

Meldrum, E., Parker, P. J., and Carozzi, A., 1991, The PtdIns-PLC superfamily and signal transduction, Biochim. Biophys. Acta. 1092:49.

Rodrigues, P. S. and Dowling, J. E., 1990, Dopamine induces neurite retraction in retinal horizontal cells via diacylglycerol and protein kinase C, Proc. Natl. Acad. Sci.. 87:9693.

Shaw, K., Montague, W., and Pallot, D. J., 1989, Biochemical studies on the release of catecholamines from the rat carotid body in vitro, Biochim. Biophys. Acta. 1013:42.

28 OPTICAL MEASUREMENTS OF MICRO-VASCULAR OXYGEN PRESSURE AND INTRACELLULAR pH IN THE CAT CAROTID BODY: TESTING HYPOTHESES OF OXYGEN CHEMORECEPTION

W.L. Rumsey [3], R. Iturriaga [1], D. Spergel [1], S. Lahiri [1] and D.F. Wilson [2]

Departments of Physiology[1] and of Biochemistry and Biophysics[2], University of Pennsylvania School of Medicine, Philadelphia PA 19104, Department of Radiopharmaceuticals[3], Bristol-Myers Squibb Pharmaceutical Research Institute, New Brunswick, NJ 08903

INTRODUCTION

The mechanism(s) of oxygen chemoreception by the carotid body is not understood. One current hypothesis states that hypoxic chemoreception is mediated by acid formation in the chemoreceptor cells (Hanson et al., 1981, Eyzaguirre et al, 1983, Fidone and Gonzalez, 1986). Using exogenously supplied lumiphors and video imaging methodology, we tested this hypothesis by monitoring changes in microvascular oxygen pressure and intracellular pH with those of chemosensory activity during oxygen deprivation. These studies were conducted using an *in vitro* preparation of the cat carotid body that was perfused and superfused with cell-free Tyrode solution (Iturriaga et al, 1990).

METHODS

Cats of either sex were anesthetized with sodium pentobarbital (35 mg/kg, IP). Excision of the carotid bifurcation and perfusion of the carotid body in the isolated state were carried out as described previously (Iturriaga et al, 1991). In brief, the carotid bifurcation was placed in a chamber for simultaneous perfusion and superfusion at 36.5 $\pm 0.5°$ C using a modified Tyrode solution containing [in mM] Na^+ [154]; K^+ [4.7]; Ca^{2+} [2.2]; Mg^{2+} [1.1]; Cl^- [110], glutamate [42.0]; glucose [5.0], and HEPES [5], pH 7.40. The media was modified by adding $NaHCO_3^-$ [21.4 mM] and deleting the same amount of NaCl for experiments on intracellular pH. Perfusion was established at a constant pressure of 80 Torr. The superfusate, equilibrated with 100% N_2 was maintained at 1.5 ml/min using a pressure gradient of 15 Torr. Measurements of PO_2, PCO_2 and pH of the media were made routinely with a blood gas analyzer (BMS3 Radiometer, Copenhagen). Chemosensory discharge was recorded from the whole carotid sinus nerve which was placed on bipolar platinum electrodes in paraffin oil.

Imaging oxygen pressure by phosphorescence quenching has been described previously in detail (Rumsey et al; 1988, 1991, Wilson 1989). The oxygen probe, Pdcoproporphyrin (3 μM, Porphyrin Products, Logan UT), and bovine serum albumin (1%, Fraction V, ICNI immunobiologicals, Lisle, IL) were added to the perfusate. A high pressure mercury lamp (100 watts) was used to illuminate (530-560 nm) the tissue. Phosphorescence (> 645 nm) was detected with an intensified CCD camera (Xybion Electronics Systems, San Diego, CA). The image was magnified with a Wild Makroskope (Leitz, Germany) positioned about 5-7 cm above the carotid body. The output from the video camera was continuously monitored (Sony PVM-12710, Japan) and recorded (HRV-3000E video recorder, All-Tronics Medical Systems, Cleveland, OH). The images were digitized and processed (Image-NAT, Universal Imaging Corp., Media, PA) via computer (Spear 386, Northbrook, IL) interfaced with the video recorder.

Changes of intracellular pH (pH^i) within the carotid body were monitored using the video system described above. The tissue was perfused with Tyrode solution containing the acetoxymethyl ester of the pH-sensitive lumiphor, 2',7'-bis(2-carboxyethyl)-5(6)-carboxyfluorescein (BCECF, 5 μM). Perfusion of the carotid body with BCECF/AM was maintained until a sufficient signal was detected using the video system. Perfusion in the absence of the fluorophor continued for about 10 min. The remaining fluorescent signal was indicative of BCECF trapped within the cells of the carotid body. The carotid body was illuminated (excitation = 420-490 nm) and emission (> 515 nm) was detected by the intensified CCD camera. The images were digitized and processed as above. The data are expressed as fluorescence intensity, i.e., the raw signal minus the background intensity, and were used to detect changes of pH_i rather than reporting the absolute pH. A decrease of intensity reflects acidification of the cytosol.

RESULTS and DISCUSSION

In the first two illustrations, changes in microvascular oxygen pressure are shown in response to lowering the oxygen pressure in the perfusate and to perfusate flow interruption. In the latter case, the chemosensory discharge is also plotted to describe the correlation between this parameter and capillary oxygen disappearance. The last two illustrations provide examples of the effects of hypoxia and hypercapnia on the relationship between intracellular pH and chemosensory discharge.

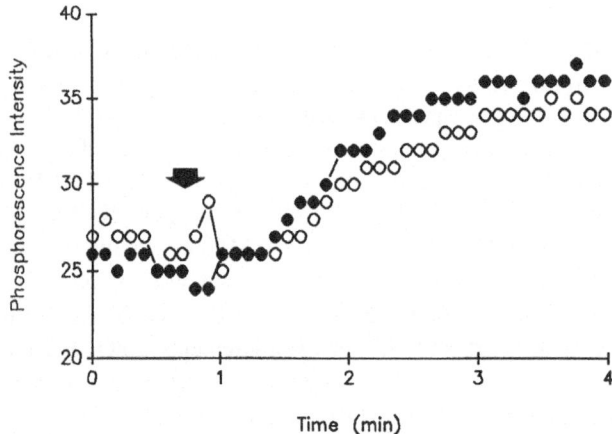

Figure 1. Effect of high and low perfusate PO_2 on phosphorescence intensity in the carotid body. Carotid body was perfused initially with high PO_2 (380 Torr) in the media and then, as indicated by the arrow, it was perfused with low PO_2 (120 Torr) in the media. The open and closed circles represent two separate regions of interest.

Figure 1 shows that the phosphorescence intensity arising from the microvasculature of the cat carotid body is increased upon transition from perfusion with media of high PO_2 (380 Torr) to one having a lower PO_2 (120 Torr, termed normoxic). In two regions of interest, the change in phosphorescence was only about 27% and 38% despite a large decrease in perfusate PO_2, i.e.. 68%. As a result of this transition, chemosensory discharge increased from 13.1 ± 1.9 imp/sec to 20.3 ± 2.2 imp/sec. As can be seen from the Stem-Volmer relationship ($I_o/I = T_o/T = 1 + k_q T_0[O_2]$; where I_o/I and T_o/T are the phosphorescence intensity and lifetime at zero oxygen pressure divided by the phosphorescence intensity and lifetime at a given oxygen pressure, respectively, and kq is a constant related to diffusion of oxygen and the oxygen probe) I_o/I and at these levels of oxygen pressure, I are directly proportional to $[O_2]$ (Vanderkooi et al, 1988). A change in perfusate PO_2 should result in a proportional change in phosphorescence intensity arising from the carotid body microvasculature. Therefore, the small change in phosphorescence, resulted likely from alterations in other physiological factors such as vascular resistance to perfusate flow.

In response to perfusate flow interruption by clamping the influent line, phosphorescence intensity from the carotid body increased several-fold. From these images, calculation of oxygen pressure within the carotid body showed that oxygen was removed from the capillaries in three phases, presumably as a result of diffusion to the mitochondria for oxidative phosphorylation, and was related to chemosensory discharge (Figure 2). It should first be noted that the PO_2 in the capillaries immediately after applying the influent clamp was about 27 Torr (higher PO_2 >200 Torr, in the perfusate resulted in levels of capillary PO_2 of about 45 Torr). Chemosensory discharge was initially about 25 imp/sec and steadily increased to slightly greater levels at about 25 seconds after the occlusion. In this first phase, oxygen disappearance was slow and steady. Chemosensory discharge then became progressively more active beginning at 30 sec and continuing to 45 sec postocclusion at which time discharges were at maximal levels . Concomitantly, oxygen disappearance became more rapid. This second phase was initiated when capillary PO_2 had decreased to about 10 Torr. The maximal level of chemosensory discharge was not maintained, rather discharge began immediately to decrease. At the same time, the rate of oxygen disappearance became progressively

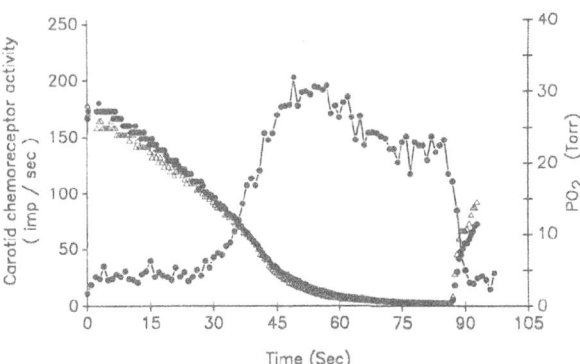

Figure 2 - *Effect of perfusate flow interruption on carotid body PO_2 and chemosensory discharge. Chemosensory discharge (ordinate) is represented by the closed circles connected by a line. Carotid body PO_2 (opposite axis) is represented by the discrete closed circles and open triangles. The PO_2 in the perfusate was 120 Torr. The influent line was clamped at time zero and flow was restored at about 87 seconds.*

sluggish, slowing to levels that were less than those found in the previous two phases of the occlusion. This last phase of oxygen disappearance began when oxygen pressure was about 3 Torr and continued until it neared zero. Restoration of perfusate flow resulted in prompt return of both capillary oxygen pressure (data not completely shown) and chemosensory activity to pre-occlusion levels.

Hypoxia was not associated with any change in fluorescence emission from the carotid body. Figure 3 shows that the transition from hyperoxic perfusate ($PO_2 = 350$ Torr) to a hypoxic one ($PO_2 = 50$ Torr, $PCO_2 = 30$ Torr, pH 7.4) results in increased chemosensory discharges, from about 7-10 imp/sec to a maximum of 50 imp/sec after about 60 seconds. Chemosensory discharge returned rapidly to baseline values when the PO2 was restored to its previous level of 350 Torr. Despite a large increase in chemosensory activity, fluorescence intensity remained essentially constant. Interruption of perfusate flow (data not shown) did not result in a decrease of intracellular pH unless the ischemia was continued for long periods of flow interruption (10 min duration). The latter change, however, occurred at a time (8-10 min post-clamp) when the levels of discharge had declined markedly from peak activity values.

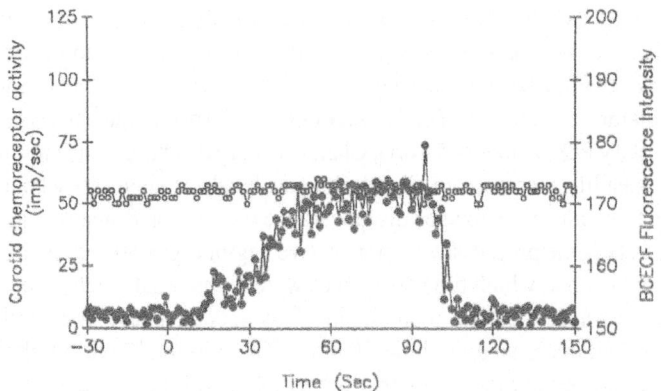

Figure 3. Effect of hypoxia on chemosensory discharges and intracellular pH. Carotid body was perfused with Tyrode; PO2 = 350 Torr, PCO2 = 30 Torr, pH 7.4. At zero time, the perfusate was switched to one with low PO2 (50 Torr). Hyperoxia was restored at about 100 seconds. Chemoreceptor discharge (ordinate) is represented by the closed circles and fluorescence intensity (opposite axis in grey levels) is represented by the open circles. A decline in fluorescence intensity indicates intracellular acidification.

In contrast to hypoxia, the transition from normocapnia ($PCO_2 = 30$ Torr) to hypercapnia ($PCO_2 = 65$ Torr) lowered intracellular pH and increased chemosensory discharge (Figure 4). Chemosensory discharge responded rapidly to hypercapnia, rising from about 10 imp/sec to a maximum of 125 imp/sec within 15 seconds and then it quickly decreased from this maximal response to about 55 imp/sec at 45-50 seconds. Chemosensory discharge remained at this level until the return to normocapnia. Normocapnia resulted in rapid return, within a few seconds, of chemosensory activity to baseline levels. When chemosensory discharge reached its plateau phase, fluorescence intensity had finally declined to its lowest level which required about 45-50 seconds to transpire. No further changes of fluorescence intensity were obtained until return to normocapnia. Restoration of normocapnia resulted in a slow increase of fluorescence intensity to levels obtained prior to the hypercapnic episode. This return to baseline levels required about 20-25 seconds. The explanation for the time lags between the responses of fluorescence intensity and chemosensory discharge were not readily apparent. The response time of the fluorophor to changes in proton concentration is rapid (less than a second). Thus other factors yet to be identified must be responsible for the transient increase in the chemoreceptor response to exogenous CO_2. It should also be noted that the changes described above were also obtained during a bolus injection of Tyrode equilibrated at a PCO_2 of 38-40 Torr into the influent line containing Tyrode buffered with HEPES, pH 7.4.

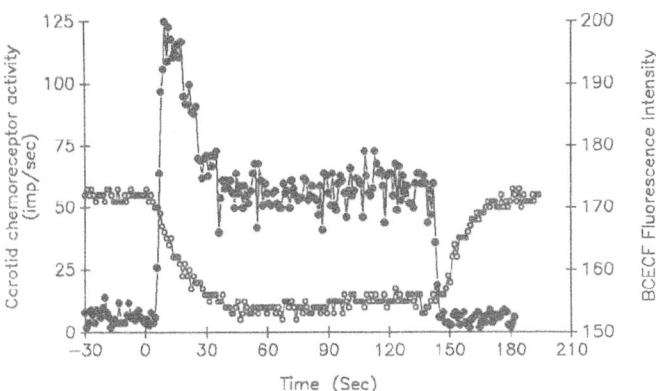

Figure 4. Effect of hypercapnia on chemoreceptor discharge and intracellular pH. Carotid body was initially perfused with Tyrode buffered with CO_2-HCO_3^-, (PCO_2 = 30 Torr, PO_2 = 350 Torr, pH 7.4). At zero time the perfusate was switched to one with increased PCO_2 (65 Torr, PO_2 unchanged, pH 7.1). The original perfusate was reinstated at about 145 seconds. Chemosensory discharges and fluorescence intensity were represented by symbols described in the previous figure.

The results presented within this report address both the "metabolic" (for recent review of the metabolic hypothesis, see Acker, 1989) and the "acid" hypotheses of oxygen chemoreception. The oxygen pressures found in the microvasculature and by accounting for an oxygen gradient from the capillaries to the consuming mitochondria are within the physiological range of values known to affect the rate of mitochondrial oxidative phosphorylation (see for example, Wilson et al, 1988, Rumsey et al 1990). The changes of microvascular oxygen pressure appear to correlate closely with the changes in chemosensory discharges during hypoxia. Moreover, an increase in carotid sinus discharge is accompanied by an increase in carotid body oxygen consumption indicated by the rate of oxygen disappearance from the microvascular compartment. On the other hand, hypoxia-induced chemosensory discharge was not associated with intracellular acidification. Contribution of ATP derived from anaerobic glycolysis and consequent production of lactic acidosis was not apparent. These results indicate, therefore, that the energy supporting the reactions involved in carotid sinus discharge is primarily derived from mitochondrial oxidative phosphorylation. Mitochondrial oxidative phosphorylation appears to play an important role in the mechanism(s) of oxygen chemoreception in the carotid body.

ACKNOWLEDGEMENTS

This work was supported by National Institutes of Health Grant HL43413.

REFERENCES

Acker, H., 1989, PO$_2$ chemoreception in arterial chemoreceptors. Ann. Rev. Physiol. 5 1 : 835- 844.

Eyzaguire, C., Fltzgerald, R.S. Lahiri, S. and Zapata, P., 1983, Arterial Chemoreceptors. In: "Handbook of Physiology. The Cardiovascular System. Peripheral Circulation and Organ Blood Flow". Bethesda, MD. Am. Physiol. Soc., 551-621.

Fidone S.J. and Gonzalez, C., 1986, Initiation and control of chemoreceptor activity in the carotid body. In: "Handbook of Physiology. The Respiratory System. Control of Breathing". Bethesda, MD: Am. Physiol. Soc., 247-312.

Fitzgerald, R.S. and Lahiri, S., 1986, Reflex responses to chemoreceptor stimulation. In: "Handbook of Physiology. The Respiratory System. Control of Breathing". Bethesda, MD: Am. Physiol. Soc., 313-362, 1986.

Hanson, M.A., P.C.G., Nye, and Torrance, R.W. 1981, The exodus of an extracellular bicarbonate theory of chemoreception and the genesis of an intracellular one. In: "Arterial Chemoreceptors", eds., Belmonte, C., Pallot, D., Acker, H., and S. Fidone. Leicester, University Press, p. 403416.

Iturriaga, R., Rumsey, W.L., Mokashi, A., Spergel, D., Wilson, D.F., and Lahiri S. 1991, In vitro perfused-superfused cat carotid body for physiological and pharmacological studies. J. Appl. Physiol. 70:1393-1400.

Rurnsey, W.L., Schlosser, C., Nuutinen, E.M., Robiolio, M. and Wilson, D.F., 1990, Cellular energetics and the oxygen dependence of respiration in cardiac myocytes isolated from adult rat heart. J. Biol. Chem. 265,15392-15399.

Rumsey, W.L., Vanderkooi, J. and Wilson, D.F., 1988, Imaging of phosphorescence: a novel method for measuring oxygen distribution in perfused tissue. Science 241, 1649-1651.

Rumsey, W.L., Iturriaga, R. Spergel, D., Lahiri, S., and Wilson, D.F. 1991, Optical measurements of the dependence of chemoreception on oxygen pressure in the cat carotid body. Am. J. Physiol. 261 (Cell Physiol. 30): C614-C622.

Vanderkooi. J.M., Maniara, G., Green, T.J. and Wilson, D.F., 1987, An optical method for measurement of dioxygen concentration based upon quenching of phosphorescence. J. Biol. Chem. 262, 5476-5482.

Wilson, D.F., Rumsey, W.L., Green, T.J. and Vanderkooi, J.M., 1988, The oxygen dependence of mitochondrial oxidative phosphorylation measured by a new optical method for measuring oxygen concentration. J. Biol. Chem. 263, 2712-2718.

29 ELEVATION OF CYTOSOLIC CALCIUM INDUCED BY pH CHANGES IN CULTURED CAROTID BODY GLOMUS CELLS

M. Sato, K. Yoshizaki and H. Koyano

Department of Physiology, Akita University School of Medicine,
Akita (Japan)

INTRODUCTION

The carotid body (CB) is a chemoreceptor organ responsive to application of cyanide (CN), decreases in environmental Po_2 and pH levels [1,2]. Hypoxia, CN and acidity increase the Ca^{2+}-dependent release of dopamine from the CB, augmenting the sensory discharge freguency of carotid nerve fibers [3,4,5] forming a synapse with the CB glomus (type I) cells[6]. There are many neurotransmitters in glomus cells [7], and application of hypoxia to cultured glomus cells elicited increases in dopamine release[8] and uptake of $^{45}Ca^{2+}$,[9]. Thus, it has been proposed that secretion of putative chemoreception-trasmitters from glomus cells will occur through an increase in cytosolic Ca^{2+} ($[Ca^{2+}]i$), although there has been some discussion about the specific role and identity of the secreted-neurotransmitters. Recently direct measurement of $[Ca^{2+}]i$ using fluorometry demonstrated that hypoxia including both anoxia and CN induced a gradual increase in $[Ca^{2+}]i$ in glomus cells[10,11]. On the other hand, a response of $[Ca^{2+}]i$ to other natural chemoreceptor stimulus such as external acidity has not yet been examined. In this study we examined the effects of changes in extracellular pH level on $[Ca^{2+}]i$ in cultured glomus cells using a fura-2 method[12].

MATERIALS AND METHODS

Dissociated cultures of CBs were prepared as previously described[11,13]. The tissues were removed from ether-anesthetized newborn rabbits and incubated for 15 min in Ca^{2+}- and Mg^{2+}-free Hank's saline containing enzymes (0.25% trypsin and 0.2% collagenase). After incubation, tissues were washed with minimum essential medium (MEM Nissui, Japan) to remove the enzymes and mechanically dissociated with a Pasteur pipette (0.5 mm) in MEM supplemented with insulin (5 µg/ml) and 10% fetal bovine serum (Nissui). Mixtures of single cells and clusters were plated on collagen-coated glass coverslips and cultured under 5% CO_2 in air for 2-10 days. Glomus cells in cluster were used for measuring $[Ca^{2+}]i$, because they showed better growth than did single glomus cells.

$[Ca^{2+}]i$ measurements in cultured glomus cells were made using a video camera-assisted microfluorometry system and a Ca^{2+}-sensitive fluorophore, fura-2, as in previous reports[11]. The coverslips carrying cells were washed with HEPES-buffered saline (HBS; 135.5 mM NaCl, 1.2 mM $CaCl_2$, 1.2 mM $MgCl_2$ •$6H_2O$, 10 mM glucose and 12 mM HEPES-NaOH, pH 7.4) and loaded with 5µM fura-2-acetoxymethylester (Dojin, Japan) for 2-4 h at 20-27 °C. After being washed with HBS, the coverslips were mounted in a measuring chamber placed on an inverted fluorescence microscope (Nikon TMD). Fluorescence was obtained by double beam excitation (340 and 360 nm) and emission (500 nm). At both the start and the end of the measurements, the cells were excited at 360 nm and, during the measurements, at 340 nm. The fluorescence image of fura-2-loaded cells was converted to digital by a video frame memory (Mitsubishi Kasei, Japan) and fed to a computer-equipped image analysis unit. The value of $[Ca^{2+}]i$ was obtained from the ratio of fluorescence intensity at 340 nm and at 360 nm by referring the value to a calibration curve. All procedures of $[Ca^{2+}]i$ measurements were made at room temperature (20-25°C). HBS was used as a control perfusate and drugs were dissolved in HBS. Acid solution was obtained by adding HCl to HBS and alkaline solution, NaOH to HBS. Sodium-free HBS contained 135.5 mM choline chloride in place of NaCl and the pH was adjusted to 7.4 with Tris (hydroxymethyl) aminomethane. Calcium-free HBS contained 3 mM EGTA. Solution changes were performed by applying the test solutions and simultaneously draining a similar volume by suction with a water flow pump. Veratridine (VRT, Sigma) and tetrodotoxin (TTX, Sankyo) were used as a voltage-dependent Na^+ channel stimulator and an ion channel inhibitor, respectively.

RESULTS

Before proceeding to an examination of the effects of changes in intracellular pH on $[Ca^{2+}]i$, glomus cells within cultured CB cells clusters were identified by the presence of sensitivity to a depolarizing agent, and the effect of its inhibitor. Stimulation by VRT, a stimulator for voltage-dependent Na^+ channels, produced a rapid increase in $[Ca^{2+}]i$, which was then inhibited by the addition of TTX, while prior application of TTX to VRT caused no increase in $[Ca^{2+}]i$ (Fig. 1).

Effects of lowering of extracellular pH on cytosolic calcium

When extracellular pH was lowered step-wise every 0.5 by changing perfusates into more acidic solutions, at pH 6.5 $[Ca^{2+}]i$ rapidly increased and then fell and relaxed within 1-2 minutes. Once the transient increase in $[Ca^{2+}]i$ was evoked, more acidity did not bring about any $[Ca^{2+}]i$ changes (Fig. 2A). But after physiological neutralization of extracellular pH the response of $[Ca^{2+}]i$ to the lowering of pH was caused repeatedly (Fig. 2A). The response to lowered pH (pH 6.0) was suppressed completely by removal of sodium (Fig. 2B) and calcium (Fig. 2C), but was unaffected by the addition of TTX (Fig. 2D).

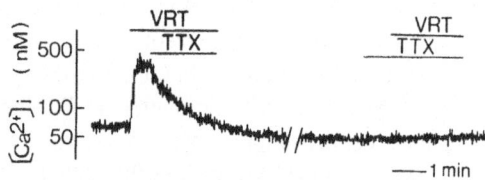

Fig. 1. *Effect of a voltage-dependent sodium channel activator, veratridine (VRT, 200 M), on $[Ca^{2+}]i$ in cultured glomus cells and inhibition of the effect by tetrodotoxin (TTX, 5 μM). Horizontal bars, time of applications of VRT and TTX.*

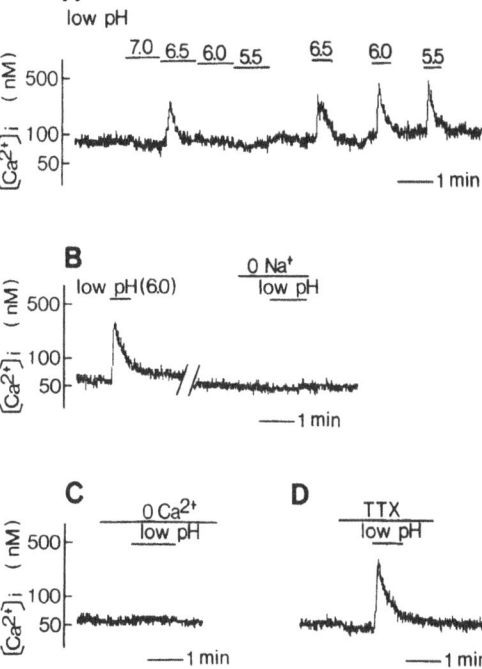

Fig. 2. Effect of lowering of extracellular pH on $[Ca^{2+}]i$ (A) and effects of removal of Na^+ (B), removal of Ca^{2+} (C) and addition of TTX (D) on the increase in $[Ca^{2+}]i$ evoked by lowering pH (pH 6.0). Horizontal bars, time of applications of acidic perfusates, TTX, Na^+-free and Ca^{2+}-free.

Effects of heightening extracellular pH on cytosolic calcium

When the extracellular pH was heightened by changing perfusates into more alkaline solutions, at pH 8.0 a gradual increase in $[Ca^{2+}]i$ was induced and enlarged with increased alkalinity (Figs. 3A and B). The response to high pH (pH 8.5) was inhibited by removal of extracellular calcium (Fig. 3D), but was unaffected by removal of sodium (Fig. 3C) and by the addition of TTX (Fig. 3E). Incidentally, TTX had still an inhibitory effect on the stimulation of voltage-dependent sodium channels with VRT even in the high pH (Fig. 3F).

DISCUSSION

The present experiments clearly demonstrated that changes in extracellular pH elicited increases of $[Ca^{2+}]i$ in cultured CB glomus cells. Specifically, lowering of extracellular pH to below pH 6.5 evoked a transient increase in $[Ca^{2+}]i$ and heightening of extracellular pH to over pH 8.0 induced a gradual increase (Figs. 2A and 3A).

Major chemoreceptor stimuli such as hypoxia and acidity increase the frequency of carotid sensory discharge and the Ca^{2+}-dependent release of dopamine from CB glomus cells[3,4,5] containing many neutrotransmitters[7]. It has been proposed that an increase in $[Ca^{2+}]i$ in the glomus cells is needed for secretion of chemoreception trasmitters, although their identity has been controversial. The proposal for hypoxic-chemotransduction is supported by the fact that hypoxia including both anoxia and CN induced a rise in $[Ca^{2+}]i$,

207

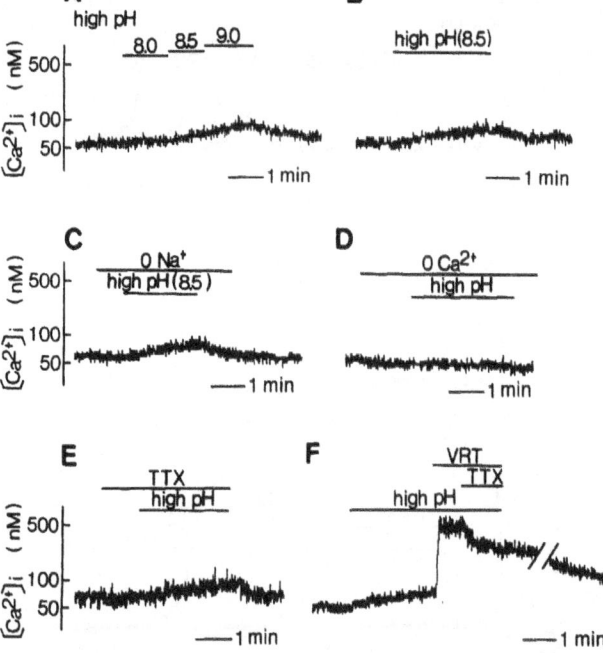

Fig. 3. *Effect of heightening extracellular pH on [Ca²⁺]i (A, B), effects of removal of Na⁺ (C), removal of Ca²⁺ (D) and addition of TTX (E) on the increase in [Ca²⁺]i induced by raising pH (pH 8.5) and effect of the high pH on inhibition of VRT-induced increases in [Ca²⁺]i by TTX (F). Horizontal bars, time of applications of alkaline perfusates, VRT, TTX, Na⁺-free and Ca²⁺-free.*

largely from intracellular pool in freshly dissociated cells from adult rabbit CB[10], in contrast largely from extracellular origin in cultured CB glomus cells of newborn rabbit[14]. Similarly the present initial finding that [Ca²⁺]i in glomus cells increased in response to low pH suggests that the transduction of the acidic stimulus may also involve an increase in [Ca²⁺]i.

However, acidity caused a rapid transient increase in [Ca²⁺]i (Fig. 2), whereas hypoxia including both anoxia and CN induced a gradual increase . The difference in the pattern of a rise in [Ca²⁺]i between the two chemostimuli may be attributable to a discrepancy in the mechanism of an increase in [Ca²⁺]i in response to them.

That an increase in [Ca²⁺]i evoked by the lowering of pH disappeared completely with removal of Na⁺ (Fig. 2B) implies that the occurrence of this response to low pH depends on the inflow of extracellular Na⁺. In recent years, glomus cells have been found to be electrically excitable[15,16]. Stimulation of TTX-sensitive voltage-dependent Na⁺ channels by VRT enhanced influx of ²²Na⁺ [13], causing depolarization, which activated Ca²⁺ entry via its own channels, leading to a rapid rise in [Ca²⁺]i, as shown in Fig. 11 . Therefore, in order to examine whether Na⁺ channels have any role in the increase in [Ca²⁺]i induced by lowered pH, we compared the effect of the stimulus on [Ca²⁺]i with and without TTX. However, TTX failed to inhibit the effect (Fig. 2D), which suggests that the Na⁺ inflow may be made via Na⁺ influx system other than TTX-sensitive voltage-dependent Na⁺ channels. Exposure of cultured glomus cells to acidity results in a decrease in intracellular pH and causes depolarization, indicating that the resting potential of glomus cells is dependent on H⁺ ions, in addition, the presence of Na⁺ dependent H⁺

extruding sytem such as Na^+-H^+ exchanger and Na^+ -dependent Cl^- - HCO_3^- antiporter exchanging external Na^+ and HCO_3^- for internal Cl^- and H^+ has recently been demonstrated in glomus cells of neonatal rat[18]. When these reports are taken into account, it seems possible that the Na^+-dependent H^+ extruding systems operate in response to low pH, and consequently alternatively-fluxed Na^+ may cause depolarization, which is inconsistent with the idea that reduction of K^+ current by low pH in glomus cells may produce depolarization[19,20].

Further, the fact that the reponse of $[Ca^{2+}]i$ to low pH also disappeared completely by removal of Ca^{2+} (Fig. 2C) implies that the source of Ca^{2+} contributing to the increase in $[Ca^{2+}]i$ is extracellular. The inhibitory effects of the removal of Na^+ and Ca^{2+} on lowered pH-induced changes in $[Ca^{2+}]i$ suggest two possible pathways of Ca^{2+} entry. One is voltage-gated Ca^{2+} channels activated by alternatively-fluxed Na^+-dependent depolarization, as described above. The other is an Na^+-H^+ exchanger-coupled Na^+ -Ca^{2+} antiporter exchanging Na^+ for extracellular Ca^{2+}, which is suggested by a report that acidic stimulus-induced release of dopamine is unaffected by Ca^{2+} agonist and antagonists, meaning no occurrence of depolarizaion accompanied acidity[21]. Either pathway may possibly be involved in the increase in $[Ca^{2+}]i$ evoked by lowered pH. At present, it is difficult to elucidate why and how lowered pH evokes a topographically increased and desensitized response.

Little is known at present about the second finding that cultured glomus cells respond to high pH with a gradual increase in $[Ca^{2+}]i$ (Figs. 3A and B). The response of $[Ca^{2+}]i$ to alkalinity, unlikely to that to acidity, seems to have no role as a signal for neurotransmitter-secretion generating action potentials at carotid sensory nerve endings, because external alkalinity depresses chemoreception discharge, while low pH augments it[17]. Further, the response of $[Ca^{2+}]i$ to external alkalinity was enhanced with higher extracellular pH (Fig. 3A). These results suggest that alkalinity-induced increases in $[Ca^{2+}]i$ may reflect cell injury, as reported in cerebral ischemia accompanied by $[Ca^{2+}]i$ accumulation in the rat hippocampus neurons[22].

We can assume that the difference in the pattern of an increase in $[Ca^{2+}]i$ between acidity and alkalinity may be due to the different mechanisms between them. In fact the response to acidity is dependent on Na^+ but that to alkalinity is not. However, extracellular Ca^{2+} is mobilized for an increase in $[Ca^{2+}]i$ evoked by changes in extracellular pH.

CONCLUSIONS

1. Changes in extracellular pH elicit increases of $[Ca^{2+}]i$ in cultured glomus cells. Specifically, decreasing extracellular pH to below pH 6.5 evoked a rapid transient rise in $[Ca^{2+}]i$, while increasing extracellular pH to over pH 8.0 induced a gradual increase. The effect of acidity on $[Ca^{2+}]i$ suggests that the transduction of the acidic stimulus involves a mechanism of Ca^{2+} dependent neurotransmitters secretion.

2. Ca^{2+} contributing to the rise in $[Ca^{2+}]i$ responding to low pH fluxes via Ca^{2+} influx system which depends on Na^+ fluxed through Na^+ influx system other than TTX-sensitive voltage-dependent Na^+ channels. It may be speculated here that the Na^+-dependent Ca^{2+} influx system involves voltage-dependent Ca^{2+} channels activated by depolarization resulting from Na^+ influx through Na^+ - H^+ antiporter and Na^+-Ca^{2+} exchanger coupled Na^+ -H^+ antiporter.

3. Ca^{2+} contributing to the increase in $[Ca^{2+}]i$ induced by high pH fluxes from extracellular solution.

REFERENCES

1. C. *Eyzaguirre and H. Koyano*, Effects of hypoxia, hypercapnia and pH on the chemoreceptor activity of the carotid body in vitro, J. Physiol (London). 178:385 (1965).

2. *U. S. von Euler and G. Liljestrand*, The effects of cyanide on respiration, Skand Archiv. Physiol. 76:27 (1937).

3. *S. J. Fidone, C. Gonzalez and K. Yoshizaki*, Effects of low oxygen on the release of dopamine from the rabbit carotid body in vitro, J. Physiol (London). 333:93 (1982).

4. *R. Rigual, E. Gonzalez, S. Fidone, and C. Gonzalez*, Effects of low pH on synthesis and release of catecholamines in the cat carotid body in vitro, Brain Res. 309:178 (1984).

5. *A. Obeso, L. Almaraz and C. Gonzalez*, Effects of cyanide and uncouplers on chemoreceptor activity and ATP content of the cat carotid body, Brain Res. 481:250 (1989).

6. *D. M. McDonald*, Peripheral chemoreceptors: structure-function relationships in the carotid body, in: "Lung Biology in Health and Disease, The Regulation of Breathing," T. F. Hornbein, ed., Dekker, New York (1981).

7. *S. Fidone and C. Gonzalez*, Initiation and control of the chemoreceptor activity in the carotid body, in: "Handbook of Physiology, The Respiratory System," A. P. Fishman, ed., American Physiological Society (1986) .

8. *M. C. Fishman, W. L. Greene and D. Platika*, Oxygen chemoreception by carotid body cells in culture, Proc Natl Acad Sci USA. 82:1448 (1985).

9. *F. Pietruschka*, Calcium influx in cultured carotid body cells is stimulated by acetylcholine and hypoxia, Brain Res. 347:140 (1985).

10. *T. J. Biscoe, M. R. Duchen, D. A. Eisner, S. C. O'Neill and M. Valdeolmillos*, Measurements of intracellular Ca^{2+} in dissociated type I cells of the rabbit carotid body, J. Physiol (London). 416:421 (1989).

11. *M. Sato, K. Ikeda, K. Yoshizaki and H. Koyano*, Response of cytosolic calcium to anoxia and cyanide in cultured glomus cells of newborn rabbit carotid body, Brain Res. 551:327 (1991).

12. *G. Grynkiewicz, M. Poenie and R. Y. Tsien*, A new generation of Ca^{2+} indicators with greatly improved fluorescence properties, J. Biol. Chem. 260:3440 (1985).

13. *M. Sato, K. Yoshizaki and H. Koyano*, Veratridine stimulation of sodium influx in carotid body cells from newborn rabbits in primary culture, Brain Res. 504:132 (1989).

14. *M. Sato, H. Iwasaki, K. Yoshizaki and H. Koyano*, Alternate carotid body glomus cells and sensory neurons pathways for increasing cytosolic Ca^{2+} in response to anoxia and cyanide, Jap. J. Physiol. 41:5667 (1991).

15. *M. R. Duchen, K. W. T. Caddy, G. C. Kirby, D. L. Patterson, J. Ponte and T. J. Biscoe*, Biophysical studies of the cellular elements of the rabbit carotid body, Neuroscience. 26: 291 (1988).

16. *J. Lopez-Barneo, J. Lopez-Lopez, J. Urena and C. Gonzalez*, Chemotransduction in the carotid body: K current modulated by Po_2 in type I chemoreceptor cells, Science. 241:580 (1988).

17. *S-F. He, J-Y. Wei and C. Eyzaguirre*, Intracellular pH and some membrane characteristics of cultured carotid body glomus cells, Brain Res. 547:258(1991).

18. K.J. Buckler, R.D. Vaughan-Jones, C. Peers and P. C. G. Nye, Intracellular pH and its regulation in isolated type I carotid body cells of the neonatal rat, J. Physiol (London). 436:107 (1991).

19. *J. Lopez-Lopez, C. Gonzalez, J. Urena and J. Lopez-Barneo*, Low pO_2 selectively inhibits K^+ channel activity in chemoreceptor cells of the mammalian carotid body, J. Gen. Physiol. 93:1001 (1989).

20. *C. Peers*, Effect of lowerd extracellular pH on Ca^{2+} -dependent K^+ currents in type I cells from the neonatal rat carotid body, J. Physiol (London). 422:381 (1990).

21. *C. Gonzalez, A. Rocher, A. Obeso, J. R. Lopez-Lopez, J. Lopez-Barneo and B. Herreros*, Ionic mechanisms of the chemoreception process in type I cells of the carotid body, in:" Arterial Chemoreception," C. Eyzaguirre, S. J. Fidone, R. S. Fitzgerald, S. Lahiri and D. M. McDonald, eds., Springer-Verlag, New York (1990).

22. *J. K. Deshpande, B. K Siesjo and T. J. Wieloch*, Calcium accumulation and neuronal damage in the rat hippocampus following cerebral ischemia, J. Cereb. Blood Flow Metab. 7:89 (1987).

30 ROLE OF CARBON DIOXIDE FOR HYPOXIC CHEMOTRANSDUCTION OF THE CAT CAROTID BODY

M. Shirahata and R.S. Fitzgerald

Departments of Environmental Health Sciences
Anesthesiology/Critical Care Medicine, Medicine, and Physiology
The Johns Hopkins Medical Institutions
615 N. Wolfe Street, Baltimore, MD 21205, USA

INTRODUCTION

The interaction between hypoxia and hypercapnia in the genesis of increased neural activity from the carotid body has been appreciated for 30 year[2,3,4,7,8]. When animals were hyperventilated, carotid chemoreceptor neural response to hypoxia was extremely reduced, and in some cases the response was almost abolished. On the other hand, the response of the carotid body to hypercapnia has never been abolished even under hyperoxic condition, although it is reduced. These observation has led us to hypothesize that the presence of carbon dioxide (CO_2) has some critical role for hypoxic chemotransduction of the carotid body. Because the carotid body is very sensitive to arterial CO_2 tension, and because CO_2/HCO_3^- is the main buffer in the biological system, regulation of intracellular pH (pHi) in the chemosensitive unit may exclusively depend on CO_2/HCO_3^- in vivo. Decrease in CO_2/HCO_3^- would increase pHi of the chemosensitive unit. Alkalinization of the unit would turn off the process of hypoxic chemotransduction. We tested this hypothesis using in vivo selective perfusion techniques of the carotid body. First, we examined the effect of the presence or absence of CO_2/HCO_3^- on the hypoxic chemotransduction of the carotid body. Second, we antagonized possible alkalinization of the chemosensitive unit during CO_2/HCO_3^--free perfusion by giving a non-volatile weak acid, butyrate. A part of this study has been reported previously[12].

METHODS

Twelve cats were anesthetized, paralyzed, and artificially ventilated. Catheters were inserted in the femoral artery and vein for measuring systemic arterial pressure and for continuous administration of glucose and saline. The carotid body area was prepared for selective and intermittent perfusion as described previously[10]. In short, a loop catheter was inserted in the common carotid artery to change the perfusion of the carotid body from natural blood supply to various kinds of cell-free perfusates. The lingual artery was catheterized. Snares were placed around the common carotid artery (proximal of the loop catheter) and around the external carotid artery. Other arterial branches except for the carotid artery were tied. The perfusion pressure was adjusted to systemic arterial pressure. Except for the short perfusion period (90 sec or 5 min), carotid body was supplied with its own natural blood flow. More than 30 min was allowed between the cell-free perfusions. Chemoreceptor neural activity was recorded from whole carotid sinus nerve in which baroreceptor component was eliminated by mechanical and thermal destruction of the baroreceptor nerve endings[5,13]. Chemoreceptor neural activity, systemic arterial pressure, and common arterial pressure were continuously monitored. Arterial blood was withdrawn periodically to measure arterial pH, Pco_2, and Po_2 .Rectal temperature was kept at 38 °C using a hot water blanket.

Experimental Protocol

In the first set of experiments seven cat were ventilated with high oxygen. The carotid body was perfused with one of the following solution for either 90 sec or 5 min. Perfusion solutions were; Krebs Ringer bicarbonate solution (KRB, composition in mM: NaCl 120, KCl 3.5, $CaCl_2$ 1.8, $MgCl_2$ 0.6, NaH_2PO_4 0.6, $NaHCO_3$ 19, glucose 20) exposed to 5 % CO_2 in 95 % N_2, HEPES buffered solution exposed to either 5 % CO_2 in 95 % N_2 (HBS+) or 100 % N_2 (HBS-). The composition of HBS+ was the same as KRB but containing 10 mM HEPES; for HBS-, the same as HBS+ except $NaHCO_3$ was replaced by NaCl.

In the second set of experiment five cats were ventilated with oxygen The carotid body was perfused with either HBS+, HBS-, or HBS- containing 5-20 mM sodium butyrate. In the solution containing sodium butyrate NaCl was reduced according to the concentration of sodium butyrate.

The responsiveness of the carotid body to sytemic hypoxia was tested before and after cell-free perfusion in three cats.

Data Analysis

Data were reported as mean ± SEM. They were analyzed with two or three way analysis of variance and Dancan's new multiple-range test.

RESULTS

Experimental Set # 1

A selective perfusion of the carotid body with hypoxic KRB increased carotid chemoreceptor neural activity as seen in the previous study [12]; the same is true with hypoxic HBS+ (Fig 1). Time lag between on the start of the perfusion and the onset of increase in neural activity is mainly due to the dead space of the perfusion setup (1.2 ml; about 10 sec). Termination of the hypoxic perfusion immediately returned neural activity to pre-perfusion level. On the contrary, perfusion with hypoxic HBS- did not increase neural activity, and re-establishment of blood supply (at normal Pco_2) in the carotid body caused a burst increased in neural activity.

In all seven cats hypoxic KRB and hypoxic HBS+ significantly elevated carotid chemoreptor neural activity. There is not statistical difference between the responses to hypoxic KRB and hypoxic HBS+. The effect of hypoxic HBS- was somewhat variable. Two showed no change in neural activity and one showed a decrease. In four cases neural activity slightly increased at the end of HBS- perfusion. However, those increases were dramatically less than the response to hypoxic KRB or hypoxic HBS+. There was no statistically significant increase in neural activity during hypoxic HBS- perfusion. The neural response to hypoxic HBS- is significantly less than the reponses to hypoxic KRB or hypoxic HBS+. Values of pH, Pco_2, and Po_2 were following, respectively: for cat, 7.414

$\pm$ 0.027, 30 $\pm$ 2 torr, 310 $\pm$ 36 torr; for KRB, 7.405$\pm$ 0.012, 32 $\pm$ 1 torr, 29 $\pm$ 2 torr; for HBS+, 7.390 $\pm$ 0.015, 30$\pm$1 torr, 29$\pm$2 torr; for HBS-, 7.403$\pm$0.013, 1$\pm$0 torr, 26$\pm$2 torr.

Since four out of even cats showed some increase in carotid chemoreceptor neural activity during hypoxic HBS- perfusion, the absence of CO_2/HCO_3^- may have simply

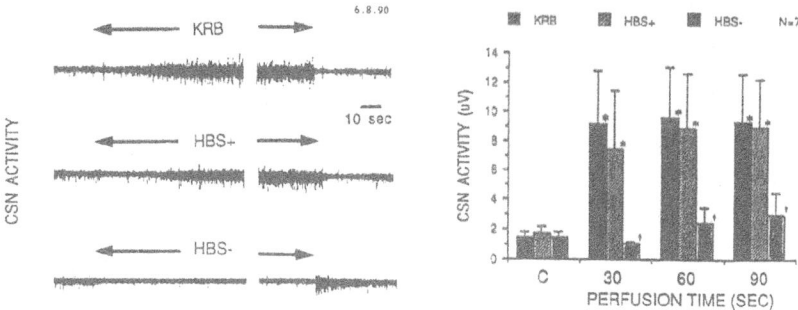

Fig. 1.   Carotid chemoreceptor neural respones to hypoxic perfusion for 90 sec. Left panel shows original
neurogram of carotid sinus nerve (CSN), during hypoxic KRB (upper trace), hypoxic HBS+
(middle), and hypoxic HBS- (lower). Right panel represents the summary of the result. C, pre-
perfusion control, *, significantly different from own control; +, significantly different from other
two.

delayed the response to hypoxia. However, prolonged perfusion (5 min) demonstrated this was not the case. Hypoxic HBS+ increased neural activity significantly from 3.3$\pm$2.5 μV of preperfusion level to 12.3$\pm$5.1 μV at 5 min perfusion. On the other hand, neural activity at 5 min of hypoxic HBS- perfusion (4.5$\pm$3.2 μV) was not significantly different from pre-perfusion level (1.7$\pm$0.8 μV).

Experimental Set # 2

Responses of the carotid body to hypoxic HBS+ (Fig 2, trace a) or hypoxic HBS- perfusion (trace b) were practically the same as above. When the carotid body was perfused with hypoxic HBS- containing 10 mM sodium butyrate chemorecptor neural activity increaed as much as during perfusion with hypoxic HBS+ (trace c). Different concentrations of sodium butyrate (5, 10 or 20 mM) caused the same level of elevation in neural activity at 90 sec of perfusion, but the onset of increase was dose-dependent (data not shown). When the carotid body was perfused with HBS- containing 10 mM butyrate exposed to 100 % O_2 neural activity did not show any change from pre-perfusion level under systemic hyperoxia (trace d). A part of the result were summarized in the right panel. The values of pH, Pco_2, and Po_2 for perfusates and cat were for HBS+, 7. 425$\pm$0.008, 30$\pm$2 torr, 22$\pm$ 9 torr; for HBS-, 7.426$\pm$0.010, 1$\pm$0 torr, 15$\pm$5 torr; for HBS- containing butyrate, 7.430$\pm$0.013, 1$\pm$1 torr, 18$\pm$4 torr; for cat, 7.400$\pm$0.005, 33$\pm$2 torr, 254$\pm$37 torr, respectively.

Responsiveness of the carotid body to systemic hypoxia

The effect of the cell-free perfusions on the function of the carotid body was examined by exposing the cats to systemic hypoxia before and at the end of experiments. Two expoures to sytemic hypoxia produced virtually the same levels of increase in neural activity in all three cats tested. The maximal neural activity during the second hypoxia was 94$\pm$4 % of that during the first hypoxia.

DISCUSSION

Our data showed that selective perfusion of the carotid body with CO_2/HCO_3^--free hypoxic perfusate did not significantly increase carotid chemoreceptor neural activity, in contrast to the "normal" response of the carotid body to CO_2/HCO_3^--containing hypoxic perfusate. No differences between the responses to hypoxic KRB and to hypoxic HBS+ suggest that HEPES itself does not have any effect on hypoxic response of the carotid body. The simplest explanation of the results would be that CO_2/HCO_3^- is necessary for hypoxic chemotransduction of the carotid body *in vivo*. Then, how necessary is $CO2/HCO_3^-$, and why have many *in vitro* studies shown a hypoxic response of the carotid body even without CO_2/HCO_3^-? Both questions could be answered in the context of pHi regulation in the chemosensitive unit.

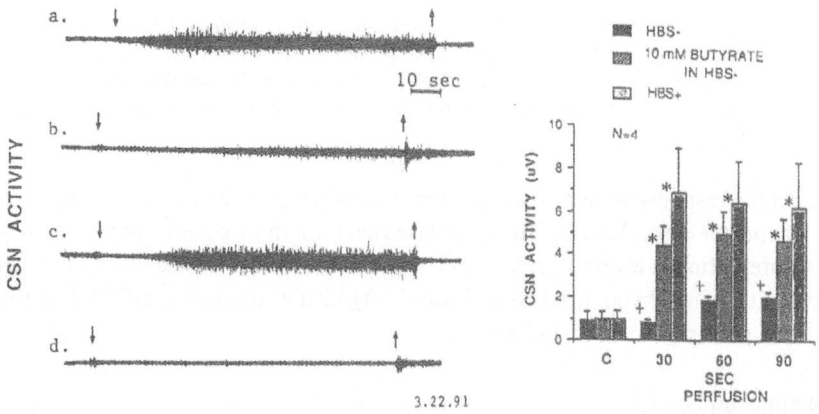

Fig. 2. *Left panel shows raw traces of carotid sinus nerve (CSN) in response to hypoxic HBS +(trace a), hypoxic HBS- (b), hypoxic HBS- containing 10 mM sodium butyrate (c), and hyperoxic HBS-containing 10 mM sodium butyrate (d). ↓ ,start of perfusion, ↑ end of perfusion. Right panel summarizes the results. C, preperfusion control, *, significantly different from own control; +, significantly different from other two.*

When the carotid body is selectively perfused with HBS -, the environment of the carotid body is acutely changed (in order of seconds) from CO_2/HCO_3^--containing blood to CO_2/HCO_3^--free solution *vice versa*. During HBS- perfusion intracellular HCO_3^- (in mM range) binds H^+ ions (in nM range) and exits cells as CO_2 leaving the cells alkalotic. If pHi of chemosensitive unit is mostly regulated by the mechanism dependent on CO_2/HCO_3^- in vivo, then, adjustment of pHi would not occur within 5 min which was the longest perfusion time in our experiments. This alkalinization of the chemosensitive unit would block the process of hypoxic chemotransduction. On the other hand, in the case of the *in vitro* experiments the carotid body is usually immersed in CO_2/HCO_3^-/-free solution. The chemosensitive unit ,then, have been exposed to CO_2/HCO_3^--free environment for hours possibly. It could well upregulate acid-loading mechanisms, and their pHi may return to a physiological level. A recent report suggested this may be the case: Buckler et al[1], found that the pHi of type I cells was 7.3 if the cells were kept in the CO_2/HCO_3^--containing solution. However, if the cells were initially kept in the CO_2/HCO_3^--

containing media and transferred to the CO_2/HCO_3^- -free solution, pHi was 7.8. If the cells were kept in the CO_2/HCO_3^- free media, intracellular pH was gradually (in hours) returned to the level in the CO_2/HCO_3^- -containing solution (7.3).

If alkalosis in the chemosensitive unit blocks hypoxic chemotransduction, we could preserve hypoxic response during HBS- perfusion by preventing alkalinization of the chemosensitive unit. Butyrate can be used for this purpose. Like any other weak acid, a butyrate-containing solution has both butyrate ions and undissociated butyric acid. Undissociated butyric acid easily and rapidly entered the cells and dissociates to H^+ and butyrate ions. The process continues until the concentration of undissociated butyrate in the cells reaches the same level as that in the extracellular fluid[14]. This process certainly would have some antagonistic effect on the alkalinization of the chemosensitive unit during CO_2/HCO_3^--free perfusion. The data have supported the hypothesis: butyrate preserved the chemoreceptor response to hypoxia during HBS-perfusion (Fig. 2). The fact that hyperoxic HBS -containing butyrate did not stimulate neural activity suggests that butyrate itself does not appear to affect carotid chemoreceptor neural activity. Thus, our results strongly suggest that certain amount of hydrogen ions is necessary for hypoxic chemotranduction.

The important role of hydrogen ions for the genesis of chemoreceptor activity was also suggested by the burst of neural activity on the reestablishment of blood supply after HBS -perfusion. When arterial blood comes into the carotid body, CO_2 in the blood imediately permeates the cell membrane and generates H^+ and HCO_3^-. Transient increase in hydrogen ions may trigger elevation of chemoreceptor neural activity.

In the physiological condition HCO_3^- seems to play a critical role for regulation of pHi in the chemosensitive unit. The presence of three transporters (Cl^-/HCO_3^- exchanger, Na^+ dependent HCO_3^- transporter, Na^+/H^+ exchanger) has been suggested as pHi regulators of type I cells[1,9]. We have previously reported the possibility that HCO_3^- -permeable chloride channels work as acid-loaders [11]. Apparently the three regulators need HCO_3^- for their activities. A balance of the activities of these putative regulators would determine the pHi of type I cells. It would be greatly affected in a CO_2/HCO_3^--free condition. The importance of CO_2/HCO_3^- has been suggested in other sytems[15]. For example, growth factors decreased pHi in CO_2/HCO_3^- containing media, but increased pHi in the CO_2/HCO_3^- -free media in several cells[6]. Since the carotid body is very sensitive to pH or CO_2 tension, the result of studies carried out in a CO_2/HCO_3^--free media must be interpreted very carefully.

It is possible that a cell- and protein- free perfusion injures the carotid body. In our preliminary studies for establishing our selective perfusion techniques, we found that the responses of the carotid body were reduced and sometimes the baseline activity increased if the perfusion period was prolonged or the perfusion was repeated too frequently without enough period of natural blood supply. Therefore, we have limited the cell-and protein-free perfusions to 90 sec in most cases; 5 min in selected cases. During more than 95 % of the experimental period the carotid body is supplied with its own blood at normal blood pressure. At the end of a experimental day (5-6 hours after the first perfusion) the carotid body is still active and responsive. In several experiments the neural response of the carotid body to hypoxia has been tested before and after the perfusions. It was virtually the same, suggesting that our perfusion methods did not affect the function of carotid body, at least in terms of the reponsiveness to hypoxia.

In summary, we have demonstrated that selective perfusion of the carotid body with hypoxic CO_2/HCO_3^--containing perfusate significantly increased carotid chemoreceptor neural activity, but not with hypoxic CO_2/HCO_3^- -free perfusion. When butyrate was added in the CO_2/HCO_3^--free perfusate, hypoxic response was restored. The results suggest that a certain level of hydrogen ions is abssolutely necessary for the carotid body to express the effect of hypoxia. Finally, CO_2/HCO_3^- plays an essential role for regulating the pHi of the chemosensitive unit in the physiological condition.

ACKNOWLEDGEMENT

This work was supported by HL 10342.

REFERENCES

1 BUCKLER, K. J. , R. D. VAUGHAN-JONES, C. PEERS, AND P. C. G. NYE. Intracellular pH and its regultion in isolated type I carotid body cells of the neonatal rat. J. Physiol. 436 :107-129, 1991

2 EYZAGUIRRE, C. , AND J. LEWIN. Chemoreceptor activity of the carotid body of the cat. J. Physiol. 159 :222-237, 1961

3 FITZGERALD, R. S. , AND D. C. PARKS. Effect of hypoxia on carotid chemoreceptor respons to carbon dioxide in cats. Respir. Physiol. 12 :218-229, 1971

4 FITZGERALD, R. S. , AND G. A. DEHGHANI. Neural responses of the cat carotid and aortic bodies to hypercapnia and hypoxia. J. Appl. Physiol. : Respirat. Environ. Exercise Physiol. 52: 596-601, 1982

5 FITZGERALD, R. S. , AND M. SHIRAHATA. The role of acetylcholine in the chemoreception of hypoxia by the carotid body. In: Arterial Chemoreception. Edited by C. Eyzaguirre, S. Fidone, R. S.Fitzgerald, S. Lahiri, D. McDonald. New York: Springer-Verlag, 1990, p 124-130

6 GANZ, M.B. , G.BOYARSKY, R.B.STERZEL, AND W.F.BORON. Arginine vasopressin enhances pHi regulation in the presence of HCO_3^- by stimulating three acid-base transport systems. Nature 337: 648-651, 1989

7 HORNBEIN, T.F. , Z.J.GRIFFO, AND A.ROOS. Quantitation of chemoreceptor activity: Interrelation of hypoxia and hypercapnia. J. Neurophysiol. 24: 561-568,1961

8 LAHIRI, S. ,AND R. G.DELANEY. Stimulus interaction in the responses of carotid body chemoreceptor single afferent fibers. Respir. Physiol. 24 : 249-266, 1975

9 ROCHER, A. , A. OBESO, C. GONZALEZ, AND B HERREROS. Ionic mechanism for the transduction of acid stimuli in rabbit carotid body glomus cells. J. Physiol. 433: 533-548, 1991

10 SHIRAHATA, M. , S. ANDRONIKOU, AND S. LAHIRI. Differential effect of oligomycin on carotid chemoreceptor responses to O2 and CO2 in the cat. J. Appl. Physiol. 63:2084-2092, 1987

11 *SHIRAHATA, M. , AND R. S. FITZGERALD*. Essential role of chloride channels on hypercapnic chemotransduction in cat carotid body (Abstract). FASEB J. 5:A1120,1991

12 *SHIRAHATA, M., AND R. S. FITZGERALD*. The presence of CO_2/HCO_3^- is essential for hypoxic chemotransduction in the in vivo perfused carotid body. Brain Research 545:297-300, 1991

13 *SHIRAHATA, M. , AND R. S. FITZGERALD*. Dependency of hypoxic chemotransduction in cat carotid body on voltage-gated calcium channels. J. Appl. Physiol. 71:1062-1069, 1991

14 *THOMAS, R. C.* Experimental displacement of intracellular pH and the mechanism of it subsequent recovery. J. Physiol. 354: 3P-22P, 1984

15 *THOMAS, R. C.* Bicarbonate and pHi response. Nature 337:601, 1989

31 METABOLIC SUBSTRATE DEPENDENCE OF CAROTID CHEMOSENSORY RESPONSES TO STOP-FLOW EVOKED HYPOXIA AND TO NICOTINE

D. Spergel

University of Pennsylvania School of Medicine
Department of Physiology
Philadelphia, PA 19104-6085, USA

INTRODUCTION

It is possible that glutamate and other amino acids enhance the responsiveness of carotid body preparations by acting as substrates for cellular energy production. Amino acids are potential metabolic substrates because they can be converted to tricarboxylic acid cycle intermediates involved in ATP generation (Altman and Dittmer, 1973). A high concentration of glutamate (42 mM) was used in the carotid body superfused in vitro (Baron and Eyzaguirre, 1977). Glutamate apparently helped to maintain responsiveness of the chemoreceptors. However, the extent to which glutamate was used as a metabolic substrate is unclear because the 5.5 mM glucose that was also present may have been an adequate substrate. Consistent with the hypothesis that glutamate and other amino acids are used as metabolic substrates by the carotid body is that vascularly isolated carotid body preparations perfused with saline solutions containing 5.5 mM glucose but lacking amino acids manifest progressive deterioration in chemosensory responsiveness (Joels and Neil, 1968; O'Regan, 1973).

The objective of this investigation was to determine whether glutamate (42 mM) or a mixture of amino acids can substitute for glucose in cellular energy production in the carotid body, enhance chemosensory responses and thus account for the maintenance of the in vitro perfused-superfused carotid body preparation (Iturriaga et al., 1991). Chemosensory responses to hypoxia evoked by perfusate stop-flow and to injection of nicotine were measured with various concentrations of glucose and glutamate, as well as with a mixture of amino acids like the one found in plasma (Altman and Dittmer, 1973).

METHODS

Cats (2-5 kg) were anesthetized with sodium pentobarbital (35 mg/kg i.p.) and the carotid bifurcation with the carotid body was excised as described by Iturriaga et al.

(1991). Chemosensory discharge from the carotid sinus nerve was recorded extracellularly while the carotid body was perfused (perfusate $PO_2 = 460 \pm 79$ Torr [mean $\pm$ S.D.], PCO_2 < 1 Torr) and superfused (superfusate $PO_2 = 32 \pm 10$ Torr [mean $\pm$ S.D.], $PCO_2 < 1$ Torr) in vitro (Iturriaga et al., 1991) at 36 °C for at least 15 min with the following substrates:

(a) 5.5 mM glucose (physiological glucose)

(b) 11 mM glucose (high glucose)

(c) 42 mM glutamate (high glutamate)

(d) 4.2 mM mixture of 19 minimal essential medium (MEM) amino acids (Sigma Chemical, St. Louis, MO) containing 100 µM glutamate plus 5.5 mM glucose (physiological glucose plus physiological glutamate)

(e) 42 mM glutamate plus 5.5 mM glucose (control), used by Baron and Eyzaguirre (1977) and Iturriaga et al. (1991).

The window discriminator was always set such that basal discharge was between 10 and 20 impulses/sec. The chemosensory response was taken to be the increase in chemosensory discharge during stop-flow or following injection of nicotine. Statistical differences were assessed using the Kruskal-Wallis test (Theodorsson-Norheim, 1986).

Chemosensory responses to multiple trials of perfusate stop-flow (cross-clamping of the perfusion line) and injection of 4 nmoles of nicotine bitartrate (British Drug House, Poole, UK) were measured. Nicotine was used as a marker because it induces an increase in chemosensory discharge similar in magnitude to that induced by stop-flow (Iturriaga et al., 1991). Yet, nicotine seems to act on a nicotinic receptor downsream from the site of O_2 chemoreception.

RESULTS

Figs. 1 and 2 illustrate the effects of metabolic substrates on carotid chemosensory discharge. Chemosensory responses to both perfusate stop-flow evoked hypoxia (Fig. 1) and to nicotine (Fig. 2) were larger ($p < 0.05$, Kruskal-Wallis test) with 42 mM glutamate plus 5.5 mM glucose, or the 4.2 mM amino acid mixture containing 100 µM glutamate plus 5.5 mM glucose, or 42 mM glutamate alone or 11 mM glucose alone than with 5.5 mM glucose.

DISCUSSION

The larger stop-flow response with 42 mM glutamate or 11 mM glucose compared to 5.5 mM glucose indicates that the latter is inadequate substrate supply and that glutamate can be used as a metabolic substrate in the carotid body. The results are consistent with the hypothesis that glutamate accounts for the longer life of the perfused-superfused carotid body in vitro preparation (Iturriaga et al., 1991) relative to other preparations perfused with 5.5 mM glucose as the only substrate (Joels and Neil, 1968;

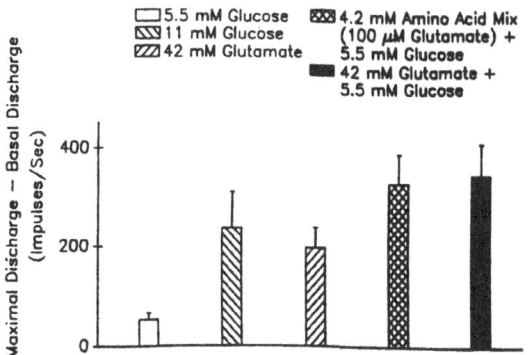

Fig. 1. Increase in carotid chemosensory discharge (Mean ± S.E.M.) during perfusate stop-flow with 5.5 mM glucose (n = 7), 11 mM glucose (n = 6), 42 mM glutamate (n = 6), 4.2 mM amino acid mixture containing 100 μM glutamate plus 5.5 mM glucose (n = 6), and 42 mM glutamate plus 5.5 mM glucose (n = 8). The value of n is the number of carotid bodies tested.

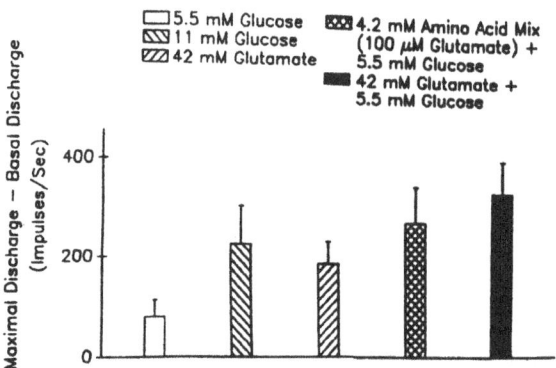

Fig. 2. Increase in carotid chemosensory discharge (Mean ± S.E.M.) in response to injection of 4 nmoles of nicotine bitartrate with 5.5 mM glucose (n = 4), 11 mM glucose (n = 4), 42 mM glutamate (n = 5), 4.2 mM amino acid mixture containing 100 μM glutamate plus 5.5 mM glucose (n = 6), and 42 mM glutamate plus 5.5 mM glucose (n = 6). The value of n is the number of carotid bodies tested.

O'Regan, 1973). The needed high concentrations of substrate indicate that the carotid body has a high metabolic rate. The superfused carotid body in vitro manifests large chemosensory responses to stop-flow with 5.5 mM glucose as the only substrate (Eyzaguirre and Lewin, 1961). However, the perfused-superfused carotid body in vitro, which was used in the present investigation, may require a larger substrate supply because of its active vasculature.

Glutamate may be used to generate ATP for chemoreception and chemosensory responses.. Glutamate is converted to the tricarboxylic acid cycle intermediate, α-ketoglutarate, according to the reaction

$$\text{glutamate} + NAD^+ + H_2O \rightarrow \alpha\text{-ketoglutarate} + NH_4^+ + NADH + H^+ \cdot$$

catalyzed by glutamate dehydrogenase (Stryer, 1981). NADH from this reaction can be used to produce ATP according to the reaction (Erecinska and Wilson, 1982)

$$NADH + H^+ + 3ADP + 3P_j + \tfrac{1}{2}O_2 \rightarrow NAD^+ + 3ATP + H_2O.$$

Moreover, α-ketoglutarate itself can be used to produce ATP, CO_2 and H_2O (Stryer,1981).

That chemosensory responses were as large with the 4.2 mM amino acid mixture containing 100 µM glutamate plus 5.5 mM glucose as with 42 mM glutamate plus 5.5 mM glucose indicates that physiological concentrations of amino acids can substitute for a high concentration of glutamate (42 mM). This finding is consistent with the observation (O'Regan, 1973) that blood, which contains a similar mixture of amino acids plus 5.5 mM glucose, restores chemosensory reponses following perfusion with 5.5 mM glucose as the only substrate. Virtually all of the amino acids in the amino acid mixture (this investigation) or in plasma are potential metabolic substrates because they can enter the tricarboxylic acid cycle through conversion to pyruvate, acetyl-CoA, α-ketoglutarate or oxaloacetate (Altman and Dittmer, 1973).

Responses to stop-flow and nicotine showed a similar dependence on substrate concentration, indicating that the effects of the various metabolic substrates were not specific to O_2 chemoreception. Metabolic substrate is required for ATP production. ATP is required for Na^+/K^+-ATPase and Ca^{2+}-ATPase activity. These ATPases may in turn be important for maintaining appropriate ionic distributions for neural activity (Hodgkin and Keynes, 1955; Gill et al., 1984).

Finally, it should be emphasized that no distinction has been made between the effects of metabolic substrate on the glomus cell versus the chemosensory nerve ending. It should also be noted that substrate utilization may not be the same in the presence and nominal absence of CO_2-HCO_3^- because the chemosensory response to stop-flow is improved in the presence of CO_2-HCO_3^- (Iturriaga and Lahiri, 1991).

ACKNOWLEDGEMENTS

This work was supported by NIH grant HL-43413. The author was a predoctoral trainee supported by NIH training grant HL-07027.

REFERENCES

Altman, P. L., and Dittmer, D. S., eds., 1973, Biology Data Book, vol. 3, FASEB, Washington, DC, p. 1544-1556 and 1791-1839.

Baron, M., and Eyzaguirre, C., 1977, Effects of temperature on some membrane characteristics of carotid body cells, Amer. J. Physiol., 233 (Cell Physiol, 2):C35-C46.

Erecinska, M., and Wilson, D. F., 1982, Regulation of cellular energy metabolism, J. Membr. Biol., 70:1-14.

Eyzaguirre, C., and Lewin, J., 1961, Effect of different oxygen tensions on the carotid body in vitro, J. Physiol. London, 159: 238-250.

Gill, D. L., Chueh, S. -H., and Whitlow, C. L., 1984, Functional importance of the synaptic plasma membrane calcium pump and sodium-calcium exchanger, J. Biol. Chem., 259:10807-10813.

Hodgkin, A. L., and Keynes, R. D., 1955, Active transport of cations in giant axons from Sepia and Loligo, J. Physiol. London, 128:28-60.

Iturriaga, R., and Lahiri, S., 1991, Carotid body chemoreception in the absence and presence of CO_2-HCO_3^-, Brain Res., in press.

Iturriaga, R., Rumsey, W. L., Mokashi, A., Spergel, D., Wilson, D. F., and Lahiri, S., 1991, In vitro perfused-superfused cat carotid body for physiological and pharmacological studies, J. Appl. Physiol., 70:1393-1400.

Joels, N., and Neil, E., 1968, The idea of a sensory neurotransmitter, in: Arterial Chemoreceptors, R.W. Torrance, ed., Blackwell, Oxford, p. 153-178.

O'Regan, R. G., 1973, Responses of the chemoreceptors of the cat carotid body perfused with cell-free solutions, Irish J. Med. Sci., 148:78-85.

Stryer, L., 1981, Biochemistry, Freeman, San Francisco, p. 407-409.

Theodorsson-Norheim, E., 1986, BASIC computer program to perform non-parametric one-way analysis of variance and multiple comparisons on ranks of several independent variables, Comp. Methods and Programs in Biomed., 23:s7-62.

REFERENCES

32   EFFECTS OF CHEMOSENSORY STIMULATI-ON MEMBRANE CURRENTS RECORDED WITH THE PERFORATED-PATCH METHOD FROM CULTURED RAT GLOMUS CELLS

A.Stea, S.A.Alexander, and C.A.Nurse

Department of Biology, McMaster University
Hamilton, Ontario, Canada. L8S 4K1

INTRODUCTION

We are using primary cultures of dissociated cells from the rat carotid body to study chemotransduction mechanisms. A major focus has been the electrophysiological characterization of the glomus (or type 1) cells which are the presumed sensors of the chemosensory stimuli, hypoxia, hypercapnia, and acidity (Biscoe and Duchen 1990; Lopez-Barneo et al 1988; Stea and Nurse 1991a; etc.). Using the giga-seal patch-clamp technique (Hamill et al 1981) we have determined that cultured rat glomus cells have membrane currents similar to those found in freshly-isolated rabbit glomus cells (Duchen et al 1988; Lopez-Lopez et al 1989; Hescheler et al 1989; see also, Peers 1990). The implementation of the novel perforated-patch recording method (Horn and Marty 1988) in many of our studies has circumvented problems associated with conventional whole cell recording such as washout of important cytoplasmic constituents, and rapid deterioration of the preparation.

In this report the effects of the three major chemosensory stimuli, low Po_2 (hypoxia), acidity, and high Pco_2 (hypercapnia), on membrane currents recorded with the perforated-patch method from cultured rat glomus cells are discussed. In several control experiments another neural crest-derivative, closely related to the glomus cell, the small intensely-fluorescent (SIF) cell of rat sympathetic ganglia, was tested to determine the specificity of the responses. Much of this work has been described recently (Stea and Nurse 1991a, 1991b; Stea et al 1991).

METHODS

The procedures for the culture of glomus and SIF cells by combined enzymatic and mechanical dissociation of the rat carotid body and superior cervical ganglion (SCG) respectively were identical to those previously described (Nurse 1990; Stea and Nurse 1991a). Perforated-patch recordings (Horn and Marty 1988) were obtained from both glomus and SIF cells using procedures described in more detail elsewhere (Stea et al 1991). Most experiments were performed using extracellular fluid of the following composition (mM): NaCl,135; KCl,5; $CaCl_2$,2; $MgCl_2$,2; glucose,10; N-2-hydroxyethylpiperazine-N'-2-ethane sulfonic acid (HEPES),10 at pH 7.4 . The stock pipette solution for most experiments contained (mM): KCl,135; NaCl,5; $CaCl_2$,0.1;

Neurobiology and Cell Physiology of Chemoreception
Edited by P.G. Data *et al.*, Plenum Press, New York, 1993

HEPES, 10 at pH 7.2. To simulate hypoxia, extracellular fluid was bubbled with 100% N_2 for 20-30 min until the $Po_2 \approx 20$ Torr prior to perfusion (see Stea and Nurse 1991a). Cytoplasmic pH was lowered by adding 3.3 - 6.6 mM of the ionophore nigericin to normal extracellular solution (see Stea et al 1991). Prior to testing the effects of a hypercapnic stimulus (10% CO_2) the perfusate was switched from a HEPES-buffered medium to a bicarbonate-buffered media equilibrated with 5% CO_2 (see Stea and Nurse 1991b).

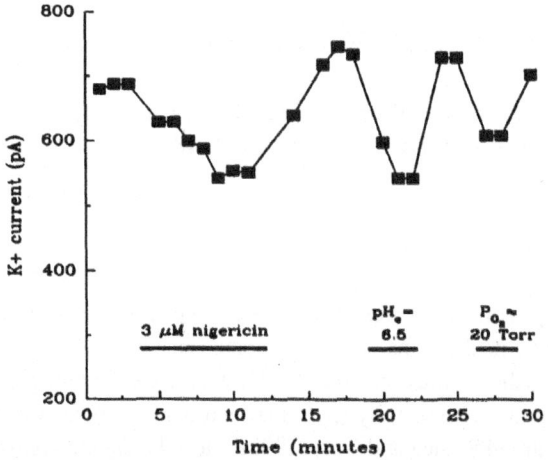

Fig. 1. Time course of the effect of various chemosensory stimuli on steady-state K^+ currents recorded from a single glomus cell. Firstly, cytoplasmic acidification induced by the addition of 3 μM nigericin to the perfusate caused a reversible decrease in the K^+ current. A similar effect was seen by reducing the external pH from 7.4 to 6.5 . Finally, hypoxia ($Po_2 \approx 20$ Torr) also caused a reversible decrease in the K^+ current in this glomus cell. Each point corresponds to the current resulting from a voltage step of -60 to +50 mV.

RESULTS

Glomus cells grow in discrete clusters in culture for several weeks and are readily identifiable and accessible for patch-clamp recording. The electrophysiological characteristics of these rat glomus cells in HEPES-buffered media (Stea and Nurse 1991a; Stea et at 1991) are similar to those reported in other laboratories (e.g. Peers 1990). In contrast to rabbit glomus cells (Duchen et al 1989; Lopez-Barneo et al 1988; see however, Hescheler et al 1989) the density of voltage-gated Na^+ channels is lower in rat glomus cells, suggesting that spiking may not be essential for chemotransduction in the latter species. The use of the perforated-patch technique has allowed recordings from glomus (and SIF cells) for well over 1 hr with minimal changes in the membrane currents. This has allowed a variety of experimental manipulations on the same cell which was not possible with conventional whole cell recording due to deterioration of the preparation.

Exposure of these cells to hypoxic stimuli (Po_2 20 Torr) caused a rapid reversible reduction of ca. 20% in the outward K^+ current (Fig. 1) while the other currents were unaffected (Stea and Nurse 1991a). This confirmed earlier findings reported for rabbit glomus cells (Lopez-Barneo et al 1988; Hescheler et al 1989). In contrast exposure of the closely related SIF cells to hypoxia resulted in no change in membrane currents (Fig.2A) suggesting that the effect was specific for glomus cells (Stea and Nurse 1991a).

Lowering the pH from 7.4 to 6.5 in the extracellular fluid caused a significant reversible reduction in both the outward K^+ (Fig. 1) and inward Na^+ currents in both glomus and SIF cells (Stea et al 1991). However, a decrease in pH_e to 7.0 did not elicit any appreciable changes in the active currents in glomus cells (cf. Lopez-Lopez et al 1989; Peers 1990). Cytoplasmic acidification of glomus cells was carried out either, by addition of the ionophore nigericin or a weak acid (acetate) to the perfusion fluid (Stea et al 1991). Nigericin is an electroneutral K^+/H^+ ionophore which will cause acidification of the cytoplasm (Fig.3) when physiological concentrations of K^+ are present inside and outside the cell (Thomas et al 1979).

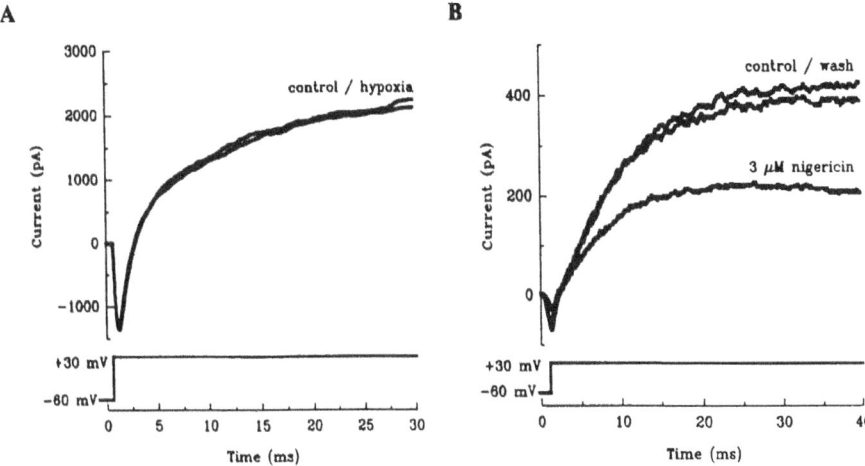

Fig.2. *Effects of hypoxia and cytoplasmic acidification on whole-cell currents in small, intensely-fluorescent (SIF) cells. A: Exposure to hypoxia ($Po_2 \approx 20$ Torr) had no obvious effects on SIF cell currents. B: Cytoplasmic acidification (3 μM nigericin) decreased both the inward Na^+ current and the outward K+ currents in this SIF cell.*

The application of 3 μM nigericin to the bathing solution caused a marked decrease in the active currents in cultured rat glomus cells (Fig. 1) and the effect was reversible upon removal of the drug. Similar results were observed when a weak acid (acetate) was applied to the bathing solution (Stea et al. 1991). Recovery of the currents was inhibited when 0.1 mM amiloride, a blocker of the Na^+/H^+ exchanger, was added to the perfusate following removal of nigericin (Stea et al. 1991). This is consistent with the involvement of the Na^+/H^+ exchanger in the mechanisms of pH_i regulation in glomus cells (Buckler et al. 1991). Unlike the effects of hypoxia, the changes elicited by an acidic pH_i were not specific for glomus cells since the active currents in SIF cells were similarly affected by nigericin (Fig. 2B).

The use of bicarbonate-based recording media has allowed us to study the response of glomus cells to hypercapnic stimuli. Interestingly, the properties of these cells are markedly different in bicarbonate-buffered media vs the typical HEPES-buffered media, used in all other patch-clamp studies on these cells. Firstly, when the perfusate was switched from a HEPES containing medium to a bicarbonate-buffered medium equilibrated with 5% CO_2 (24 mM $NaHCO_3$; $pH_e \approx 7.4$) a marked depression in outward K^+ current was seen in glomus cells (Stea and Nurse 1991b). Presumably this was due to cytoplasmic acidification caused by the hydration of CO_2 in the presence of carbonic anhydrase which has been localized to these cells (Nurse 1990).

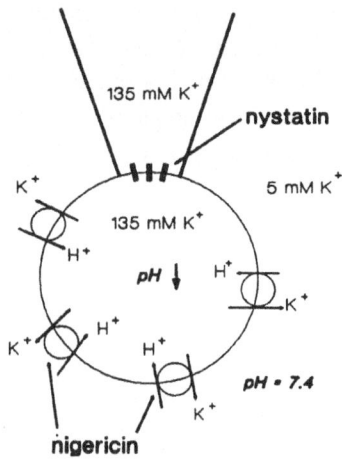

Fig. 3. Schematic representation of a perforated-patch recording from a glomus cell indicating the action of the ionophore nigericin in inducing cytoplasmic acidification.

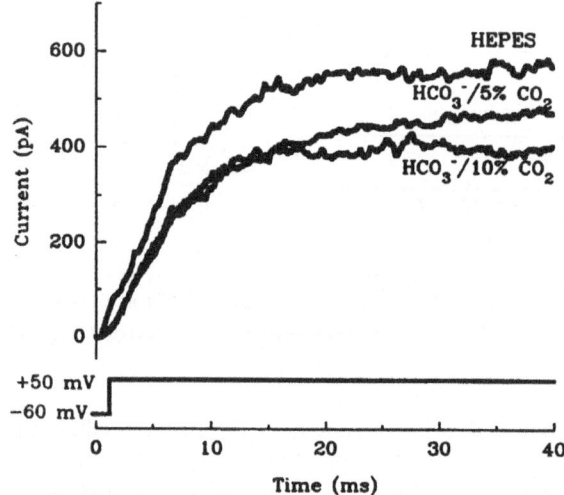

Fig.4. Effect of varying CO_2 levels on whole -cell currents in a glomus cell. Exposure to bicarbonate-buffered media (BBM) equilibrated with 5% CO_2 ($pH_e \approx 7.4$) caused a rapid decrease in the K^+ current in this glomus cell. BBM equilibrated with 10% CO_2 ($pH_e \approx 7.2$), which simulates hypercapnia, caused a further decrease in the K^+ current.

This interpretation was confirmed in preliminary experiments where the decrease in K^+ current upon exposure to bicarbonate-buffered media was delayed in the presence of 10 μM acetazolamide, an inhibitor of carbonic anhydrase (see also Buckler et al. 1991).Secondly, an obvious decrease in the input resistance was seen upon switching from HEPES to bicarbonate-buffered media (Stea and Nurse 1991b) possibly due to the opening of large conductance Cl⁻ channels previously seen in inside-out patches from these cells (Stea and Nurse 1989).Glomus cells exposed to a hypercapnic stimulus (10% CO_2 equilibrated in bicarbonate-buffered media; pH ≈7.2)showed a further decrease in the outward K^+ current (Fig.4) compared to cells exposed to 5% CO_2; this is likely due to an even greater drop in pH_i in the presence of 10% CO_2.

DISCUSSION

Recent studies employing the patch-clamp technique have indicated that the glomus (or type 1) cell of the carotid body is the likely sensor of chemosensory stimuli in arterial blood (Biscoe and Duchen 1990; Lopez-Barneo et al 1988; Hescheler et al 1989; Peers 1990; Stea and Nurse 1991a). We have now shown that cultured rat glomus cells respond to hypoxia, acidity, and hypercapnia (Stea and Nurse 1991a, 1991b; Stea et al 1991), which are known to excite the carotid body *in vivo*, by suppression of voltage-activated currents.

Hypoxia causes a subset of O_2-sensitive K^+ channels to close and this is likely to cause the glomus cell to depolarize. This depolarization would cause an influx of Ca^{2+} through voltage-activated Ca^{2+} channels known to be present in these cells (Hescheler et al 1989, Lopez-Lopez et al 1989; Stea and Nurse 1991a) and trigger transmitter release onto the apposed sensory nerve terminals. Extracellular acidity may also cause a similar influx of extracellular Ca^{2+} following cell depolarization caused by K^+ channel closure. Hypercapnic stimuli (10% CO_2) also causes a depression of K^+ currents. It is likely that the effects of extracellular acidity and hypercapnia are mediated throught a similar cytoplasmic acidification mechanism.Indeed studies with fluorescent pH dyes and pH microelectrodes have shown that the cytoplasmic pH follows the external pH changes (Buckler et al 1991, He et al 1990).Thus chemosensory stimuli known to excite the carotid body may act through, or be modulated by, changes in pH_i. The changes in input resistance seen during recording in bicarbonate -buffered media suggest that Cl^- ions may play a key role in governing the resting potential of these cells (Oyama et al 1986; Stea and Nurse 1991b). Since this model of chemotransduction is based on the opening of voltage-activated currents when the membrane depolarizes to -30mV or higher then a resting potential governed by Cl^- may help keep the membrane potential near this level. This would make the cell very sensitive to any changes in K^+ conductance caused by chemosensory stimuli.

ACKNOWLEDGEMENTS

This work was supported by a grant from the NIH # 1 RO1 HL 43412.In addition A. Stea was the recipient of a NSERC post-graduate scholarship.

REFERENCES

Biscoe T.J., Duchen M.R. (1990) Responses of type I cells dissociated from the rabbit carotid body to hypoxia. J. Physiol. 428:39-59.

Buckler K.J., Vaughan-Jones R.D., Peers C., Nye P.C.G. (1991) Intracellular pH and its regulation in isolated type I carotid body cells of the neonatal rat. J. Physiol. 436:107-129.

Duchen M.R., Caddy K.W.T., Kirby G.C., Patterson D.L., Ponte J., Bisooe T.J. (1988) Biophysical studies of the cellular elements of the rabbit carotid body. Neuroscience 26:291-311.

Hamill O.P., Marty A., Neher E., Sackmann B., Sigworth F.J. (1981) Improved patch-clamp techniques for high-resolution current recordings from cells and cell-free membrane patches. Pflugers Arch. 391:85-100.

He S.F., Wie J.Y., Eyzaguirre C. (1990) Intracellular pH of cultured carotid body cells. In: Eyzaguirre C., Fidone S.J., Fitzgerald R.S., Lahiri S., McDonald D.M. (eds) Arterial Chemoreceptors. Springer Verlag, New York, pp 18-23.

Hescheler J., Delpiano M.A., Acker H., Pietruschka F. (1989) Ionic currents on type I cells of the rabbit carotid body measured by voltage-clamp experiments and the effect of hypoxia. Brain Research 486:79-88.

Horn R., Marty A. (1988) Muscarinic activation of ionic currents measured by a new whole-cell recording method. J. Gen. Physiol. 92:145-159.

Lopez-Barneo J., Lopez-Lopez J.R., Urena J., Gonzalez C. (1988) Chemotransduction in the carotid body: K^+ current modulated by PO_2 in type I chemoreceptor cells. Science 241:580-582.

Lopez-Lopez J., Gonzalez C., Urena J., Lopez-Barneo J. (1989) Low Po_2 selectively inhibits K channel activity in chemoreceptor cells of the mammalian carotid body. J. Gen. Physiol. 93: 1001-1015.

Oyama Y., Walker J.L., Eyzaguirre C. (1986) The intracellular chloride activity of glomus cells in the isolated rabbit carotid body. Brain Res. 368:167-169.

Nurse C.A. (1990) Carbonic anhydrase and neuronal enzymes in cultured glomus cells of the carotid body of the rat. Cell Tissue Res. 261:65-71.

Peers C. (1990) Effect of lowered extracellular pH on Ca^{2+}-dependent K^+ currents in type I cells from the neonatal rat carotid body. J. Physiol. 422:381-395.

Stea A., Alexander S.A., Nurse C.A. (1991) Effect of pH_e and pH on membrane currents recorded with the perforated-patch method from cultured chemoreceptors of the rat carotid body. Brain Research. 567: 83-90.

Stea A., Nurse C.A. (1989) Chloride channels in cultured glomus cells of the rat carotid body. Am. J. Physiol. 257:C174-C181.

Stea A., Nurse C.A. (1991) Whole-cell and perforated-patch recordings from O_2-sensitive rat carotid body cells grown in short and long-term cultures. Pflugers Arch. 418:93-101.

Stea A., *Nurse C.A.* (1991) Contrasting effects of HEPES vs HCO_3^- buffered media on whole-cell currents in cultured chemoreceptors of the rat carotid body. <u>Neuroscience Letters.</u> 132:239-242.

Thomas J.A., Buchsbaum R.N., Zimniak A., Racker E. (1979) Intracellular pH measurements in ehrlich ascites tumor cells utilizing spectroscopic probes generated in situ. <u>Biochemistry</u> 18:2210-2218.

33 CARBONIC ANHYDRASE NEAR CENTRAL CHEMORECEPTORS

R. W. Torrance

University laboratory of Physiology
Parks Road
Oxford OX1 3PT, UK

INTRODUCTION

Our findings on carbonic anhydrase (CA) near to central chemoreceptors arose rather by accident. We knew that an injection of high PCO_2 excites arterial chemoreceptors quickly and that one of low PCO_2 and high PO_2 abolishes their discharge, and we had used the reflex effects of abolishing discharge to elaborate von Euler's model of the respiratory centre (Nye, Hanson & Torrance, 1981). High PO_2 saline is a good stimulus for this purpose because it suddenly completely abolishes discharge whereas CO_2 saline, an excitatory stimulus, is less valuable because it gives a less clear cut and reproducible response with a huge and unphysiological peak and an uncertain time course of decay.

CLOSE ARTERIAL INJECTIONS

When we had finished this work on the peripheral receptors, we studied whether the central chemoreceptors have similar reflex effects. To do so we had to develop a method for exciting them quickly with a close arterial injection. The essential point about our method for giving close arterial injections at the carotid body to excite it quickly is that the injections are given into a cannula in the external carotid artery pointing centrally. The cannula opens distal to the carotid sinus and the injection is made towards the aorta without clamping the common carotid and it instantly (on the time scale of respiration) fills the carotid sinus and enters the artery to the carotid body. Such an injection changes blood pressure in the lingual artery little and the chemoreceptor response starts about 100 ms from the start of an injection of CO_2 saline. We adapted this method for the central receptors (Hanson, Nye & Torrance, 1981), by making use of the anatomy of their arterial supply. The two vertebral arteries join on the ventral surface of the medulla to give the single midline basilar artery, and if one of the vertebrals is occluded, their junction serves as an anastomosis and the open artery takes over the supply of the whole medulla. We therefore cannulated one vertebral artery of cats, approaching the artery from ventrally at the level of the first intercostal space. As with the external carotid, a sudden injection pushed blood back towards the aorta down the open vertebral artery which could be seen to blanche transiently at the first intercostal space. Again with the central receptors we found that reducing the stimulus gave the simplest results to analyse and so we used 150 mM alkaline Tris for the stimulus, and we found that central reflex responses are similar to those from the periphery (Nye, Hanson and Torrance, 1983).

Neurobiology and Cell Physiology of Chemoreception
Edited by P.G. Data *et al.*, Plenum Press, New York, 1993

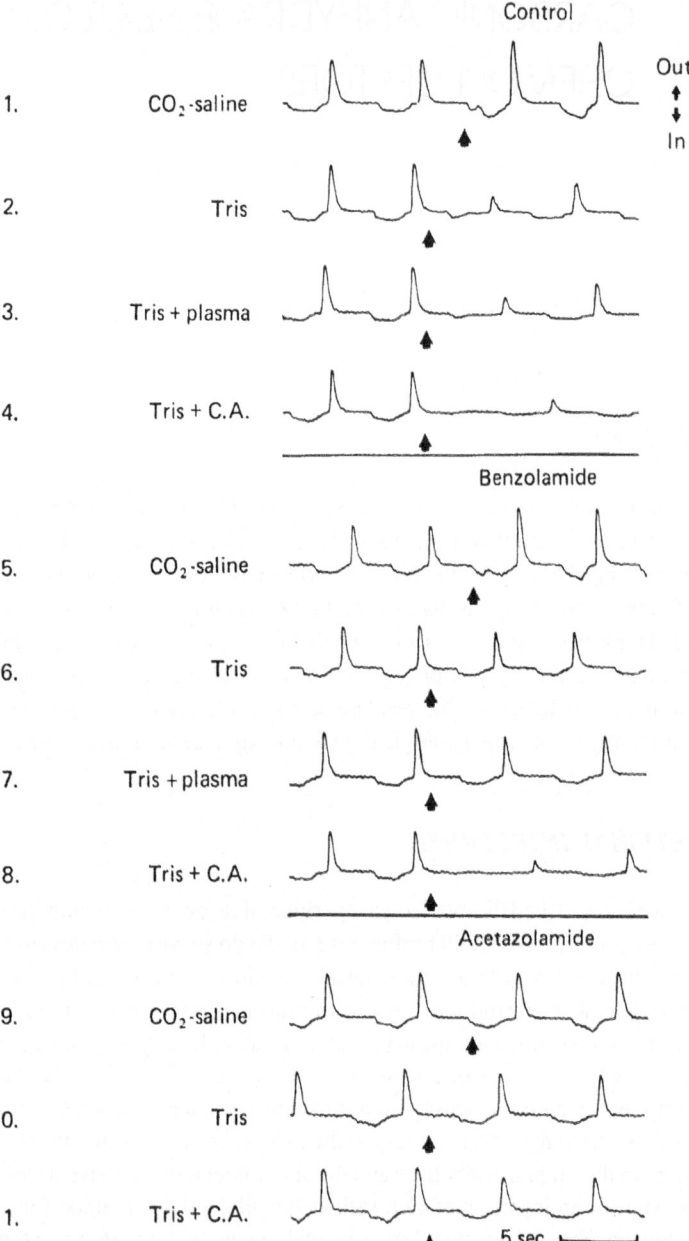

Fig. 1. *Pneumotachograph traces from pentobarbitone anaesthetized, vagotomized cat. Expiration upwards. 0.5 ml of specified injectate given cranially into right vertebral artery at arrows. C.A.: carbonic anhydrase. CO2 (1) unaffected by benzolamide (5) but abolished by acetazolamide (9). Tris (2) abolished by benzolamide (6) restored by C.A. (8) abolished by acetazolamide (11). (From Hanson, Nye & Torrance, 1981).*

CENTRAL CARBONIC ANHYDRASE

But the method also gave very interesting observations on the significance of CA in the medulla (Hanson, Nye & Torrance, 1981). Because a sharp close arterial injection presents the stimulus to the receptors from the blood for barely a second and the blood brain barrier (BBB) allows only fat soluble substances to cross it quickly, the CO_2 alone in a normal range of injections has any effect on central chemoreceptors. Figure 1 gives the essential findings. They reveal three distinct lots of CA in response to CO_2.

1) HEADACHE RECEPTORS (Hanson, Nye & Torrance, 1982)

The immediate effect of a CO_2 injection (Fig. 1.1 and 1.5), is inhibition of whatever phase of respiration is in progress and the inhibition of breathing lasts for the duration of the injection, a fraction of a second in our work but 5-10 seconds in some of Arita's. This response is abolished (Fig. 1.9), over about an hour by acetazolamide i.v. in a large dose of 50 mg/kg. It comes from pain receptors in the walls of cerebral arteries which contain CA and the pH at or in the receptors is pushed very acid by hydration of CO_2 from the injectate and it remains acid so long as there is a very high PCO_2 in the arterial lumen. Vertebral injections proved a useful method for Arita when he was studying pathways for headache in the brainstem since the method can be used many times over for exciting headache receptors and it is not necessary to open the skull to expose them.

2) CARBONIC ANHYDRASE ON THE CAPILLARY LUMEN (Hanson, Nye & Torrance, 1981)

This second and quite distinct lot of CA serves to bring the plasma buffers of the blood to equilibrium with PCO_2 whilst the blood is still in the capillaries of the chemoreceptors, as does the luminal CA of the pulmonary capillaries. In the lung, we used changes in the alveolar PCO_2 to reveal this process: in the medulla we used changes in breathing. Tris has proved a useful stimulus here. It takes up a lot of CO_2 by buffering it to give Tris bicarbonate, but first the CO_2 must be hydrated. In the unpoisoned animal Tris draws CO_2 out of the chemoreceptors (Fig. 1.2), but acetazolamide stops this and so does benzolamide even though it does not cross the BBB (Fig. 1.6 and 1.7). An inhibitor of CA will stop Tris even if it is injected mixed in with the Tris and not before it. Adding CA to the Tris increases the response in the unpoisoned animal (Fig. 1.4) and restores it when it has recently been abolished by an inhibitor of CA (Fig. 1.8). Plasma does not: it contains little CA (Fig. 1.7). This CA must be very accessible to an inhibitor from the lumen of the capillary. A position stuck onto the wall of the capillary is the only conceivable site for it to be in.

This component of the medullary CA brings the plasma and red cells of the blood passing through the central chemoreceptors to CO_2 equilibrium before the blood leaves the capillaries and thus avoids the problem of disequilibrium that Forster (Bidani et al., 1978) discussed for blood after it leaves the lungs.

3) CARBONIC ANHYDRASE AT THE MEDULLARY CHEMORECEPTORS

Tris quickly reduces breathing (Fig. 1.2). Airflow in inspiration is reduced within half a second of the start of an injection and then respiration returns to control over about half a minute (Nye, Hanson & Torrance, 1983). The response to CO_2 - bubbled saline is not simply the reverse of that to Tris for it is complicated by the presence of the headache inhibition we have already discussed (Hanson, Nye & Torrance, 1982). With a sudden injection, the inhibition lasts only about half a second but it lasts throughout a slow injection made steadily over several seconds. The chemoreceptors respond maximally within a second with a sudden injection, but with a slow one, the stimulus at the receptors must build up steadily throughout the injection, since the PCO_2 of the injection is far above any tissue PCO_2 it might conceivably produce. The transient peak of ventilation is equal to that produced in the steady state by a change of $PACO_2$ of the order of only 5 torr. This excitation of chemoreceptors depends on CA beyond the BBB. If the luminal CA has been knocked out by a small dose of benzolamide, CO_2 saline still excites the chemoreceptors (Fig. 1.5) but now Tris only affects them if CA is added to it to replace the inhibited luminal CA (Fig. 1.8). The injections still act respectively as a source and as a sink for CO_2. But if a further large dose of acetazolamide, 50 mg/kg is given, it gradually blocks both stimuli, taking an hour or more to do so. This is so because the acetazolamide only slowly penetrates the BBB. Benzolamide which never penetrates the BBB, never abolishes these effects of CO_2 saline.

SITE OF CA AT RECEPTOR

Thus we have shown that there is a third lot of CA beyond the BBB, and it hydrates CO_2 near to the receptors themselves or within their cytoplasm. To localize this CA further, Hanson et al. (1984) have applied CA inhibitors of various diffusibilities from the CSF and the inhibitors suggest, as they do for the arterial chemoreceptors, that CO_2 is hydrated by CA within a cell when it excites chemoreceptors (Torrance, 1991) .

Thus central chemoreceptors may be like arterial ones in responding to an intracellular pH (Hanson, Nye & Torrance, 1981), and not to the pH of the cerebral extracellular fluid around them (Loeschcke, 1982).

The results of Arita et al. and Ichiwara et al. (1989) can be used to argue in favour of this view. They have used pH-sensitive microelectrodes to record the course of pH changes in the extracellular fluid of the medulla in response to vertebral injections and have tried to correlate the pH changes with the timing of excitation of chemoreceptors. To do this one needs to know how the excitation of chemoreceptors does develop in response to an injection that alters PCO_2 in the CISF. This is best given by Tris because the respiratory response to it comes simply from a reduction in the excitation of chemoreceptors whereas the response to CO_2 is more complex, being the resultant of an early prepotent but shortlasting inhibition from headache receptors and a longer lasting excitation from chemoreceptors. The response to Tris is much faster than the pH changes Arita et al. (1989) recorded which are slower than the respiratory responses to CO_2 saline even if no allowance is made for headache receptors inhibition. It is not a pH such as Arita et al. (1989) have recorded that sets off the response but rather a pH where an ample supply of CA makes pH change in parallel with PCO_2 as is shown by the dependence of the response on CA. This pH could well be the intracellular pH of a cell that contains CA but it might be the pH at the surface of a cell with a membrane-bound CA at its external surface: if CA acts on the CISF and it is not uniformly distributed as Ichikawa et al. (1989) show, it must be restrained from diffusing in some way such as this, and so the problem of the pH near to fixed CA has to be considered.

The respiratory responses to a quick injection of Tris (and even a very quick one of CO_2) do seem to follow the likely change of tissue pCO_2, but it would be valuable to have this determined directly during a transient, perhaps by injecting a little CA beforehand down a micropipette barrel beside the one from which pH CISF is picked up (Kogo & Arita, 1990).

238

Whether or not this alternative interpretation of results is correct, Arita and his colleagues are now applying techniques that are appropriate for studying the transients of excitation at a sudden change of stimulus, the importance of the CA involved in the transients and its location inside or outside cells.

REFERENCES

Arita, H., Ichikawa, K., Kuwana, S. & Kogo, N. (1989). Possible locations of pH-dependent central chemoreceptors: intramedullary regions with acidic shift of extracellular fluid pH during hypercapnia. Brain Res., 485, 285-293.

Bidani, A., Crandall, E.D. & Forster, R.E. (1978). Analysis of postcapillary pH changes in blood in vivo after gas exchange. J.Appl. Physiol., 44, 770-781.

Hanson, M.A., Nye, P.C.G. & Torrance, R.W. (1981). The location of carbonic anhydrase in relation to the blood-brain barrier at the medullary chemoreceptors of the cat. J.Physiol., 320, 113-125.

Hanson, M.A., Nye, P.C.G. & Torrance, R.W. (1982). The effects on respiration in the cat of sudden excitation of cerebral vascular nociceptors by carbondioxide. Clin.Sci., 63, 505-511.

Hanson, M.A., Holman, R.B. & McCooke, H.B. (1984). Further studies on carbonic anhydrase at the medullary chemoreceptors of the cat. In: 'The Peripheral Arterial Chemoreceptors', ed. Pallot, D.J. Croom Helm, London, pp.409-414.

Ichikawa, K., Kuwana, S. & Arita, H. (1989). ECF pH dynamics within the ventrolateral medulla: A microelectrode study. J.Appl.Physiol., 67, 193-198.

Kogo, N. & Arita, H. (1990). In vivo study on medullary H+-sensitive neurons. J.Appl.Physiol., 69, 1408-1412.

Loeschcke, H.H. (1982). Central chemosensitivity and the reaction theory. J.Physiol., 332, 1-24.

Nye, P.C.G., Hanson, M.A. & Torrance, R.W. (1981). The effect on breathing of abruptly stopping carotid body discharge. Resp. Physiol., 46, 309-326.

Nye, P.C.G., Hanson, M.A. & Torrance, R.W. (1983). The effect on breathing of abruptly reducing the discharge of central chemoreceptors. Resp.Physiol., 51, 109-118.

Torrance, R.W. (1990). The action of carbon dioxide in central and peripheral chemoreceptors. In: 'Chemoreceptors and Chemoreceptor Reflexes', ed. Acker, H., Trzebski, A. & O'Regan, R.G. Plenum Press, New York, pp.43-48.

UPDATE ON THE BICARBONATE HYPOTHESIS

R.W. Torrance, E. M. Bartels and A.J.McLaren

University laboratory of Physiology
Parks Road
Oxford OX1 3PT, UK

INTRODUCTION

A single chemoreceptor fibre is excited by both hypoxia and hypercapnia. The bicarbonate hypothesis provides an explanation of how CO_2 and O_2 converge to give impulses in the same chemoreceptor fibre. Many other explanations are conceivable (R.W. Torrance, 1977), but no other has been put out in the same detail as the bicarbonate one. Here we will outline the hypothesis and the observations any hypothesis has to account for. We will consider how the bicarbonate hypothesis accounts for them and how it has stood up over the last ten years.

The bicarbonate hypothesis is a particular example of the general idea that the carotid body is basically an acid receptor and that a pH becomes more acid in hypoxia. It abandons the old idea that the acidity is produced in hypoxia by the release of an acid and supposes instead that a mechanism stabilising pH becomes less effective (in hypoxia). Thus CO_2 and O_2 still converge at a pH but CO_2 changes the pH by-being hydrated to carbonic acid by carbonic anhydrase (CA), whilst O_2 now somehow determines the activity of a membrane transport mechanism.

This idea needed a short name and it has come to be called a bicarbonate hypothesis, but that should not be taken to imply that the P_{O_2} component of H^+ management is achieved by active movement of HCO_3^-, nor even that some CO_2/HCO_3^- buffer must be present.

BACKGROUND

The P_{CO_2} throughout the carotid body is nearly constant, for CO_2 diffuses freely; if then the pH in the intracellular space of a cell is different from that of another cell or from the pH of the extracellular space, the CO_2/HCO_3^- buffer in these spaces can differ only in respect of its (HCO_3^-). The isohydric principle states that each buffer in a mixture of buffers will have its components at concentrations appropriate for the single common pH. Movement of H^+ into or out of a cell leads to a change in all of its buffers. A 'bicarbonate hypothesis' as the name has come here to be used, is not tested by showing that the carotid body responds to hypoxia in the absence of CO_2/HCO_3^- buffer as Iturriaga et al. (1991) has supposed, since pH_i can be managed in its complete absence.

BEHAVOUR OF CHEMORECEPTORS

Any hypothesis has to account for the behaviour of the carotid afferents, both in the steady and in the transient state. The steady responses have been reported by Fitzgerald

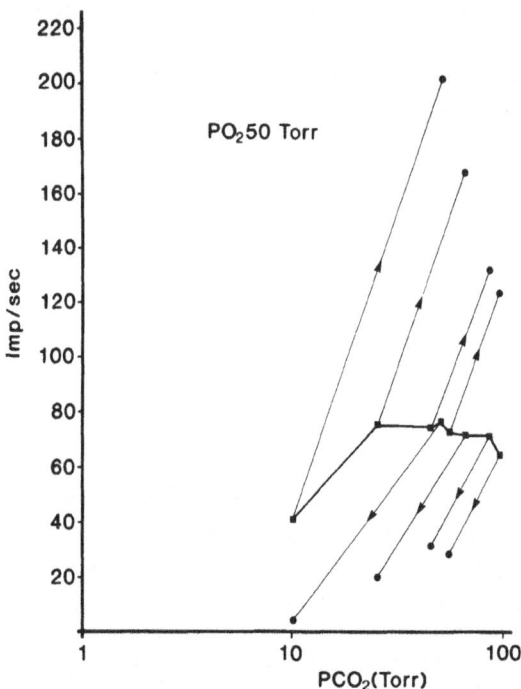

Fig. 1. Semilogarithmic plot of the steady-state (-■-) and the transient responses to sudden P_{CO2} steps, all of 40 torr, from different initial steady P_{CO2} levels, in hypoxia (P_{O2} 50 torr). The flattening of the steady-state response line is seen above P_{CO2} 20 torr. The transient rises or falls in response to the step P_{CO2} changes precede adaptation to the steady-state value at the new P_{CO2}. When plotted logarithmically they are parallel as the hypothesis requires. (McLaren, Nye & Paterson, unpublished data).

and Parks (1971) and later workers and are decribed by an equation by Nye and Painter (1989). The response to O_2 at a constant P_{CO2} is hyperbolic, being proportional to $1/P_{O2}$-C. The response to CO_2 is linear and steeper at a lower P_{O2} with a 6-fold range of slopes or more. This fan of CO_2 response lines converges to intersect beneath the CO_2 axis near to the origin. Thus the thresholds for O_2 and CO_2 are at the points where the lines cut the P_{CO2} axis. The P_{O2} and P_{CO2} interact at threshold. The response to CO_2 is not great, changing by less than 5% per torr P_{CO2} at a resting P_{CO2}, in great contrast to the central receptors which are about ten times as sensitive.

At a change of P_{CO2} the response increases quickly and then adapts. At a fall of P_{O2} it rises more slowly and adapts little. There is a marked undershoot at the off of CO_2 and there may be some undershoot at the off of hypoxia. A transient response curve describes the immediate response when P_{CO2} suddenly changes. It is steeper than the steady-state curve. Above some P_{CO2} the steady-state curve (Fig. 1) suddenly becomes parallel to the P_{CO2} axis and the P_{CO2} at which it does this is lower in hypoxia. This phenomenon was described by Fitzgerald and Dehghani (1982). Within this range of the responses a sudden rise of P_{CO2} still makes discharge rise suddenly and a fall makes discharge fall but in each case it adapts back completely to its initial level.

Any hypothesis of the convergence of P_{O2} and P_{CO2} must account for these characteristics of the responses of chemoreceptors.

THE INTRACELLULAR HYPOTHESIS

The intracellular version of the HCO_3^- hypothesis (Hanson, Nye & Torrance, 1981) supposes that when P_{CO2} rises, CO_2 is hydrated by CA within a cell and changes pH_i, the pH within the cell, but that pH_i is stabilized by a membrane mechanism which is P_{O2} dependent. At a sudden change of P_{CO2} the immediate change of pH_i is determined simply by the buffering power of the cell cytoplasm and it is independent of P_{O2}. The membrane pump then moves pH_i back towards its initial value, almost completely in hyperoxia but very little in hypoxia. pH_i somehow sets off nerve impulses. The response to pH_i is instant, linear and without adaptation. CO_2 is hydrated quickly by CA and so pH_i quickly becomes acid and the nervous response starts quickly. The membrane pump is slower and so it gradually reduces the response: the response adapts. And since the pumps action is greater in hyperoxia, adaptation is greater in hyperoxia and slight in hypoxia. Thus it is not that hypoxia calls into play some quite distinct mechanism of excitation such as the release of an acid. It is rather that O_2 switches off the immediate response to CO_2, and that in hypoxia the response is not switched off.

THE REFERENCE HCO₃ LEVEL

This idea was first put out in 1974. It was then most confidently believed that it was the extracellular pH that sets off impulses in medullary chemoreceptors, and so it was supposed, in imitation, that an extracellular pH was similarly important in the carotid body (Torrance, 1974). The extracellular spaces of the carotid body communicate quite freely with the blood: peroxidase can enter them from the blood (Woods, 1975). Thus bicarbonate pumping was supposed to alter the local bicarbonate relative to the blood bicarbonate, and therefore

$$pH_L = pK + \log \frac{(HCO_3^-)_a + \Delta HCO_3^-}{\alpha P_{CO2}}$$

P_{CO2} was not significantly different from that in the blood. ΔHCO_3^- was supposed to equal βP_{O2} where β was a Michaelis-Menten type function of P_{O2} and the term included P_{CO2} to make the reponse curves of the model fan. It also made the pumping stabilize pH (Torrance, 1976).

When it emerged that it is an intracellular pH that is important and not an extracellular one, the blood bicarbonate was no longer the appropriate reference HCO_3^- level but rather the intracellular HCO_3^- that would be present if pumping stopped entirely and H^+ came to be distributed across the cell membrane according to a Donnan equilibrium with K^+. Then pH_i is about 6.2 and at P_{CO2} of 40 torr (HCO_3^-) will be only about 1 mM. Then

$$pH_i = pK + \log \frac{1 + \Delta HCO_3^-}{\alpha P_{CO2}}$$

If pumping raises (HCO_3^-) to about 20 mM, pH_i will be alkaline.

How does (HCO_3^-) vary with P_{CO2} at any P_{O2}? In a CO_2 response curve at some P_{O2} the response of chemoreceptors is linearly related to P_{CO2} and the hypothesis suppose that the nerve endings respond linearly to pH_i. The pH_i must then be linearly related to P_{CO2}. Thus the (HCO_3^-) rises with CO_2 but with a decreasing slope. On entering the flat region discharge holds steady and so pH_i does also and $(HCO_3^-)/P_{CO2}$ must also remain constant.

WHY AN INTRACELLULAR HYPOTHESIS?

The hypothesis was originally proposed for an extracellular space that leaked to extracellular fluid and plasma, but it was known that CA is necessary for the rapid response of the carotid body (Black, McCloskey & Torrance, 1971), and so it had to be supposed that the CA acted in this extracellular space and it did so because it was attached to the walls of the space as it is in the kidney tubule or on the lumen of pulmonary capillaries (Hanson et al., 1981). If that is so, all inhibitors of CA should have easy access to the CA from the blood. But it was found that an inhibitor which does not penetrate cells easily slows the response to CO_2 much less than one which does penetrate cells. This suggested that CA acts intracellularly and that a change in pH_i is all important in excitation rather than a change of pH in an extracellular space.

DNP also has provided arguments that a pH_i is important in exciting chemoreceptors. It has long been known to excite them (Shen & Hauss, 1939) and when it was recognized that it depletes ATP by uncoupling oxidative phosphorylation, it was supposed that it excites chemoreceptors in the same way. Its action became an important piece of evidence for the metabolic hypothesis that hypoxia excites by depressing ATP. DNP does however excite very quickly and intensely, rather as does CO_2 (Nye & Torrance, 1980), and this would hardly be so if ATP had first to be run down before DNP could cause nerve impulses. But DNP, like CO_2, is an acid that is readily diffusible across membranes when it is unionized. In a cell, its diffusibility can affect pH_i in two ways:
1) It diffuses into the cell in an unionized form and then ionizes, 2) It acts as a protonophore across the cell membrane and so makes the cell acid. Thus it may make the pH_i acid in three ways as Boron and de Weer (1976) suggested for the squid axon. (1) As a diffusible acid, (2) as a protonophore or (3) as a metabolic inhibitor.

Now Obeso et al. (1989) and Rocher et al. (1991) have shown that it excites the carotid body without reducing its ATP content, and regard it as acting simply by altering pH_i. Certainly it gives a massive discharge of the afferents and also a great release of dopamine. Whether this heralds the complete demise of the metabolic hypothesis remains to be seen.

ADAPTATION IN RELATION TO OXYGEN

The hypothesis does then bring together quite a large number of observations on the carotid body. Also it makes assertions about the transient responses. At a sharp change of P_{CO2}, the immediate change of pH_i is caused by CO_2 hydration before any change in the rate of HCO_3^- pumping has had any effect. It is given by $logP_{CO2}$ and this is independent of P_{O2}. The slope of the transient response curves at any initial P_{CO2} should be independent of P_{O2}. This was suggested by the finding that the oscillations in the discharge of chemoreceptors with respiration have an amplitude which is independent of P_{O2} (Band &

Wolff, 1978; Goodman, Nail & Torrance, 1974) and that they are principally due to the oscillations of P_{CO_2} (Nye & Marsh, 1982). Now Kumar, Nye and Torrance (1988) have stepped P_{CO_2} between two levels to show better that transient response curves are parallel, independent of P_{O_2}. Thus at a step of P_{CO_2} the response rises up one of the parallel transient lines and adapts down to lie on one of the fan of steady-state lines. It does adapt more in high O_2 as the hypothesis asserts for the range of the fan. But what happens outside this range when the steady response becomes independent of P_{CO_2}? The receptor is not saturated: the response to CO_2 is still prompt but now it adapts completely (McLaren, Nye & Paterson, unpublished data, Fig. 1). In hypoxia at a low P_{CO_2} adaptation is slight. At a high P_{CO_2} it is complete and a range of P_{CO_2} for the step can be chosen so that adaptation is anywhere between these extremes. A single figure for adaptation at any P_{O_2} is then meaningless unless the initial and final stimulus levels are given. It is simpler to think of the well-established steady-state response lines and parallel transient lines.

What then does the flat part of the steady-state line in hypoxia mean? If the response of nerve impulses to a change in pH_i is indeed 'instant linear and without adaptation', then the pH_i, at a change of P_{CO_2} into the flat range, swings far to acidity but is brought sharply right back. This determines the flatness. Why should this be so? It is as if a very acid pH_i would somehow damage something within the cell and a massively effective nocifensor membrane channel is called into play when its threshold pH_i is exceeded and the channel clamps the membrane within a limited range of pH_i.

Fidone and Gonzalez (1986) summarize criticism of this hypothesis. Those deriving from observations on adaptation have already been commented upon and the arguments about the intensity of excitation that hypoxia can produce seem to have been abandoned in view of the reassessment of how DNP excites.

EFFECT OF INHIBITION OF CA ON STEADY RESPONSES

If H^+ is pumped into a system of buffers which includes CO_2/HCO_3^-, the pH depends on whether the system is open for CO_2 as the carotid body is and whether it contains active CA to accelerate CO_2 hydration within it. The hypothesis therefore requires that inhibition of the CA of the carotid body should change its responses in the steady-state. If it does, then the direction of the change should help to develop the hypothesis.

We (Hayes, Maini & Torrance, 1976) found that inhibition of CA reduces the steady-state reponses of the carotid body as well as the transient ones. We argued from this that H^+ is being steadily manipulated.in the carotid body. H^+ is being removed from some space which contains CO_2/HCO_3^- buffer and is open for CO_2. When the CA within the space is inhibited, a disequilibrium pH towards alkalinity results, and so discharge is reduced. It is the reverse of the well known acid disequilibrium pH of the kidney tubule into which H^+ is being pumped. Were it quite strictly HCO_3 that is pumped in the model, the disequilibrium produced by CA inhibition would be towards acidity since the escape of CO_2, an acid, would be impeded. The model then does not require the CO_2/HCO_3^- buffer be present for it to work, though its responses are altered if some is present and if the speed of buffering is altered by inhibiting CA.

Work on isolated cells (Buckler et al., 1991) gives no so simple answer. Acetazolamide does slow strikingly the pH_i change at a change of P_{CO_2} and also it makes pH_i more alkaline, which all fits with its effects on transient and steady-state responses.

But some of the cell's mechanisms of pH control simply involve Na^+-H^+ exchange, as has been suggested, whilst in others it is said that the HCO_3^- ion itself is quite strictly moved. And when it comes to the control of a cell's pH_i it must all be less simple than the headlong secretion of an acid into a kidney tubule. Nevertheless it can be said that suddenly blocking CA would increase the effect on pH_i of steady OH- generation in the cell but should reduce the effect of steady HCO_3^- generation.

SITE OF pH_i: CELLS OR NERVE ENDINGS ?

The type I cells are necessary for the nerve endings to be excited by natural stimuli but it has not been shown that they are phasically active whenever the nerve endings are excited. The space in which pH_i is controlled could be in the large 'en calyce' endings or it could be in the type I cells. The endings are large enough to accommodate the bits and pieces to do the job. They also have a high r^2:r^3 ratio with their punctured tennis ball, en calyce, shape, and that would be good for quick active responses for this hypothesis does present the problem that pH_i is changed throughout the buffered cytoplam of the cell and not just at its surface as are the ions which give membrane potentials. The common sense answer does of course remain that it is the type I cells that taste the blood and manage their pH_i as de Castro showed in 1928. And they are the cells most confidently said to contain CA.

The central assertion of this hypothesis is however that the change in pH_i produced by a change of P_{CO2} is reduced by an active O_2 dependent mechanism. Such a response has seldom been found in isolated cells. It must however be borne in mind when the type I cells of the carotid body are studied in isolation, that isolation in saline away from the normal cell environment, for example away from type II cells and often not even in clusters of type I cells, will be at least as damaging as perfusion with saline. In perfusion the responses are reduced (O'Regan, 1979), and Gray (1968) did not see adaptation of the response to CO_2 steps during perfusion with bicarbonate saline.

SPECULATIONS ABOUT THE NEONATE AND THE HIGH ALTITUDE NATIVE

The hypothesis postulates two things, a receptor mechanism for pH_i and a mechanism which stabilizes pH_i. There are some abnormalities or immaturities of chemoreceptor behaviour. Could these be explained by abnormalities or immaturities of one or other of these two mechanisms? If there were no pH_i receptor mechanism there would be no response. If the pH_i control mechanism were absent, the nerve fibre would respond to CO_2 as if there were intense hypoxia or the receptor were stuck on a transient response curve and could not adapt. The responses would not be altered by changes of P_{O2}, but could be altered by P_{CO2}.

Natives of high altitude have little or no response to a change in P_{O2} and also their alveolar P_{CO2} (P_{ACO2}), even after adaptation to sea level, is low (Monge & Leon-Velarde, 1991). The P_{ACO2} of the altitude native at sea level is equal to the P_{ACO2} of a sea level native when he is adapted to the altitude at which the altitude native was born (Hackett et al., 1980; Fitzgerald, 1913). It is as if the native of high altitude develops soon after birth a capacity to switch off his chemoreceptors which is adequate for the degree of hypoxia of the altitude of his birth and at which he is then living. If he does not develop early a full capacity to switch off he never does, and he remains with vigorously discharging chemoreceptors which he cannot switch off. This refers to the neonate, who starts to respond to a changing P_{O_2} over the first week of his life (Williams et al., 1991). What is then happening to his CO_2 sensitivity is not known. If each set of subjects had 'lack of switch off' as a single lesion it would leave the CO_2/H^+ responses intact and there would be a brisk reponse to a changing P_{ACO_2}.

One of the characteristics of the high altitude native is that his carotid body is hypertrophied. A carotid body that cannot be switched off could well show work hypertrophy as a result of the sustained activity that is caused by failure of development of a mechanism for switching it off. And the neonate may not be so extraordinarily illogical if what he does learn after birth is how to switch off his chemoreceptors and not how to switch them on. They would now help to cope with the danger from hypoxia at and around birth by providing a reflex drive to breathe. It is easy to invent a teleological tale for a chemoreceptor that only slowly develops an ability to switch off after birth so that it is maximally active at birth but it is not so easy for one that slowly learns to switch on.

ENVOY

It seems to be accepted

1) That a change of pH_i sets off the CO_2/H^+ response.
2) That there is a mechanism for pH_i control.
3) That pH_i control contributes to adaptation in response to CO_2.
4) That adaptation to CO_2 is huge in high P_{O_2} and declines with hypoxia.

Does this leave any room for PO_2 to act in any way but by setting the degree of pH_i control?

ACKNOWLEDGEMENTS

This work was supported by the MRC and the Wellcome Trust.

REFERENCES

Band, D.M. & Wolff, C.B. (1978). Respiratory oscillations in discharge frequency of chemoreceptor afferents in sinus nerve of anaesthetized cats at normal and low arterial oxygen tensions. J.Physiol., 182, 1-6.

Black, A.M.S., McCloskey, D.I. & Torrance, R.W. (1971). The responses of carotid body chemoroceptors in the cat to sudden changes of hypercapnic and hypoxic stimuli. Respir.Physiol., 13, 36-49.

Boron, W.F. & De Weer, P. (1976). Intracellular pH transient in squid giant axons caused by CO_2, NH_3, and metabolic inhibitors. J.Gen. Physiol., 67, 91-112.

Buckler, K.J., Vaughan-Jones, R.D., Peers, C. Nye, P.C.G. (1991). Intracellular pH and its regulation in isolated type I carotid body cells of the neonatal rat. J.Physiol., 436, 107-130.

De Castro, F. (1928). Sur la structure et innervation du sinus carotidienne de l'homme et des mammiferes. Nouveaux faits sur l'innervation et la fonction du glomus carotidien. Trab.Lab.Invest. Biol.Univ. Madrid, 25, 331-384.

Fidone, S.J. & Gonzalez, C. (108). Initiation and control of chemoreceptor activity in the carotid body. In: 'Handbook of Physiology III: 'The Respiratory System', Vol.2, Part 1, ed. Fishman. A.P., Cherniack, N.S., Widdicombe, J.G. & Geiger, S.R. American Physiological Society, Bethesda, MD, USA, pp.242-312.

Fitzgerald, M. P. (1913). The changes in the breathing and the blood at various high altitudes. Phil.Trans.Roy.Soc.Lond., 203, 351-371.

Fitzgerald, R.S. & Dehghani, G.A. (1982). Neural responses of the cat carotid and aortic bodies to hypercapnia and hypoxia. J.Appl. Physiol., 52(3), 596-601.

Fitzgerald, R.S. & Parks, D.C. (1971). Effect of hypoxia on carotid chemoreceptor response to carbondioxide in cats. Respir.Physiol., 12, 218-222.

Goodman, N.W., Nail, B.S. & Torrance, R.W. (1974). Oscillations in the discharge of single carotid chemoreceptor fibers of the cat. Respir.Physiol., 20, 251-269.

Gray, B. (1968). Response of the perfused carotid body to changes in pH and P_{CO_2}. Respir.Physiol., 4, 229-245.

Hackett, P.H., Reeves, J.T., Reeves, C.D., Grover, R.F. & Rennie, D. (1980). Control of breathing in sherpas at low and high altitude. J.Appl.Physiol., 49, 374-379.

Hanson, M.A., Nye, P.C.G. & Torrance, R.W. (1981). Studies on the localization of pulmonary carbonic anhydrase in the cat. J.Physiol., 319, 93-109.

Hanson, M.A., Nye, P.C.G. Torrance, R.W. (1981). The exodus of an extracellular bicarbonate theory of chemoreception and the genesis of an intracellular one. In: 'Arterial Chemoreceptors'. Ed.Belmonte, C. Pallot, D.J., Acker, H. & Fidone, S. Leicester University Press, Leicester pp.403-416.

Hayes, M.W., Maini, B.K. & Torrance, R.W. (1976). Reduction of the responses of carotid chemoreceptors by acetazolamide. In: Morphology and Mechanisms of Chemoreceptors', ed. Paintal, A.S. Vallabhbhai Patel Chest Institute,Delhi, pp.36-45.

Iturriaga, R., Lahiri, S. & Okashi, A. (1991). Carbonic anhydrase and chemoreception in the cat carotid body. Am.J.Physiol., 261, C565-C573.

Kumar, P., Nye, P.C.G. & Torrance, R.W. (1988). Do oxygen tension variations contribute to the repiratory oscillations of chemoreceptor discharge in the cat? J.Physiol., 395, 531-552.

McLaren, A.J., Nye, P.C.G. & Paterson, D.J. (unpublished data).

Monge, C. & Leon-Velarde, F. (1991). Physiological adaptation to high altitude: Oxygen transport in mammals and birds. Physiol.Rev., 71, 1135-1172.

Nye, P.C.G. & Marsh, J. (1982). Ventilation and carotid chemoreceptor discharge during venous CO_2 loading via the gut. Resp.Physiol., 50, 335-350.

Nye, P.C.G. & Painter, R. (1989). Quantifying the steady-atate discharge of the carotid body of the anaesthetized cat. J.Physiol., 417, 177P.

Nye, P.C.G. & Torrance, R.W. (1980). A breath of carbon monoxide reduces chemoreceptor discharge in the cat. J.Physiol., 307, 47P.

Obeso, A., Almaraz, L. & Gonzalez, C. (1980). Effects of cyanide and uncouplers on chemoreceptor activity and ATP content of the cat carotid body. Brain Res., 481, 250-257.

O'Regan, R.G. (1979). Reponses of the chemoroceptors of the cat carotid body perfused with cell-free solution. Irish J.Med.Sci., 148, 78-85.

Rocher, A., Obeso, A., Herreros, B. & Gonzalez, C. (1991). Involvement of $Na^+:H^+$ and $Na^+:Ca^{++}$ antiporters in the chemotransduction of acidic stimuli. In: 'Chemoreceptors and Chemoreceptor Reflexes', ed. Acker, H., Trzebski, A. & O'Regan, R.G. Plenum Press New York, pp.35-41.

Shen, T.C.R. & Hauss, W.H. (1939). Influence of dinitro-phenol 1-2-4, dinitro-orthocresol 1,2,4 and paranitro-phenol upon the carotid sinus chemoreceptors of the dog. Arch.int.Pharmacodyn. LXIII fasc.2, pp.251-258.

Torrance, R.W. (1974). Arterial chemoreceptors. In: 'Respiratory Physiology', ed. Widdicombe, J.G. MTP Physiology Series One, Butterworth. Vol.2, pp.247-271.

Torrance, R.W. (1976). A new version of the acid receptor hypothesis of carotid chemoreceptors. In: 'Morphology and Mechanism of Chemoreceptors', ed. Paintal, A.S. Vallabhbhai Patel Chest Institute Delhi. pp.131-137.

Torrance, R.W. (1977). Convergence of stimuli in arterial chemoreceptors. In: 'Tissue Hypoxia and Ischemia', ed. Reivitch, M., Coburn, R., Lahiri, S. & Chance, B. Adv.exp.med.biol., 78. Plenum Press, pp.203-207.

Williams, B.A., Smyth, J., Boon, A.W., Hanson, M.A., Kumar, P. & Blanco, C.E. (1991). Development of respiratory chemoreflexes in response to alternations of fractional inspired oxygen in the newborn infant. J.Physiol., 442, 81-90.

Woods, R.I. (1975). Penetration of horseradish peroxidase between all elements of the carotid body. In: 'Peripheral Arterial Chemoreceptors', ed. Purves , M.J. Cambridge University Press pp.195-205.

35 REGULATION OF INTRACELLULAR pH IN TYPE I CELLS OF THE NEONATAL RAT CAROTID BODY

R.D.Vaughan-Jones, K.J.Buckler, C.Peers* and P.C.G.Nye

University Laboratory of Physiology
Parks Road, Oxford, OX1 3PT, U.K.
*Department of Pharmacology, Worsley Medical
and Dental Building, Leeds, LS2 9JT, U.K.

INTRODUCTION

Although the carotid body is the principal arterial pH receptor, little is known about its ionic mechanisms for pH-detection. A prime candidate for the H^+-chemosensory element is the type I (glomus) cell, also believed to be the site of O_2-chemoreception. This cell contains a rich variety of neuroactive compounds, most notably catecholamines, which are secreted in response to extracellular acid/base challenges and to hypoxia (Fidone & Gonzalez, 1986). In this way, local endings of the carotid sinus nerve are believed to be elicited. Indirect evidence has suggested that, in order to get chemoreception of extracellular acidosis/alkalosis there must first be a change of intracellular pH in the type I cell (Hanson, Nye & Torrance, 1981; Rigual, Lopez-Lopez Gonzalez, 1991). Until recently, however, there was no accurate way of directly measuring pH in this cell. Consequently, it was not known if pH_i was regulated by acid equivalent efflux and influx mechanisms as it is in many other cells (Roos & Boron, 1981; Thomas, 1984 & Vaughan-Jones, 1988). In the present work we have therefore used the intracellular fluorescent probe SNARF-1 to examine intracellular pH and its regulation in type I cells isolated enzymically from the neonatal rat carotid body.

METHODS

Type I cells were enzymically isolated from the carotid bodies of 7-11 day old rat pups (Peers & O'Donnell, 1990). Only the most commonly found spherical, phase bright cells of approximately 10µm diameter, or clumps of such cells, were used in experiments.

HCO_3^--buffered solution contained NaCl 117, KCl 4.5, $NaHCO_3$ 23, $MgCl_2$ 1, $CaCl_2$ 2.5, glucose 11 and was equilibrated with 5% CO_2/95% air, pH at 37°C was 7.4-7.45. For HEPES solution, $NaHCO_3$ was replaced by 20mM NaHEPES, pH 7.40 at 37°C. Na-free solution was made by replacing NaCl with N-methyl-D-glucamine chloride (NMG). Chloride-free solution was made by replacing NaCl and $NaHCO_3$ by equivalent amounts of Na glucuronate and K gluconate and by replacing $CaCl_2$ by 12mM Ca gluconate (elevated to compensate for the Ca^{2+}-binding properties of glucuronate/gluconate anions). Ethyl isopropyl amiloride (EIPA) was a gift from Drs.Scheibli & Fuhrer, Ciba Geigy, Switzerland. DIDS (diisothiocyanatostilbene disulphonic acid) from SIGMA chemicals.

Intracellular pH was measured using the dual emission fluoroprobe carboxy SNARF-1. Cells were loaded with SNARF-1 by incubation in a 5μM solution of the acetoxymethyl ester (SNARF-1-AM) for 8 min. For further details of the use and calibration of this dye, see Buckler & Vaughan-Jones, 1990.

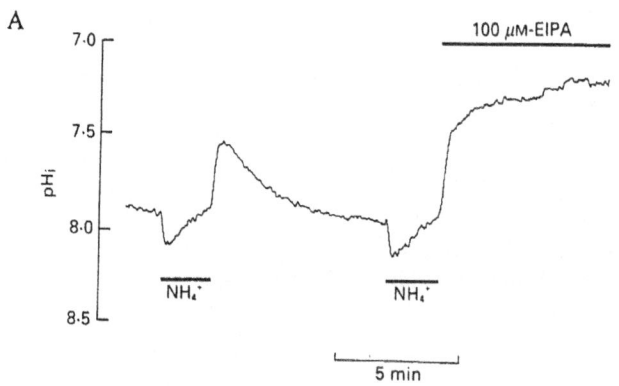

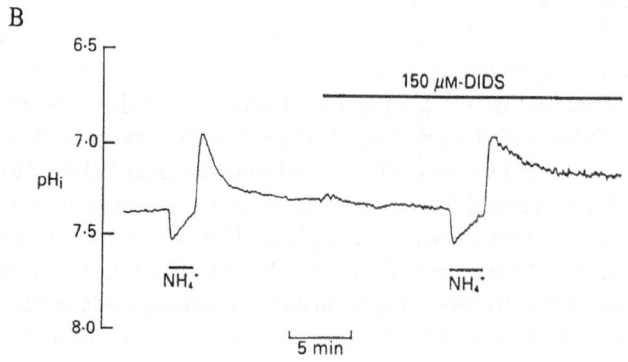

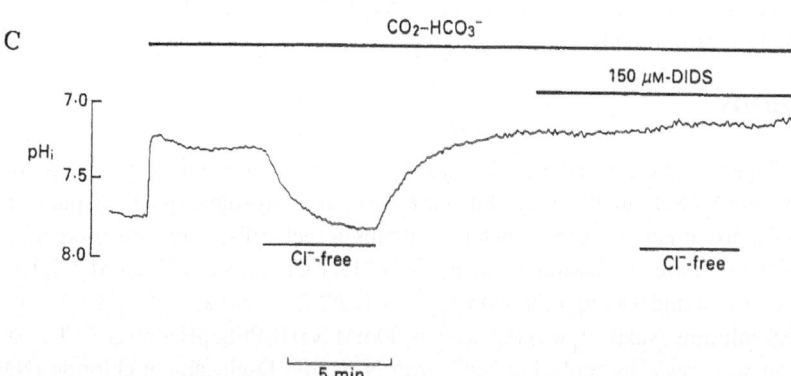

Fig. 1 Carboxy SNARF-1 recording of pH$_i$ in neonatal rat type I cells; pH$_o$ 7.40 throughout. **A** Single cell superfused with air equilibrated HEPES-buffered medium; bars at bottom indicate addition of 10mM NH$_4$Cl. **B** Clump of cells superfused with 5%CO$_2$/HCO$_3^-$ buffered solution; bars at bottom indicate periods of addition of 10mM NH$_4$Cl. **C** Single cell superfused with CO$_2$/HCO$_3^-$ buffered solution for period shown by upper bar (note that recording commences with superfusion of HEPES solution); bars at bottom indicate periods in chloride-free solution.

RESULTS

Fig.1A shows a recording of pH_i in a clump of type I cells bathed in HEPES-buffered medium (nominally free of CO_2/HCO_3^-). Note that resting pH_i was particularly alkaline (~7.80; see also Buckler et al., 1991a). Acid-loading the cell (using the ammonium-rebound, technique, Roos & Boron, 1981) resulted in a fall of pH_i by ~0.3 units followed by a recovery which was totally inhibited (n= 10) by 100μM EIPA (ethyl isopropyl amiloride), a high-affinity inhibitor of Na^+/H^+ exchange. A similar inhibition of recovery occurred if all Na^+_o was replaced isosmotically by N-methyl-D-glucamine (NMG; n=6). Thus in HEPES media, Na^+/H^+ exchange must be the principal acid extrusion system following intracellular acidosis. When type I cells were switched from HEPES to 5% CO_2/HCO_3^-,-buffered media, intracellular pH fell rapidly (entry and hydration of CO_2 within the cell) and stabilised at a lower pH_i (see initial part of Fig. 1C, pH_i stabilised finally at ~7.25 cf. Buckler et al., 1991a). When cells under these conditions were acid-loaded (ammonium-rebound), pH_i-recovery rate was slowed only ~~50% by 100μM EIPA (n=3, data not shown; see Buckler et al., 1991a). The recovery rate, was also slowed by 150μM DIDS (an inhibitor of anion movements) as shown in Fig. 1B. Indeed, DIDS-pretreatment (150μM for 5 min.) of an acid-loaded cell (ammonium rebound) with subsequent replacement of the DIDS by 100μM EIPA (the effects of DIDS are only partially reversible after about 10 min.) inhibited pH_i-recovery (not shown, n=4). Thus, in effect, application of DIDS plus EIPA inhibited all of the pH_i recovery observed in CO_2/HCO_3^- medium. Moreover, pH_i-recovery from an ammonium rebound in 5% CO_2-medium was also completely inhibited by replacement of Na^+_o with NMG (n=4, not shown). From this we conclude that, in CO_2/HCO_3^- -buffered media, acid extrusion is mediated by at least two Na^+_o-requiring, ionic mechanisms, one needing both Na^+ and HCO_3^--ions and inhibitable by DIDS (probably configured as a co-influx into the cell, see Discussion), the other being Na^+/H^+ exchange (inhibitable by EIPA).

Finally, there is evidence for the existence of a separate Cl^-/HCO_3^--exchanger in the type I cell. Fig.1C shows that removal of extracellular chloride produced a reversible alkalosis which was DIDS-inhibitable (n=4). This reversible alkalosis in Cl^--free solution was unaffected by removal of extracellular sodium for periods of up to 15 minutes (n=6) and must therefore be independent of the two other mechanisms identified above since the others failed to work in Na^+-free solution. We suggest that the alkalosis is caused by Cl^- efflux and HCO_3^- ion influx on a DIDS inhibitable chloride/bicarbonate exchanger.

DISCUSSION

Of the three ionic mechanisms for transporting acid or its ionic equivalent across the surface membrane of the type I cell, only the two sodium-dependent systems (Na^+/H^+ exchange and $Na^+-HCO_3^-$-dependent mechanism) seem capable of limiting an intracellular acidosis by extruding acid equivalents (recall that all pH_i recovery from intracellular acidosis was blocked by Na^+_o-removal even though this manoeuvre did not block Cl^-/HCO_3^- -exchange). We have yet to characterize fully the ionic dependence of the $Na^+-HCO_3^-$ mechanism. In order to produce pH_i recovery from an acid load the carrier must produce bicarbonate influx, so it seems reasonable to propose that this is coupled with Na^+ influx (the possibility that the mechanism also needs and transports chloride ions has not yet been tested). The role of the third mechanism, anion exchange, is uncertain at present although it would seem reasonable to propose that it normally functions as it does in many other cells, i.e. as an acid-equivalent influx mechanism (HCO_3^- ion efflux coupled to Cl^-

ion influx; Vaughan-Jones, 1982; Ganz, Boyarsky, Sterzel & Boron, 1989). We have indeed observed that DIDS-application can sometimes raise resting pH$_i$ by 0.1-0.2 units in cells bathed in CO_2/HCO_3^- solution (seen in 6 out of 9 cells, not shown). This would be consistent with inhibition of a resting bicarbonate efflux mediated through the anion exchanger (note that DIDS-inhibition of the Na$^+$/HCO$_3^-$-dependent mechanism would be expected to acidify pH$_i$, thus ruling out this mechianism). The reversal of anion exchange observed in Cl$^-$ free solution in Fig.lC (Cl$^-$ efflux for HCO$_3^-$ ion entry) would therefore be an unusual and non-physiological mode for the anion exchanger in the type I cell, observed only when the chemical driving force for chloride ions becomes outward (or, at least, when the usually inwardly-directed chloride gradient becomes < the inwardly-directed chemical gradient for bicarbonate ions).

We propose the model illustrated in Fig.2 for pH$_i$-control in the type I cell. There is a dual system for acid extrusion activated by a fall of pH$_i$ and a single system for acid equivalent influx, perhaps activated by a rise of pH$_i$ (cf. Vaughan-Jones, 1982, 1988; Ganz et al., 1989). The net effect of these mechanisms is that pH$_i$ is stabilised at ~7.2 in CO_2/HCO_3^- -buffered solution (pH$_o$ 7.4). It is notable that, in HEPES media when Na$^+$ / H$^+$ exchange only will be able to operate, resting pH$_i$ can become extremely alkaline (Fig.lA). The reason for this is yet to be established but the result stresses the importance of working with CO_2/HCO_3^-, media when experimenting on isolated type I cells, to ensure that pH$_i$ is not abnormally high.

In conclusion, there are at least three carriers that move acid equivalents across the membrane of the type I cell. These will serve to regulate pH$_i$ but, presumably, they may

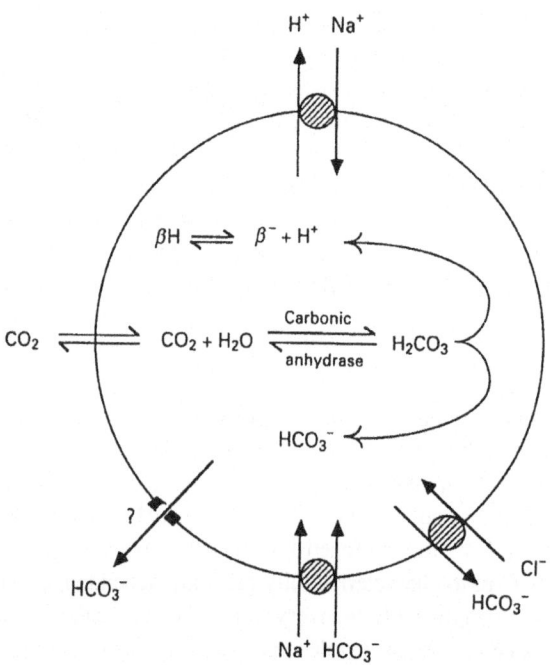

Fig.2 Model of pH$_i$-regulation in the carotid body type I cell. The three identified acid equivalent carriers are indicated (note that the full ionic dependence of the Na$^+$-HCO$_3^-$ system has not been established). Also shown are a possible bicarbonate efflux via an anion (chloride) channel and an intracellular buffer system (β) which includes the catalysis of CO_2-hydration by carbonic anhydrase (see Buckler et al. 1991a).

also subserve the role of mediating changes of pH_i in response to extracellular acid/base challenges ie. changes in pH_o (cf. Buckler, Vaughan-Jones, Peers, Lagadic-Gossmann & Nye, 1991b). They may therefore play a role in the mechanism of arterial pH-chemotransduction.

ACKNOWLEDGEMENTS

This worlk was supported by the Wellcome Trust (all authors) and the British Heart Foundation (R.D.V-J.; K.J.B.).

REFERENCES

Buckler, K.J. Vaughan-Jones, R.D., 1990, Application of a new pH-sensitive fluoroprobe (carboxy SNARF-1) for intracellular pH measurement in small, isolated cells. Pflugers Archive., 417:234.

Buckler, K.J., Vaughan-Jones, R.D., Peers, C. & Nye, P.C.G., 1991a, Intracellular pH and its regulation in isolated type I carotid -body cells of the neonatal rat. J.Physiol., 436:107.

Buckler, K.J., Vaughan-Jones, R.D., Peers, C., Lagadic-Gossmann, D. & Nye, P.C.G., 1991b, Effects of extracellular pH, PCO_2 and HCO_i on intracellular pH in isolated type-I cells of the neonatal rat carotid body. J.Physiol., 444:703.

Fidone, S.J. & Gonzalez, C., 1986, Initiation and control of chemoreceptor activity in the carotid body. In: Handbook of Physiology III: The Respiratory System, vol.2, part 1; American Physiological Society, Bethesda, MD., USA., ed. Fishman, A.P., Cherniack, N.S.,Widdicombe, J.G. & Geiger, S.R. :247.

Ganz, M.B., Boyarsky, G., Sterzel & Boron, W.F., 1989, Arginine vasopressin enhances pH_i regulation by stimulating three acid-base transport systems. Nature, 337:648.

Hanson, M.A., Nye, P.C.G. and Torrance, R.W., 1981, The exodus of an extracellular bicarbonate theory of chemoreception and the genesis of an intracellular one. In: Arterial Chemoreceptors; Leicester University Press, Leicester; ed. Belmonte, C., Pallot, D.J., Acker, H. & Fidone, 5.: 403.

Peers,C. & O'Donnell, J., 1990, Potassium currents recorded in type-I carotid body cells from the neonatal rat and their modulation by chemoexcitatory agents. Brain Research, 522:259.

Rigual,R., Lopez-Lopez,J.R. & Gonzalez, C., 1991, Release of dopamine and chemoreceptor discharge induced by low pH and high PCO_2 stimulation of the cat carotid body. J.Physiol., 433:519.

Roos, A & Boron, W.F., 1981, Intracellular pH. Physiological Reviews, 61:296.

Thomas, R.C., 1984, Review Lecture. Experimental displacement of intracellular pH and the mechanism of its subsequent recovery. J. Physiol., 354: 'P.

Vaughan-Jones, R.D., 1982, Chloride-bicarbonate exchange in the sheep cardiac Purkinje fibre. In: Intracellular pH: Its Measurement. Regulation and Utilization in Cellular Functions, A.R.Liss, New York, ed. Nuccitelli, R. & Deamer, D.:239.

Vaughan-Jones, R.D., 1988, Regulation of intracellular pH in cardiac muscle. In: Proton passage across cell membranes. Wiley, Chichester (Ciba Foundation 139:23.

SECTION III

Neurotransmitter

36 NORADRENERGIC INHIBITION OF THE GOAT CAROTID BODY

G. Bisgard, M. Warner, J. Pizarro, W. Niu and G. Mitchell

Department of Comparative Biosciences
University of Wisconsin
Madison, Wisconsin 53706

INTRODUCTION

Circulating catecholamines are considered to be respiratory stimulants in humans. This conclusion is based on studies in which intravenous infusion of norepinephrine or epinephrine produced ventilatory stimulation (Cunningham, et al., 1963; Heistad et al., 1972). Because the respiratory stimulation was diminished during hyperoxia, it was concluded that the carotid bodies were responsible for the reflex effect. Studies of norepinephrine and epinephrine in non-human species have revealed complex and sometimes conflicting effects on the carotid body and ventilation. Joels and White (1968) found tht in anesthetized and decerebrate cats, these agents stimulate ventilation, an effect eliminated by carotid sinus nerve transection. Folgering et al. (1982) found that norepinephrine and epinephrine produced transient inhibition followed by excitation of carotid chemoreceptor afferent neural discharge frequency in cats. Although similarly complex effects were found in rabbits (Matsumoto et al., 1981), Bisgard et al. (1979) found primarily inhibition but also excitation of the carotid body in dogs. More recently, the possibility of a prominent alpha-2 adrenergic inhibitory mechanism in the cat carotid body has been described (Kou et al., 1991).

We wished to determine the ventilatory effects of elevated circulating norepinephrine in goats. This relates to our interest in mechanisms of ventilatory stimulation in states in which plasma catecholamines might be elevated, e.g. in exercise and in hypoxia.

METHODS

Under general anesthesia induced by 15 mg/kg thiamylal sodium IV and maintained by 1% Halothane, 50% nitrous oxide and oxygen, five adult goats were prepared for study. The preparation included bilateral translocation of both common carotid arteries to a subcutaneous location, and denervation of one carotid body. After a minimum of one week of recovery, the animals were studied awake. A catheter was place percutaneously in each of the translocated carotid arteries for infusion of norepinephrine, arterial blood gas sampling and for monitoring of arterial blood pressure. With one carotid body denervated, we were able to use infusions on the denervated side as control. Doses as low as 0.2 µg/kg/min and as high as 7.0 µg/kg/min were infused by the intracarotid route in preliminary studies. All doses produced inhibition of ventilation, but the higher doses produced excessive arterial hypertension. Therefore, a nominal dose of 1 g/kg/min for three minutes was selected for further studies as it produced a substantial effect on ventilation without excessive hypertension.

One goat was studied under general anesthesia (15 mg/kg thiamylal sodium, IV, followed by IV chloralose, 5 to 15 mg/kg/hr). The animal was paralyzed with gallamine (2.0 mg/kg followed by 1.5 mg/kg/0.5h) and ventilated artificially. In this study the carotid sinus nerve was exposed for recording of chemoreceptor discharge frequency.

Norepinephrine solution was prepared fresh each day. The norepinephrine was diluted in sterile physiological saline and this solution was infused at the rate of 5 ml/min. Infusion of saline without norepinephrine had no effect on ventilation.

RESULTS

Norepinephrine infusion at the rate of 1 µ/kg/min produced a marked decrease in ventilation which was maximal at approximately 30-60 seconds into the infusion fig. 1). The inhibition of ventilation was not maintained throughout the period of infusion. This was attributed, at least in part, to a rising $Paco_2$ which would stimulate central chemoreceptors and overcome the effects of carotid body inhibition. Blood gas sampling at one minute into the infusion showed that $Paco_2$ was, on average , elevated by 4 Torr. In some goats, upon termination of the infusion, ventilation increased above the control, pre-infusion level until $Paco_2$ recovered to a normal level. The decrease in breathing during infusion of norepinephrine began to occur before arterial pressure increased. Arterial pressure began to rise at approximately 45 soconds into the infusion and increased by an average of 20 to 30 Torr. Infusion of norepinephrine on the carotid body

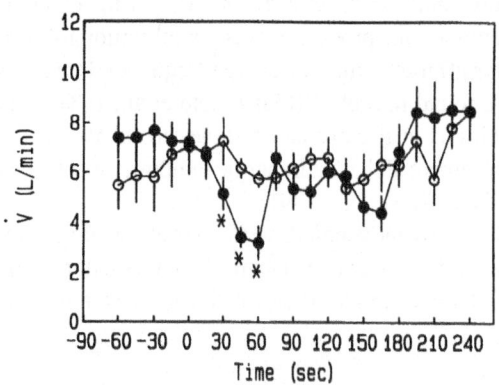

Fig. 1 Minute ventilation (V̇) during infusion of 1µg/kg/min of norepinephrine intra-arterially into the common carotid artery on the carotid body intacts side (closed circles) and on the carotid body denervated side (open circles). Infuson was initiated at time 0.

denervated side had little or no offect on ventilation (fig. l) even though arterial pressure rose to the same degree and with the same time course as during infusions on the intact side. Therefore, it was concluded that increased arterial pressure was not the cause of ventilatory inhibition. Furthermore, these findings suggested that the ventilatory inhibition was mediated by a carotid body effect.

In the goat prepared for carotid body afferent nerve recording intracarotid infusion of norepinephrine (1 µg/kg/min) produced abrupt inhibition of afferent discharge during the entire 2 minute infusion (fig. 2). The inhibition ended promptly on termination of the infusion. While no firm conclusion can be made on the basis of one study, these data support the hypothesis that exogenous norepinephrine infusion produces inhibition of carotid body afferent discharge frequency. In this animal, guanabenz, an alpha-2 adrenergic agonist, was also infused (fig. 3). This drug was selected because it was used extensively by Kou et al. (1991) to demonstrate the presence of alpha-2 adrenergic inhibition of the carotid body in cat. We found that guanabenz also produced inhibition of carotid body afferent discharge frequency, but the dose to obtain inhibition was approximately 10 times that for norepinephrine (fig. 3).

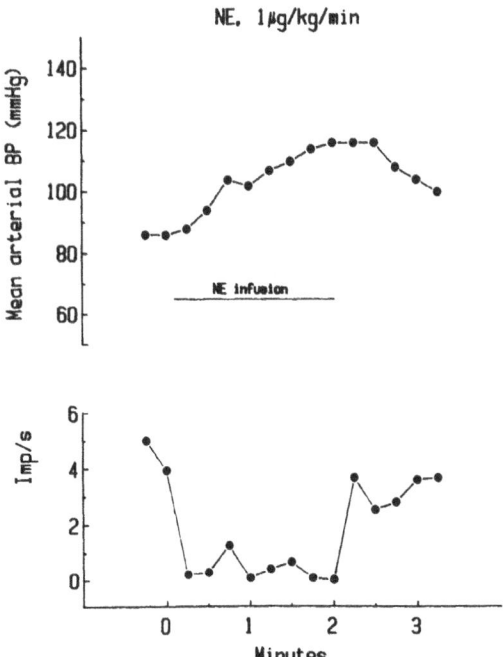

Fig. 2. *Afferent discharge frequency (Imp/s) of a paucifiber carotid body afferent and mean arterial blood pressure turing intra-arterial infusion of norepinephrine (1 µg/kg/min). Infusion was started at time 0 and ended at 2 minutes.*

Preliminary studies using adrenergic pharmacologic antagonists were carried out. The tentative conclusions from thes studies were: (1) that beta-adrenergic receptors were not responsible for CB inhibition as propranolol (1 mg/kg IV) was ineffective in preventing the inhibition; (2) alpha-adrenergic blockade with phenoxybenzamine (1 mg/kg IV) attenuated the inhibition; and, (3) dopaminergic blockade wlth domperidone (1 mg/kg IV) effectively prevented the norepinephrine mediated inhibition of ventilation.

DISCUSSION

These studies provide strong evidence that exogenously administered norepinephrine causes inhibition of ventilation that is mediated by the carotid body chemoreceptors. Inhibition followed by stimulation of breathing was not noted during infusion, although ventilation approached pre-lnfusion control values before termination of the norepinephrine infusion. It is possible that some of this recovery of ventilatory drive could have been

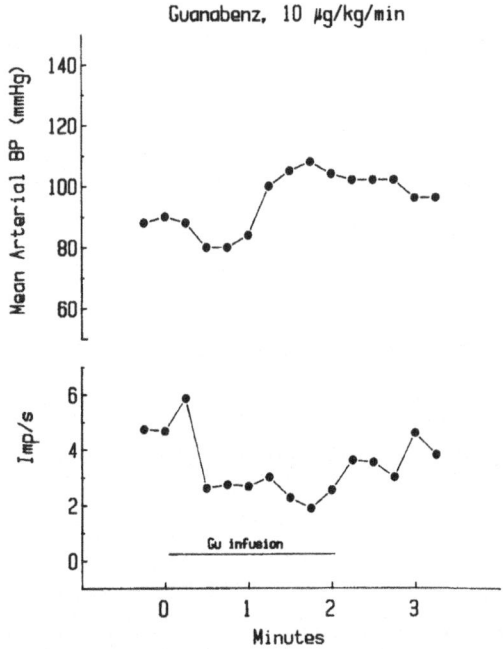

Guanabenz, 10 μg/kg/min

Fig. 3. Afferent discharge frequency (Imp/s) of a paucifiber carotid body afferent and mean arterial blood pressure during intra-arterial infusion of guanabenz (10 μg/kg/min) just downstream from the carotid body. Infusion was started at time 0 and ended at 2 minutes.

due to an excitation of the carotid body; however, no such excitation was noted in the single animal in which chemoreceptor discharge was measured. The increase in ventilation and overshoot after termination of infusion is likely to be due to the stimulation of medullary chemoreceptors by elevated $Paco_2$. These data indicate that the predominant effect of exogenous norepinephrine is inhibitory to the goat carotid body.

Based on preliminary pharmacologic evidence, the norepinephrine inhibitory effect may be mediated by involvement of both alpha-adrenergic and dopaminergic receptors. Alpha-2 adrenergic receptors may be involved based on preliminary findings with guanabenz, an alpha-2 agonist, in the anesthetized sinus nerve recording study. Such an effect would be compatible with the findings of Kou et al.(1991). We have not used alpha-2 adrenergic antagonists because of the behavioral effects of these agents in awake goats. We speculate that norepinephrine may release dopamine from carotid body type 1 cells. The dopamine may then depress carotid body afferent discharge. Dopamine is known to inhibit carotid body activity in goats (Bisgard et al., 1980; Kressin et al., 1986).

These studies suggest that elevated circulating norepinephrine is unlikely to be an important component in the stimulation of ventilation during hypoxia or exercise in goats. Rather, an inhibitory role is suggested in both physiologic states.

ACKNOWLEDGEMENTS

This work was supported by grants from the National Institutes of Health, USPHS,(HL15473 and HL36780). The authors thank Gordon Johnson for excellent technical assistance.

REFERENCES

Bisgard, G.E., Mitchell, R.A., and Herbert, D.A., 1979, Effects dopamine, norepinephrine and 5-hydroxytryptamine on the carotid body of the dog, Respir. Physiol., 37:61-80.

Bisgard, G.E., Forster, H.V., Klein, J.P., Manohar, M. and Bullard, V.A., 1980, Depression of ventilation by dopamine in goats: effects of carotid body excision, Respir. Physiol., 41:379-392.

Cunningham, D.J.C., Hey, E.V., Patrick, J.L., and Lloyd, B.B., 1963. The effect of noradrenaline infusion on the relation between pulmonary ventilation and the alveolar Po_2 and Pco_2 in man, Ann. N.Y. Acad. Sci.,109:756-771.

Folgering, H, Ponte, J, and Sadig, T, 1982, Adrenergic mechanisms and chemoreception in the carotid body of the cat and rabbit, J. Physiol. (London), 325:1-21.

Heistad, D.D., Wheeler, R.C., Mark, A.L., Schmid P.G., and Abboud, F.M., 1972, Effects of adrenergic stimulation of ventilation in man, J. Clin. Invest, 51:1469-1475.

Joels, N., and White, H., 1968, The contribution of the arterial chemoreceptors to the stimulation of respiration by adrenaline and noradrenaline in the cat, J. Physiol. (London), 197:1-23.

Kou, Y.R., Ernsberger, P., Cragg, P.A., Cherniack N.S., and Prabhakar, N.R., 1991, Role of a_2-adrenergic receptors in the carotid body response to isocapnic hypoxia, Respir. Physiol, 83:353-364.

Kressin, N.A., Neilsen, A.M., Laravuso, R., and Bisgard, G.E., 1986, Domperidone induced potentiation, of ventilatory responses in awake goats, Respir. Physiol, 65:169-180.

Matsumoto, S., Ibi, T., and Nakajima, T., 1981, Effects of carotid body chemoreceptor stimulation by norepinephrine, epinephrine and tyramine on ventilation in the rabbit, Arch. int. Pharmacodyn, 252:152-161.

37 ROLE OF SUBSTANCE P IN RAT CAROTID BODY RESPONSES TO HYPOXIA AND CAPSAICIN

P.A. Cragg, Y.R. Kou and N.R. Prabhakar

Department of Medicine, Pulmonary Division
Case Western Reseve University
Cleveland, Ohio 44106, U.S.A.

INTRODUCTION

Substance P (SP) belongs to a family of structurally related peptides called *tachykinins*. In the peripheral nervous system, SP is associated with the processing of sensory signals[1]. In recent years, several studies have examined the role of SP in the chemosensory function of the cartid body. SP-like peptides have been demonstrated in the glomus tissue [2,3,4] and exogenous administration of the peptide stimulates the carotid body's afferent activity [4,5,6,7]. Futhermore, antagonists specific for tachykinin peptides, when given in doses sufficient to block the excitatory effects of SP, attenuate or abolish the chemoreceptor response to hypoxia, but not that caused by CO_2[7]. Thus far, studies on the cat carotid body suggest that SP is associated with the transmission of the hypoxic signal.

Carotid bodies of other species, e.g., rabbits and rats, have also been shown to contain SP-like peptides. The significance of SP in the chemoreceptor function of species other than cats, however, has not been investigated. If SP is associated with transmission of the hypoxic stimulus, then the stimulatory effects of the peptide should be uniform across species, and the tachykinin antagonist should attenuate the carotid body response to hypoxia in species other than cats. Therefore, we investigated the significance of SP in the carotid body function of rats.

METHODS

General. Adult Sprague-Dawley rats (250 to 400 g) were anaesthetized by intraperitoneal injection of urethane (1.8 g/kg) and rectal temperature was maintained at 37 to 38°C with a heat lamp. Intravenous administration of fluids and drugs was accomplished via a femoral vein catheter. Cannulation of a femoral artery permitted measurement of arterial blood pressure (ABP) and arterial sampling for blood gases. After tracheal intubation, the trachea and oesophagus were sectioned and retracted rostrally to expose the carotid bifurcation. The intracarotid cannula was a PE-10 catheter inserted retrogradely into the external carotid artery until its tip was positioned just caudal to the origin of the lingual artery.

Recording of carotid body nerve activity. The carotid sinus nerve was separated from the connective tissue and cut with a small piece of glossopharygneal nerve still attached for ease of handling. The baroreceptors were eliminated by crushing or cutting the tissue on the rostral margin of the entire carotid sinus enlargement of the internal carotid artery. The rat was then paralyzed with pancuronium (0.8 mg/kg) and artificially ventilated. Whole nerve activity was recorded with conventional electrophysiological equipment using a unipolar platinum-iridium decode with a silver electrode in nearby tissue. The mass discharge rate was measured and also a window discriminator fed to a rate meter was used to count the action potentials above the baseline noise. An audio signal in which there was no discharge in phase with the ABP pulse confirmed carotid baroreceptor denervation as did the lack of a decrase in nerve activity when the common carotid artery was clamped for 10s during hyperoxic ventilation.

Neonatal capsaicin treatment. Two-day old rat pup were injected subcutaneously at three sites (dosal neck and both axillae) with either 80 mg/Kg of capsaicin (5 mg/ml) or just the vehicle solution. The blepharospasm response of the adult to a capsaicin (0.1 mg/ml) eye drop was found to be abolished by neonatal s.c. capsaicin treatment.

Drug Solution. Stock solutions of SP and its antagonist, spantide, (1 nmo/μl and 1 μg/ml respectively) were prepared in sterile 0.01 M acetic acid. Capsaicin was dissolved in a 1:1:8 ratio of Tween-80, ethanol and saline and the stock solutions were 200 μg/ml for i.v. and 5 mg/ml for neonatal s.c. All drugs were stored at -50°C. During each experiment they were kept on ice and before i.c. or i.v. injections were diluted in saline to the required dose.

Data analysis. All variables were recorded on an oscilloscopic recorder. Chemoreceptor activity during air ventilation was counted over 5 or 10 s intervals and the peak response to SP or capsaicin was compared with the average count over the proceding 1 min control. Hyperoxic ar hypoxic challenge was achieved by connecting the inspiratory port of the ventilator to 100%, 12% or 6% O_2. Arterial blood samples were taken during the 4th min of each gas challenge and chemoreceptor activity is reported for the 5th min. Statistical significance was evaluated by ANOVA coupled with a Tukey's test (dose responses to SP and capsaicin), paired t-tests (effect of SP-antagonist) and unpaired t-tests (effect of neonatal capsaicin treatment).

RESULTS

I. Effects of intracarotid administration of SP on carotid chemoreceptor activity. Control saline injecons had no discernable effects on chemoafferent activity, whereas, injection of various doses of SP consistently and significantly augmented the carotid body sensory discharge. Average results with four different doses of SP are summarized in Table 1. There was a dose-dependent excitation between 1 and 3 nmoles of SP. The magnitude of chemoreceptor stimulation by 10 and 30 nmoles of SP was the same as with 3 nmole, suggesting the response reached a plateau. However, as the doses of SP increased, chemoreceptor excitation commenced more rapidly. For instance, latency for chemoreceptor excitation with 1 nmole of SP averaged 10 ± 1.8 s, whereas with 30 nmoles it was 4.7 ± 1.5 s ($p<0.05$;n=7).

Arterial blood pressure (ABP) changes were variable in response to intracarotid injections of SP. Low doses of SP, i.e. 1 and 3 nmoles, decreased arterial blood pressure by 15 ± 4 mmHg. On the other hand, higher doses of SP, i.e. 10 and 30 nmoles, increased blood pressure by 28 ± 5 mmHg.

II. Effect of systemic administration of capsaicin on carotid body activity. Capsaicin has been shown to release endogenous SP from neuronal tissues[10,11]. Whether systemic administration of capsaicin stimulates the carotid body activity by releasing SP was examined in 7 rats. Intravenous injections (0.1 ml) of capsaicin stimulated the chemoreceptor activity in a dose-dependent manner. In response to 0.5, 1.0, 2.5, and 5.0 μg/kg of capsaicin, chemoreceptor acvity increased by $\Delta 33 \pm 11, \Delta 100 \pm 25, \Delta 166 \pm 44$ and $\Delta 252 \pm 34$ respectively (p < 0.01) from a baseline of 58 ± 16 impulses/s.

Table 1. Effect of intracarotid injections of SP on carotid body activity.

SP Dose (nmole)	% Change Carotid Body Activity (100% = control)	P Values
Saline	106 ± 3	N.S.
1	126 ± 5	<0.01
3	155 ±15	<0.01
10	170 ± 8	<0.01
30	150 ± 11	<0.01

Anaesthetized rats (n = 7) were ventilated with room air: PaO_2 74 ± 2 mmHg, $PaCO_2$ 24 ± 1 mmH, pH 7.35 ± 0.01 and ABP =78 ± 7 mmHg. Volume of injections was 0.2 ml given over 10 s. Data given as mean ± S.E.M. N.S. = not statistically significant.

III. Effect of SP-antagonist on carotid body response to capsaicin. The experiments described above demonstrated that capsaicin stimulates the carotid body activity. Whether capsaian-induced chemoreceptor stimulation is due to release of endogenous SP was examined using spantide, a tachykinin receptor antagonist. Spantide was infused close to the carotid body at a rate of 25-30 μg/Kg/min in 1 ml for 5 minutes. This dose of spantide not only reduced the carotid body response to 10 nmoles of SP from $\Delta 181 \pm 15$ to $\Delta 10 \pm 4$ impulses/s (p < 0.05), but also significantly attenuated the capsaicin evoked chernoreceptor stimulation (2.5 μg/kg from $\Delta 166 \pm 44$ to $\Delta 6 \pm 3$, and 5 μg/kg from $\Delta 252 \pm 34$ to $\Delta 12 \pm 5$ impulses/s; p < 0.01, n = 7).

IV. Effect of SP-antagonist on carotid chemoreceptor response to hypoxia. Chemoreceptor responses to three levels of arterial PO_2 ($216 \pm 12, 73 \pm 3$, and 45 ± 2 mmHg) while maintaining the $PaCO_2$ at 24 ± 3 mmHg were examinated before and after intracarotid infusion of SP-antagonist (n = 6 rats). The dose of the antagonist was 40 μg/kg/min and was infused close to the carotid body over a period of 10 minutes. Before the antagonist, there were large increases in chemoreceptor activity as PaO_2 decreased (Table 2). SP-antagonist had no effect on the carotid body activity recorded under hyperoxic ventilation. As can be seen from Table 2, SP-antagonist significantly attenuated the carotid body response to decreases in arterial PO_2.

In order to determine whether attenuation of hypoxic rcsponse by SP-antagonist was due to non-specific blockade of carotid body activity, the following experiment was performed in the same animals as above. Before administration of SP-antagonist, intraveous administration of $NaHCO_3$ (1 ml of 4.2% strength) stimulated the carotid body activity, probably by releasing CO_2 from the arterial blood. This augmentation of carotid body activity by $NaHCO_3$ was not affected by SP-antagonist.

Table 2. Effect of SP-antagonist on carotid chemoreceptor responses to hypoxia.

PaO2 mmHg	Before SP-Antagonist	After SP-Antagonist	P Values
216 ± 12	100%	100%	-
73 ± 3	160 ± 7	107 ± 6	<0.01
45 ± 2	332 ± 15	133 ± 14	<0.01

Data expressed as % of hyperoxic response (mean $\pm$ S.E.M. from 6 rats).

V. Effect of neonatal capsaicin treatment on carotid chemoreceptor responses to hypoxia. It has been reported that neonatal capsaicin treatment results in a blunted ventilatory response to hypoxia[12,13]. We examined whether neonatal capsaicin (80 mg/Kg s.c.) affects the carotid body response to hypoxia. The results showed that the carotid body nerve activity of capsaicin-treated rats (n = 7) had a significantly attenuated response compared with 7 control rats to an arterial PO_2 of 30 ± 2 mmHg caused by 6% O_2 (254 ± 47 cf 447 ± 34 impulses/s; p < 0.01). In contrast the activity in air (32 ± 6 cf 33 ± 7 impulses/s) and in 100% O_2 (17 ± 5 cf 14 ± 2 impulses/s) was unaffected by neonatal capsaicin treatment.

DISCUSSION

Close carotid body administration of SP stimulated the chemoreceptor activity in the rats. These findings are consistent with the previous studies on in vivo cat carotid bodies[4,5,6]. Under the present experimental conditions, the doses required to stimulate the rat cartid bodies ranged between 1 and 30 nanomoles. The reported doses for cat carotid bodies were between 10 and 100 nanomoles[4]. Thus, rat chemoreceptor tissue appears to be relatively more sensitive to SP than cats. In the present study, intracarotid administration of SP in low doses decreased, whereas higher doses increased arterial blood pressure. Despite the variability in blood pressure responses, all doses of SP consistently stimulated the carotid body. Therefore, we believe that chemoreceptor excitation by SP is due to direct action of the peptide on the glomus tissue, rather than secondary to blood pressure changes.

Several investigators have shown that acute capsaicin administration releases endogenous SP from neuronal tissues[10,11]. In the present study, systemic administration of capsaicin stimulated the carotid body in a dose-dependent manner. The fact that SP-antagonist abolished the capsaicin-induced carotid body stimulation suggests that the stimulating effects are mediated by endogenous SP. The stimulatory effects of exogenous SP and capsaicin, when taken together, make it clear that SP is excitatory to the carotid body.

Intracarotid administration of SP-antagonists attenuated the chemoreceptor response to hypoxia, but not the excitation caused by $NaHCO_3$, suggesting that the actions of the antagonist were not due to non-specific blockade of nerve conduction. Moreoever, lower doses of antagonist produced only a partial blockade, whereas higher doses completely abolished the hypoxic response. The dose-dependent effects of the antagonist can be attributed to its action on SP-receptors in the glomus tissue. Thus, the present observations in rats and previous studies on cat[4,7] support the notion that SP and/or a related peptide is associated with transmission of the hypoxic stimulus at the carotid body.

Previous studies have demonstrated a blunted hypoxic ventilatory drive in adult rats treated with capsaicin neonatally[12,13]. In the present study, we found that the carotid body responses to hypoxia were markedly attenuated in capsaicin-treated rats, compared to controls. Therefore, the blunted hypoxic response could conceivably be due to impaired hypoxic chemotransmission at the carotid body. Neonatal administration of capsaicin results in degeneration of unmyelinated sensory fibers[14] and depleted SP in the sensory ganglion[15]. Attenuation of hypoxic response in the capsaicin-treated rats could be due to either degeneration of unmyelinated carotid body afferents and/or depletion of SP from the carotid body.

ACKNOWLEDGEMENTS

This work was supported by National Institutes of Health grants: HL-38986 and Research Career Development Award to Nanduri R. Prabhakar: HL-02599.

REFERENCES

B. Pernow, Substance P, Pharmacol Rev. 35:85 (1983).

A.C. Cuello and D.S. McQueen, Substance P: A cat carotid body peptide, Neurosci. Lett. 17:215 (1980).

I.V. Chen, R.D. Yates, and J.T. Hansen, Substance P-like immunoreactivity in rat and cat carotid bodies: Light and electron microscopic studies, Histol. Histopathol. 1:203 (1986).

N.R. Prabhakar, S.C. Landis, G.K. Kumar, D. Mullikan-Kilpatrick, N.S. Cherniack, and S.E. Leman, Substnce P- and neurokinin A in the cat carotid body: Localization, exogenous effects, and changes in content in response to arterial PO_2. Brain Res. 481:205 (1989).

D.S. McQueen, Effects of substance P on carotid chemoreceptor activity in cats, J. Physiol. (London) 302:31(1980).

L. Monti-Bloch and C. Eyzaguirre, Effects of methionine-enkephalin and substance P on the chemosensory discharge of the cat carotid body, Brain Res. 338:297 (1985).

N.R. Prabhakar, J. Mitra, an N.S. Cherniack, Role of substance P in hypercapnic excitation of carotid chemoreceptor, J. Appl. Physiol. 63: 2418 (1987).

G. Hansen, L. Jones, and S.J. Fidone, Physiological chemoreceptor stimulation decreases enkephalin and substance P in the carotid body. Peptides 7:767 (1986).

N.R. Prabhakar, M. Runold, N.S. Cherniack, and G.K Kumar, Analysis of chemoreceptor responses to tachykinins in rats, cats, and rabbits, in: "Chemoreceptors and Chemoreceptor Reflexes," H. Acker, A. Trzebski, and R. O'Regan, eds., Plenum Press, New York (1990).

A. Bucsics and F. Lembeck, In vitro release of substance P from spinal cord by capsaicin congeners, Euro. J. Pharmacol. 71:71(1981).

R. Gamse, A. Molnar, and F. Lembeck, Substance P release from spinal cord slices by capsaicin, Life Sci, 25:629 (1979).

S.M. Bond, F. Cervero, and D.S. McQueen, Influence of neonatally administered capsaicin on baroreceptor and chemoreceptor reflexes in adult rat, Br. J. Pharmacol. 77:517 (1982).

G.T. De Sanctis, F.H.Y. Green, and J.E. Remmers, Ventilatory responses to hypoxia and hypercapnia in awake rats pretreated with capsaicin, J. Appl. Physiol. 70:1168 (1991).

S.H. Buck and T.F. Burkes, The neuropharmacology of capsaicin: Review of some recent observations, Pharmacol. Rev. 38:179 (1986).

R. Gamse, S.E. Leeman, P. Holzer, and F. Lembeck, Differential effects of capsaicin on the content of somatostatin, substance P, and neurotension in the nervous system of the rat, Naunyn-Schmiedeberg's Arch. Pharmacol. 317:140 (1981).

38 CAROTID SINUS NERVE INHIBITION MEDIATED BY ATRIAL NATRIURETIC PEPTIDE

S.J. Fidone, Z.-Z. Wang, J. Chen, L. He, W.-J. Wang, B. Dinger and L.J. Stensaas

Department of Physiology, University of Utah School of Medicine, Salt Lake City, Utah, USA

INTRODUCTION

DeBold first described the correlative relationship between the incidence of distinct storage granules in atrial cardiocytes and body water/electrolyte balance[5]. In later studies, DeBold and colleagues[6] showed that the injection of atrial granular extracts initiated a rapid and extensive diuresis and natriuresis, and they identified the active substituent as the 28 amino acid peptide, atrial natriuretic peptide (ANP), which is released from the atria by distension or local hypoxia[see2,8,10]. Receptors for ANP have been localized in the renal glomerular vasculature, suggesting that the circulating peptide directly modulates kidney function[3]. The unique physiologic role of ANP as a regulator of volume and electrolyte levels in the periphery led others to postulate parallel functions for this peptide in the central nervous system (CNS). In fact, later studies revealed the presence of ANP[15,25] and ANP receptors[19,20] in specific CNS sites associated with cardiovascular control. In addition, central administration of ANP was shown to decrease the release of arginine vasopressin (AVP) in dehydrated animals[21,22] and to decrease salt-water intake following salt deprivation. Thus, ANP appears to act globally at specific targets both in the CNS and periphery to initiate coordinated adjustments in response to changes in systemic hydration and electrolyte balance.

Reflexes involved in the regulation of respiration and of blood pressure are initiated by chemoreceptors of the carotid body and petrosal ganglia. Although classical views do not include a direct role for this arterial chemoreceptor organ in water and solute balance, an early study by Eyzaguirre and his colleagues[9] demonstrated that carotid sinus nerve (CSN) fibers sensitive to hypoxia were also excited by increased osmolarity. A recent review by Honig[14] has also documented a relationship between arterial chemoreceptor activity and solute metabolism[14]. These data suggest that reflexes initiated by hypoxic stimulation of the carotid body produce an increase in the renal excretion of sodium. Interestingly, this response appears to be mediated by a hormonal mechanism, although the possible involvement of ANP in these effects is uncertain[14].

In the studies summarized here, we have utilized a multidisciplinary approach to investigate the possible role of ANP in chemoreceptor processes in the carotid body. Immunocytochemical experiments have localized ANP-like immunoreactivity in the chemosensory tissue. We have further shown that submicromolar concentrations of the synthetic ANP analog, atriopeptin III (APIII), profoundly modulate chemoreceptor activity evoked by low O2 stimuli. Because in other tissues the transmembrane receptor for ANP incorporates guanylate cyclase activity[4], we have also attempted to elucidate the role of cGMP in this chemoreceptor modulation. Our findings demonstrate that APIII activates guanylate cyclase in the carotid body, which in turn initiates a rapid synthesis of cGMP. Immunocytochemical studies have further shown that the cGMP response occurs in type I (glomus) cells in a dose-response relationship. Although the physiological significance of ANP in the carotid body remains obscure, collectively our data suggest a possible role for this peptide in the modulation of chemoreceptor activity via the "efferent" inhibitory pathway. Results of these studies have been published elsewhere[29,31].

EXPERIMENTAL PROCEDURES

The experiments were conducted on carotid bodies surgically removed from adult cats anesthetized with sodium pentobarbital (40 mg/kg). Detailed descriptions of the methods have been published elsewhere, including those for immunocytochemical localization of ANP[29] and cGMP[31], radioimmunoassay of cGMP[28], and the techniques for recording carotid sinus nerve (CSN) activity from the in vitro superfused carotid body[29].

RESULTS

ANP-like immunoreactivity was widespread in type I parenchymal cells but was absent in type II parenchymal cells of lobules within the cat carotid body (Fig. 1). Although quantification of the incidence of stained cells was not attempted, multiple sections from 5 carotid bodies indicated that at least 80% of type I cells contained ANP immunoreactivity. Reaction product for ANP was uniform amongst the type I cells, and occured throughout the cytoplasm in accord with the known distribution of large, dense-core cytoplasmic vesicles[27]. Staining was absent from sections in which normal preimmune antiserum was substituted for the primary antibody, or when the ANP antibody was preabsorbed with APIII.

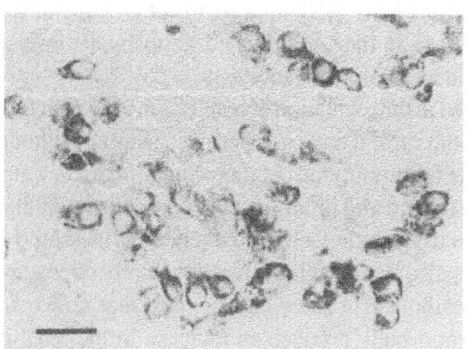

FIG. 1. ANP-like immunoreactivity in cat carotid body. Reaction product is restricted to lobules of type I (glomus) cells. Staining is absent in interlobular connective tissue and vascular elements. Scale bar, 20 μm.

In each of 8 superfused preparations of the carotid body with its attached CSN, the effects of APIII on basal and stimulus-evoked CSN activity were evaluated in a series of 3 stimulus trials consisting of 1), a control response to hypoxia in the absence of APIII; 2), the basal and hypoxia-evoked CSN activity recorded in the prosence of a selected concentration of APIII; and finally, 3), evoked activity recorded after removal of the drug. The quantitative effect of APIII in each case was determined by comparing the average of the first and third responses with the second response. In 100% O2-equilibrated superfusion medium, the mean CSN discharge rate was 632+238 impulses/10 sec (X ± SEM). The discharge rate during a 5 min low O2-stimulus period (20% O2 medium) was elevated by an average of 1660 + 222 imp/10 sec. The chemosensory discharge rapidly subsided when 100% O2 medium was reintroduced to the preparation. Administration of APIII at concentrations up to 100 nM did not alter basal CSN activity (100% O2 medium), but the drug significantly inhibited the response to hypoxia (20% O2 medium). The response typically recovered completely following superfusion for 30-40 min in the absence of the drug. Table I summarizes the dose-response relationship for APIII inhibition of CSN chemosensory activity evoked by media equilibrated with 20% O2. APIII was an effective inhibitor of the hypoxic response at 1, 10 and 100 nM, and at the lattermost concentration up to 93% inhibition was observed.

TABLE I   CSN Inhibition Produced by Atriopeptin III (APIII)

APIII (nM)	1	10	100
% CSN	26.74 (3)	69.27 (8)	86.05 (6)
Inhibition	± 2.49	± 1.27	± 1.87

Values are % ± SEM. Numbers in parentheses = n

Virtually no cGMP immunoreactivity occurred in tissue incubated in control medium (Fig. 2A), a result consistent with the low levels of cGMP detected by RIA (cGMP = 75.3 ± 17.4 pmole/g tissue). Changes in the levels of cGMP reaction product following _in vitro_ incubation with 0.01 µM, 0.1 µM and 1.0 µM APIII for 10 min are shown in Figures 2B, 2C and 2D. A significant, dose-dependent increase in the number of immunopositive type I cells was observed. The intensity of the reaction product ranged from lightly stained elements to those which were very dark, and was observed with all concentrations of APIII. Type II cells lacked immunoreactivity following exposure to APIII as did nerve, vascular and connective tissue elements. The RIA measurements of cGMP accorded well with the immunocytochemical observations, revealing a significant increase in second messenger content following a 10 min exposure to 0.1 µM APIII (cGMP was elevated to 774.3 ± 193.1 pmole/g tissue).

DISCUSSION

Our findings suggest that ANP is present in type I (glomus) cells of the cat carotid body where it may act to modulate CSN activity evoked by natural stimuli. The results presented here do not, however, establish type I cells as a site of synthesis for ANP, and

273

it is not known whether ANP is released by these cells under physiological conditions. We have addressed these further issues in preliminary experiments where hypoxic exposure both in vivo and in vitro was found to dramatically reduce the immunocytochemical staining for ANP in type I cells. Furthermore, ANP staining in vitro recovered within 2hr following a return to normoxia, suggesting that type I cells are capable of de novo synthesis of ANP in the absence of circulating peptide[26]. In separate experiments, we have used RIA techniques to show that calcitonin gene-related peptide (CGRP)

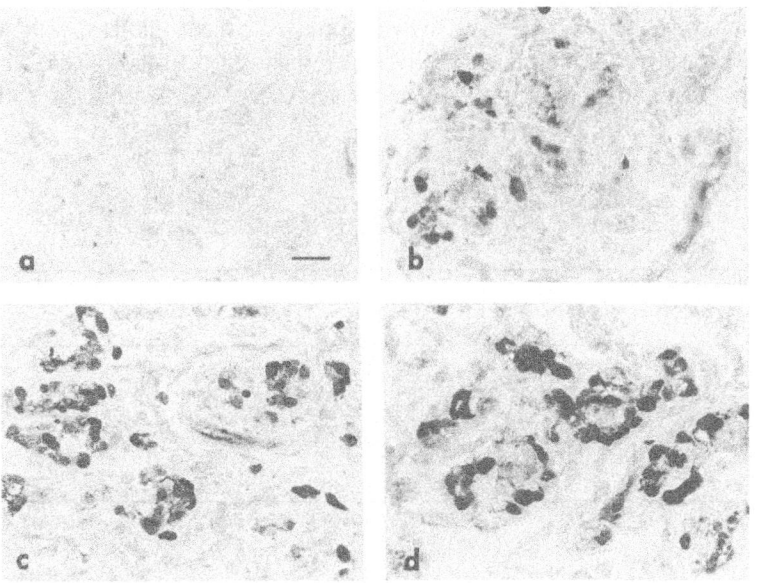

Fig. 2. Immunocytochemical localization of cyclic GMP in the in vitro cat carotid body and its response to increasing concentrations of APIII. (A) In control carotid bodies, cGMP reaction product was undetectable. (B) 0.01 μM APIII produced low levels of staining in some type I cells. (C and D) 0.1 μM and 1.0 μM APIII resulted in moderate to high levels of cGMP in a majority of the type I cells. Scale bar, 20 μm.

also evokes the release of ANP from cat carotid bodies superfused in vitro. These preliminary findings suggest that ANP metabolism in the carotid body is regulated in a manner similar to that for ANP storage in the atria, where hypoxia [2] and CGRP [32] have been shown to evoke the release of this natriuretic peptide.

ANP has also been reported to occur in adrenalin (A) and noradrenalin (NA) cell types of the rat adrenal medulla, another peripheral catecholaminergic organ. However, mRNA for ANP was found in only 15% of the cells, all of which were presumed to be of the NA-type[18]. Autoradiographic studies have demonstrated that A-cells preferentially bind and internalize circulating ^{125}I-ANP. Ultrastructural immunocytochemical data indicate that ANP in NA cells is stored in secretory vesicles, while in contrast A cells contain the peptide in the cytoplasmic matrix and plasma membrane[13]. Although our preliminary findings suggest that most type I cells are capable of synthesizing and releasing ANP[26], the morphological and biochemical similarities between catecholamine containing cells of the carotid body (type I cells) and adrenal medulla (A and NA cells) suggest that a complex pattern of ANP metabolism may be characteristic of both structures. Whatever the dynamics of ANP in the carotid body, its localization to type I cells portends an important role for this peptide in carotid body physiology.

Our observation that submicromolar concontrations of ANP inhibit hypoxia-induced CSN discharge suggests that this natriuretic peptide may modulate chemoreceptor function under physiological conditions. Although the mechanism of this ANP-mediated inhibition is not known, the dose dependency of this effect points to the involvement of specific ANP receptors[see 13,23,24] which our preliminary findings suggest are localized to type I cells in the carotid body[30]. It is significant in this regard that the ANP receptor has been shown to consist of a single transmembrane protein subunit which incorporates guanylate cyclase[4]. Consequently, our finding that nanomolar concentrations of ANP inhibit chemoreceptor activity suggests the possible involvement of cyclic GMP (cGMP) in this response. Interestingly, we have found that the cGMP analogs, 8-bromoc-GMP and dibutyryl-cGMP, produce a similar depression of the hypoxic response in vitro[11]. Moreover, previous studies in our laboratory have shown that hypoxia decreases the cGMP content of the carotid body[28], which is likewise consistent with an inhibitory role for this putative second messenger in chemoreception. A role for cGMP in ANP-induced CSN inhibition is also supported by our findings that 1), ANP stimulates cGMP production by the type I cells in a dose-dependent fashion and in the same dose-range as the inhibitory chemoreceptor effect[11,31], and 2), ANP immunoreactivity co-exists in the same type I cells which produce cGMP in response to APIII stimulation[31]. This latter finding suggests that ANP released from type I cells may act on autoreceptors to induce the inhibitory effects.

The possible neurotransmitter mechanisms which mediate ANP induced inhibition have not yet been clarified. In preliminary experiments, however, we have attempted to assess the involvement of the two transmitter agents most frequently proposed for inhibitory roles in chemosensation, namely, dopamine (DA) and met-enkephalin (ME). Studies of catecholamine release have shown that APIII does not alter the amount of DA and NE released from carotid bodies superfused in either normoxic or hypoxic conditions[11]. In contrast, CSN recordings from the in vitro carotid body demonstrate that the opiate receptor antagonist, naloxone, as well as the highly specific delta receptor antagonist, naltrindole, reverse the CSN inhibition produced by APIII or cGMP analogs, suggesting that ME is released from type I cells in response to these agents[12].

Collectively, the available data suggest that endogenous and/or circulating ANP modulates stimulus-evoked CSN activity. The actions of ANP in the carotid body appear to be mediated by autoreceptors on type I cells which are coupled to guanylate cyclase. These findings are consistent with the notion that ANP, via cGMP generated in type I cells, evokes the release of an inhibitory neurotransmitter, perhaps ME. Finally, because hypoxia and CGRP evoke the release of ANP from type I cells, our data support the possibility that CGRP containing unmyelinated nerve fibers, previously described within and adjacent to carotid body glomeruli[16,17], may participate in the ANP mediated inhibition.

REFERENCES

1. *Antunes-Rodrigues, J., McCann, S.M. and Samson, U.K.* Central administration of atrial natriuretic factor inhibits saline preference in the rat. Endocrinology 118: 1726-1728, 1986.

2. *Baertschi, A.J., Hausmaninger, C., Walsh, R.S., Mentzer, R.M., Wyatt, D.A. and Pence, R.A.* , Hypoxia-induced release of atrial natriuretic factor (ANF) from the isolated rat and rabbit heart. Biochem. Biophys. Res. Commun. 140: 427-433, 1986.

3. *Bianchi, C., Gutkowska, J., Thibault, G., Garcia, R., Genest, J. and Cantin, M.* Distinct localization of atrial natriuretic factor and angiotensin II binding sites in the glomerulus. Am. J. Physiol 251 *(Renal Fluld Electrolyte Physlol. 20)*: F594-F602, 1986.

4. *Chinkers, M., Garbers, D.L., Chang, M.-S., Lowe, D.G., Chin, H., Goeddel, D.V. and Schulz, S.* A membrane form of guanylate cyclase is an arterial natriuretic peptide receptor. Nature Lond. 338: 7883, 1989.

5. *DeBold, A.J.* Heart atria granularity. Effects of changes in water and electrolyte balance. Proc. Soc. Exu. Biol. Med. 161: S08-511, 1979.

6. *DeBold, A.J., Borenstein, H.B., Veress, A.J. and Sonnenberg, H.* A rapid and potent natriuretic response to intravenous injection of atrial myocardial extract in rats. Life Sci. 28: 89-94, 1981.

7. *De Vente, J., Steinbusch, H.W.M. and Schipper, J.* A new approach to immunocytochemistry of 3',5'-cyclic guanosine monophosphate: preparation, specificity and initial application of a new antiserum against formaldehyde-fixed 3',3'-cyclic guanosine monophosphate. Neuroscience 22: 361,373, 1987.

8. *Flynn, T.G. and Davies, P.L.* The biochemistry and molecular biology of atrial natriuretic factor. Biochem. J. 232: 313-321, 1985.

9. *Gallego, R., Eyzaguirre, C. and Monti-Bloch, L.* Thermal and osmotic responses of arterial receptors. J. Neurophysiol. 42: 665-680, 1979.

10. *Genest, J., Cantin, M., Anand-Srivastava, M.B., Cusson, J.R., De Lean, A., Garcia, R., Gutkowska, J., Hamet, P., Kuchel, O., Larochelle, P., Nemer, M., Schiffrin, E.L., Schiller, W., Thibault, G. and Tremblay, J.* The atrial natriuretic factor: its physiology and biochemistry. Rev. Physiol. Biochem. Pharmacol. 110: 1-145, 1988.

11. *He, L., Chen, J., Wang, W.-J., Dinger, B. and Fidone, S.* Cyclic GMP release and carotid chemoreceptor inhibition produced by atrial natriuretic peptide. Soc. Neurosci. Abstr. 16: 877, 1990.

12. *He, L., Chen, J., Dinger, B. and Fidone, S.* Mechanism of action of atrial natriuretic peptide (ANP) in rabbit carotid body. Soc. Neurosci. Abstr. 17: 117, 1991.

13. *Heisler, S. and Morrier, E.* Bovine adrenal medullary cells contain functional atrial natriuretic peptide receptors. Biochem. Biophys. Res. Commun. 150: 781-787, 1988.

14. *Honig, A.* Peripheral arterial chemoreceptors and reflex control of sodium and water homeostasis. Am. J. Physiol. 257 (Regulatory Integrative Comp. Physiol. 26): R1282-R1302, 1989.

15. *Kawata, M., Nakao, K., Morii, N., Kiso, Y., Yamashita, H., Imura, H. and Sano, Y.* Atrial natriuretic polypeptide: topographical distribution in the rat brain by radioimmunoassay and immunocytochemistry. Neuroscience 16: 521-546, 1985.

16. *Kondo, H. and Yamamoto, M.* Occurrence, ontogeny, ultrastructure and some plasticity of CGRP (calcitonin gene-related peptide)immunoreactive nerves in the carotid body of rats. Brain Res. 473: 283-293, 1988.

17. *Kummer, W. and Fischer, A.* Tachykininergic axons in the guinea pig carotid body: origin, ultrastructure and co-existence with other peptides. In: C. Eyzaguirre, S.J. Fidone, R.S. Fitzgerald, S. Lahiri and D.M. McDonald (eds.) Arterial Chemoreception. SpringerVerlag, New York, 1990, pp. 229-234.

18. *Morel, G., Chabot, J.-G., Garcia-Caballero, T., Gossard, F., Dihl, F., Belles-Isles, M. and Heisler, S.* Synthesis, internalization and localization of atrial natriuretic peptide in rat adrenal medulla. Endocrinology 123: 149-158, 1988.

19. *Quirion, R., Dalpe, M. and Dam, T.-V.* Characterization and distribution of receptors for atrial natriuretic peptides in mammalian brain. Proc. Natl. Acad. Sci. USA 83: 174-178, 1986.

20. *Saavedra, J.M., Correa, lF.M., Plunkett, L.M., Israel, A., Kurihara, M. and Shigematsu, K.* Binding of angiotensin and strlal natriuretic peptide in brain of hypertensive rats. Nature Lond. 320: 758-760, 1986 .

21. *Samson, W.K.* Atrial natriuretic factor inhibits dehydration and hemorrhage - induced vasopressin release . Neuroendocrinology 40: 27 7 279, 1985.

22. *Samson, W.K., Aguila, M.C., Martinovic, J., Antunes-Rodrigues, J. and Norris, M.* Hypothalamic action atrial natriuretic factor to inhibit vasopressin secretion. Peptides 8 : 449-454, 1987 .

23. *Schenk, D.B., Johnson, L.K., Schwartz, K., Sista, H., Scarborough, R.H. and Lewicki, J.A.* Distinct atrial natriuretic receptor sites on cultured bovine aortic smooth muscle and endothelial cells. Biochem. Biophys. Res. Commun. 127: 433-442, 1985.

24. Shionoiri, H., Hirawa, N., Takasaki, I., Ishikawa, Y., Minamisawa, K., Miyajima, E., Kinoshita, Y., Shimoyama, K., Shimonaka, M., Ishido, M. and Hirose, S. Presence of functional receptors for atrial natriuretic peptide in human pheochromocytoma. Biochem. Biophys. Res. Commun. 148: 286 - 291, 1987.

25. Skofitsch, G., Jacobowitz, D.M., Eskay, R.L. and Zamir, N. Distribution of atrial natriuretic factor-like neurons in the rat brain. Neuroscience 16 : 917-948, 1985.

26. Stensaas, L.J., Wang, Z.-Z., Dinger, B. and Fidone, S. Alteration of atrial natriuretic peptide (ANP) immunostaining in the rat carotid body evoked by hypoxia. Soc. Neurosci. Abstr. 17 : 118, 1991 .

27. Verna, A. Ultrastructure of the carotid body in the mammals. Int. Rev. Cytol. 60: 271-330, 1979.

28. Wang, W.-J., Cheng, G.-F., Dinger, B.G. and Fidone, S.J. Effects of hypoxia on cyclic nucleotide formation in rabbit carotid body in vitro. Neurosci. Lett. 105 : 164-168, 1989 .

29. Wang, Z.-Z., He, L., Stensaas, L.J., Dinger, B. and Fidone, S.J. Localization and action of atrial natriuretic peptide in the carotid body of the cat. J. Appl. Physiol. 70: 942-946, 1990.

30. Wang, Z.-Z., Hirose, S., Dinger, B., Fidone, S. and Stensaas, L.J. Localization of atrial natriuretic peptide receptors in the carotid body. Soc. Neurosci. Abstr. 17 : 118, 1991.

31. Wang, Z.-Z., Stensaas, L.J., Wang, W.-J., Dinger, B., de Vente, J. and Fidone, S.J. Atrial natriuretic peptide increases cyclic GMP immunoreactivity in the carotid body. Neurosci. 49: 479-486, 1992.

32 . Yamamoto, A., Kimura, S ., Hasui, K. , Fuj isawa, Y. , Tamaki, T. , Fukui, K., Iwao, H. and Abe, Y. Calcitonin gene-related peptide (CGRP) stimulates the release of atrial natriuretic peptide (ANP) from isolated rat atria. Biochem. Biophys. Res. Comm. 155: 1452-1458, 1988.

39 NEUROTRANSMITTERS AND SECOND MESSENGER SYSTEMS IN THE CAROTID BODY

M.T. Pérez-García, A. Gómez-Niño, L. Almaraz and C. González

Depto. Bioquímica y Biología Molecular y Fisiologia. Facultad de Medicina. Universidad de Valladolid. 47005 Valladolid. Spain

INTRODUCTION

It is well known that multiple cellular events are regulated by second messenger systems (Greengard and Costa, 1970; Sekar *et al.*, 1986; Hockberger *et al.*, 1987). Among second messengers, cAMP have been shown to modulate several ionic channels and neurotransmitter synthesis and release in different structures (Joh *et al.*, 1978; Hockberger *et al.*, 1987; Morita *et al.*, 1987; Santiago *et al.*, 1990). It has been reported that several stimuli of the carotid body (CB) like hypoxia, hypercapnic acidosis, high extracellular K^+ and cyanide, stimulate adenylate cyclase activity, and that hypoxia was the only one that induces cAMP increases in a calcium-free medium, this accumulation being proportional to the intensity of the hypoxic stimulation (Pérez-García *et al.*, 1990). Therefore it was proposed a primary role for cAMP in CB chemotransduction (Pérez-García *et al.*, 1990, 1991). It was further suggested that the specific activation of adenylate cyclase by hypoxia could reflect a direct effect of low pO_2 on the cAMP biosynthetic machinery.

cAMP would affect chemoreception by modulating the release of neurotransmitters from type I cells (Pérez-García *et al.*, 1991; Wang *et al.*, 1991) and probably also by a direct effect on the O_2 transducing mechanism (Pérez-García *et al* ., 1990) . Treatment of the CBs with different substances that increase cAMP levels in the organ such as forskolin (an adenylate cyclase activator), dBcAMP (a permeant analogous of cAMP) and isobutylmethyl-xantine (IBMX; an inhibitor of cAMP degrading phosphodiesterase) induced an important potentiation of the low pO_2-evoked catecholamines (CA) release and increased the frequency of action potentials recorded in the carotid sinus nerve (Pérez-García *et al.*, 1990, 1991; Wang *et al.*, 1991). These findings show that in the CB, cAMP is a possible mediator of the effects of substances that modify the chemosensitivity of the organ.

In other structures, the receptors for several neurotransmitters are linked to the adenylate cyclase system. Chemoreceptor cells possess a great variety of neurotransmitters (Fidone and González, 1986) but, for most of them, little is known about the mechanisms by which they exert their regulatory actions in the organ.

In the present work we have explored the existence of a possible link between the receptors for several neurotransmitters and the cAMP system in the CB. For some of them, we have also studied their effects on low pO_2-induced CA release. The possible relationships between these two actions of the neurotransmitters are discussed. In

Neurobiology and Cell Physiology of Chemoreception
Edited by P.G. Data *et al.*, Plenum Press, New York, 1993

addition we provide preliminary evidence to suggest that other second messenger system, the araquidonate derived prostaglandins (PGs), modulates the chemoreception process.

MATERIAL AND METHODS

Adult New Zealand white rabbits (1.8-2.5 Kg) of both sexes were used in these experiments. The animals were anesthetized with sodium pentobarbital (40 mg/Kg, i.v.) and the carotid bifurcations rapidly removed. The CBs were cleaned of surrounding conective tissue in a chamber filled with ice-cold 100% O_2-equilibrated modified Tyrode solution (in mM: NaCl 140; KCl 5; CaCl$_2$ 2; MgCl$_2$ 1.1; HEPES 5; glucose 5; pH 7.42).

For cAMP measurements the CBs were incubated in a metabolic shaker bath in 100% O_2-equilibrated Tyrode for 30 minutes at 37°C. At this time the solution was changed for a new one containing 0.5 mM IBMX (control CBs), or IBMX plus different receptor agonists (experimental CBs). After 10 minutes the organs were placed in ice-cold 5% trichloroacetic acid (0.3 ml) for 10 minutes and thereafter quickly weighted in an electrobalance (Sartorius Supermicro). In the experiments in which forskolin was used, the diterpene was present in the incubation solution for 30 minutes (control CBs) and bethanecol was added during the last 10 minutes of incubation (experimental CBs). cAMP levels in the CBs were determined using a radioimmunoassay kit (Amersham) and are expressed as picomole/mg of wet tissue.

For the release experiments, the CA stores of the CB were labelled by incubating the organs for 2 hours in the presence of 20µM ^{3}H-tyrosine ([3,5-^{3}H tyrosine], 30 Ci/mmol). Then, the CBs were transfered to vials containing 4 ml of precursor-free medium equilibrated with 100% O_2. The media were renewed every 30 minutes and discarded, and after 2 hours the changes were performed every 10 minutes and the solution was collected for ^{3}H-CA analysis (Pérez-García et al., 1991). The general protocol followed in the experiments is shown in figure 2 in the Results Section.

Data are expressed as means ± SEM . Statistical significance of the differences was assessed using a Student's t test for unpaired data.

RESULTS

Effects of several receptor agonists on cAMP levels. Figure 1A shows the effect of dopamine (10 µM), isoproterenol (50 µM), serotonin (10 µM), substance P (10 µM) and prostaglandin E$_2$ (PGE$_2$; 300 nM) on cAMP levels of CBs incubated in the presence of 0.5 mM IBMX. Control cAMP levels averaged 14.10 ± 1.08 pmol/mg tissue (mean ± SEM, n = 21). Maximal accumulation (about 4 times control levels) was obtained treating the CBs with low dosis (0.3 µM) of PGE$_2$. In the presence of isoproterenol, cAMP was 150% of that found in control CBs and the effect of this β agonist was completely prevented by 10µM propranolol (data not shown). The addition of dopamine, substance P or serotonin to the incubating solutions increased the CB cAMP levels to values between 200 and 300% of that found in control conditions. In figure 1B it is shown that the muscarinic agonist bethanecol, at a dosis of 100 µM, almost halved the cAMP levels reached in presence of 1 µM forskolin, a specific activator of adenylate cyclase. As previously reported (Pérez-García et al., 1990), this agent increased the cAMP content of the CB 10 times above basal levels. These data suggest that in the CB multiple receptors are linked, positively or negatively, to the cAMP system.

Effect of IBMX and isoproterenol on the ^{3}*H-CAs release induced by low* pO_2. Figure 2 shows the ^{3}H-CA release obtained in three different CBs exposed to two cycles of hypoxic stimulation. The empty bars represent the release obtained during incubation of the CBs in a control, 100% O$_2$ equilibrated solution, and the dashed ones that obtained during incubation in a solution equilibrated with 5% O$_2$. The cpm above

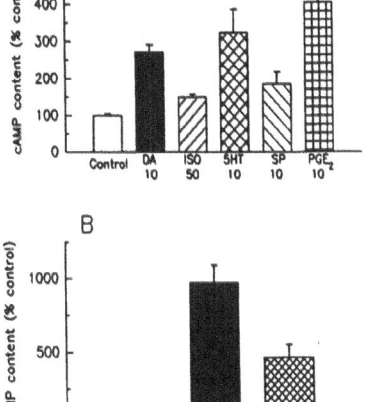

Fig. 1. Effect of different receptor agonists on the CB cAMP levels. A.This part of the figure shows cAMP levels in IBMX (0.5mM) treated CBs incubated in the presence of the indicated receptor agonists. Data are expressed as percentage of the levels found in control (IBMX treated) organs. DA (dopamine), Iso (isoproterenol), 5HT (serotonin), SP (substance P) and PGE$_2$ (prostaglandin E$_2$); concentrations are in mM. B. Effect of forskolin and forskolin plus bethanecol on the cAMP levels of CBs. Data are expressed as percentage of cAMP levels found in control, untreated CBs. Mean ±SEM of four or more values.

the horizontal lines correspond to the ^{3}H-CA evoked release for each stimulus (S$_1$ and S$_2$) in the three CBs. Left panel of the figure shows that the release of ^{3}H-CA obtained during the second stimulation (S$_2$) in control CB is lower than that obtained during the first presentation of the stimulus, a finding that probably reflects a transient emptying of the releasable labelled CA pool. In the middle pannel it is shown that in the presence of IBMX (0.5 mM) the ^{3}H-CA released in S$_2$ is even greater than that obtained during the first

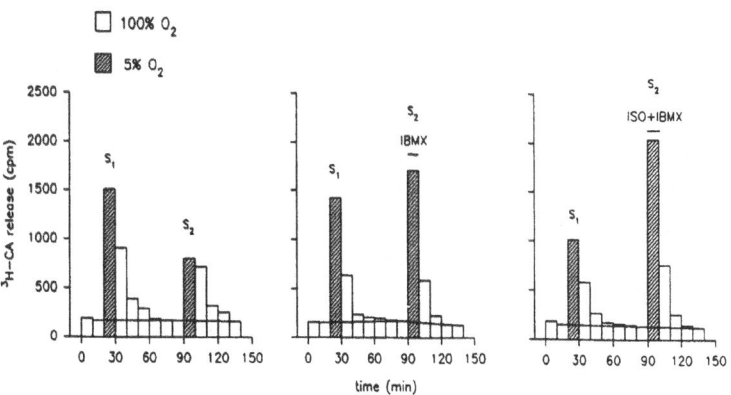

Fig. 2. *Effects of IBMX and isoproterenol on the release of* ^{3}*H-CAs evoked by low* pO_2 *from rabbit carotid body (CB). Each pannel represent the* ^{3}*H-CAs release from one CB subjected to two cycles of hypoxic stimulation (5% O$_2$-equilibrated medium; dashed bars). Left pannel, control CB. Middle pannel, the CB was incubated with 0.5 mM IBMX during the second stimulation. Right panel, the CB was incubated with 0.5 mM IBMX plus 50 μM isoproterenol during the second stimulation period. Horizontal lines through the colums separate the stimulus evoked release (cpm above the lines) for the first (S$_1$) and second (S$_2$) stimulus, from the basal release. See text for additional explanations.*

stimulation (S_1). When IBMX and isoproterenol were applied simultaneusly as in the third CB (right pannel), the evoked ^{3}H-CA release in S_2 doubled that obtained in S_1 in absence of drugs.

Figure 3 sumarizes the effects of isoproterenol on the release of ^{3}HCA induced by low pO_2. Part A of the figure shows the results obtained in four experiments as the one presented in figure 2. It may be seen that, as previously reported (Pérez-García et al., 1991), IBMX increased by a 70% the release response evoked by low pO_2 (5% O_2-equilibrated solution) . Additionally the Figure shows that isoproterenol, a selective β agonist , produced a further 50% increase in the release of ^{3}H-CA. Figure 3B represents the effect of different concentrations of isoproterenol on the release of ^{3}H-CAs evoked by low pO_2 (5%-O_2 equilibrated media) in the absence of IBMX. Low dosis of the agonist (2 and 5 μM) were inefective, but a significant potenciation of the hypoxic response was obtained with 10 and 50 μM isoproterenol. In both parts of the figure data are expressed as the ratio of the evoked release in the second stimulus to that obtained in the first stimulus (S_2/S_1) These results indicate that the activation of β receptors in the CB increases the sensitivity of type I cells to hypoxia, an effect that is accompanied by an elevation in the CB cAMP content (see Fig.1).

Effect of indomethacin on the ^{3}H-CA release induced by low pO_2. It is known that indomethacin, at low concentrations, is a specific inhibitor of the prostaglandin synthesis (Kulmacz et al., 1978), and that PG are potent modulators of the release of CA in adrenal medulla and sympathetic terminals (Stjärne, 1989). Indomethacin at a concentration of 2 μM induced a 250% increase in the release of ^{3}H-CA induced by exposure of the CB to a solution equilibrated with 10% O_2 without modifying the basal release (data not shown). This observation indicates that endogenous PGs are modulating the release process.

DISCUSSION

In the present work a link between the receptors for several neurotransmitters and the cAMP system is defined in the CB. Dopamine, isoproterenol, serotonine, substance P and PGE_2 increased cAMP levels in CBs treated with an inhibitor of the phosphodiesterase (IBMX). These results suggest that the receptors for these agents are possitively coupled to adenylate cyclase in the CB. Additionally, we report that the muscarinic agonist bethanecol, decreases the cAMP levels of CBs incubated in presence of forskolin, a specific activator of the catalytic subunit of adenylate cyclase. Since the type I cells possess muscarinic receptors (Dinger et al., 1991) at least a part of this cAMP reduction probably takes place in these cells. If this is the case, it is tempting to propose a link among the reduction in cAMP, the decrease in the evoked ^{3}H-CA release (Dinger et al., 1991) and the inhibition of carotid sinus nerve action potential frecuency (Monti-Bloch and Eyzaguirre, 1980) produced by muscarinic agonist in the rabbit CB. The nature of the intimate mechanisms involved in this proposed link is presently unknown.

Dopamine (DA) is a neurotransmitter whose metabolism is well characterized in the CB. DA is stored in chemoreceptor cells and released upon stimulation of the organ (Fidone and González, 1986). From a biochemical and pharmacological point of view, dopaminergic receptors have been classified in D-1 and D-2. While the D-2 receptors are negatively or not coupled to adenylate cyclase, D-1 receptors are positively coupled to this enzyme (Creese and Fraser, 1987). Electrophysiological and radioligand binding studies have established the presence of D-2 receptors in the CB (Llados and Zapata, 1978; Dinger et al., 1981; Mir et al., 1984; Fidone and González, 1986). Some of these receptors seem to be located on type I cells (Dinger et al., 1981; Mir et al., 1984) where they inhibit the release of CA (Gómez-Niño et al., 1989). Our data showing cAMP accumulation in rabbit CBs incubated in the presence of dopamine, reveals the presence of D-1 receptors in this animal species. Supporting this idea we have found that SKF38393, a specific D-1 agonist, reproduces the effect of dopamine on CB cAMP levels and that this effect may be prevented by a specific D-1 antagonist, SCH23390 (Almaraz et al., 1991). The vascular localization of this type of receptors in other peripheral

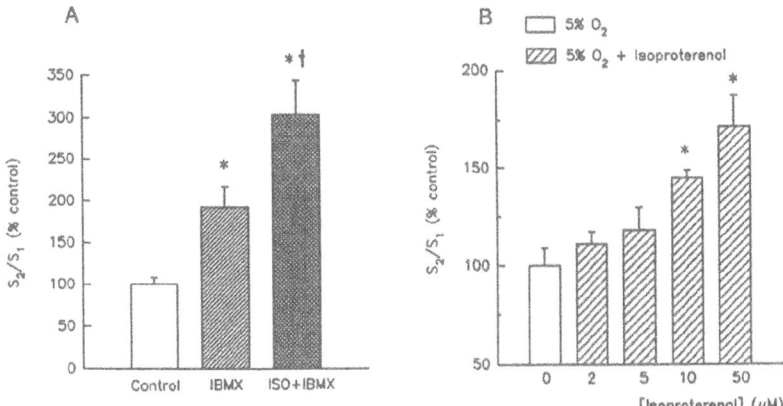

Fig. 3. *Effect of isoproterenol on the ^{3}H-CA release induced by low pO_2 from carotid bodies (CB). Experimental protocols as in fig. 2. A. Effect of IBMX 0.5 mM and IBMX 0.5 mM plus isoproterenol (Iso) 50mM. B: Effect of different concentrations of isoproterenol (Iso). S_2/S_1 represents the ratio between the ^{3}H-CAs-evoked release in the second stimulation to that obtained in the first one (S_2/S_1; see Fig 2). S_2/S_1 obtained in CB incubated in a drug free solution were taken as 100%. Data are means ± SEM of four or more individual values. *p < 0.05 comparing with control values, + < 0.05 comparing with IBMX treated CB.*

structures, where they mediate vasodilation (Creese and Fraser, 1987), gives a new putative modulatory role to the dopamine released during the activation of the CB. Vasodilation and indeed increases in blood flow through the CB could be one of the mechanisms by which dopamine inhibits chemosensory discharges in the rabbit "in vivo" (Docherty and McQueen, 1979), an effect that is replaced by excitation in the vascularly-isolated preparation (Monti-Bloch and Eyzaguirre, 1980). A reinterpretation of the role of dopamine in chemoreception, taking into consideration the diversity and multiple locations of the dopaminergic receptors found in the CB, has been presented elsewhere (Almaraz et al., 1991). It should be noted that previous studies performed "in vivo" could not detect any effect of DA on rat CB cAMP levels (Mir et al., 1983).

The effect of isoproterenol increasing CB cAMP levels parallels that found by Mir and coworkers (1984) in the same species. Several laboratories have reported an increase in the chemosensory discharges after isoproterenol administration (Lahiri et al., 1981; Folgering et al., 1982; Gonsalves et al., 1984). Additionally, radioligand binding studies using (±) cyanopindolol as β agonist have shown the existence of β receptors in the CB of the rat and the rabbit (Mir et al., 1984), however there were no indications of the intraglomic location of these adrenergic receptors. Our data showing that isoproterenol increases the release of ^{3}H-CA induced by low pO_2 suggest that, at least a part of the effects of isoproterenol are mediated through type I cells because it is well known that the low pO_2 induced release of ^{3}H-CA from the CB is a specific parameter of these cells. Since this effect of isoproterenol is similar to that described for substances such as forskolin, IBMX or dBcAMP (Pérez-García et al., 1991) our results also suggest that β receptor activation by isoproterenol will produce its effects on the secretory response by increasing cAMP synthesis.

The β receptor subtype activated by isoproterenol and mediating their effects on the release of ^{3}H-CA in the rabbit CB has not been investigated in the present work. There is only a study in which the potency of specific β_1 and β_2 antagonist in blocking the effects of isoproterenol has been performed in the rat CB (Mir et al., 1983; 1984). In this species the cAMP accumulation induced by isoproterenol disappeared after treatment with ICI 118.551, a β_2 selective antagonist whereas betaxolol, a β_1 selective antagonist was ineffective. Preliminary experiments (not shown) performed in our laboratory suggest that the effects of isoproterenol on cAMP levels are also mediated through β_2 receptors in the rabbit CB.

We found also an activation of CB adenylate cyclase by serotonin and substance P, two agents that excite the CB (Fidone and Gonzalez, 1986). The parallelism described for these drugs increasing cAMP and chemosensory discharges simultaneously, applies also for forskolin and for β receptor agonists. In the latter cases a potenciation of the release of ^{3}H-CA evoked by hypoxia is also present (Pérez-García et al., 1991; Wang et al., 1991; this paper). A correlation between cAMP levels, ^{3}H-CA release and activity in the carotid sinus nerve is also found for the muscarinic agonist bethanecol; in this case a decrease in the three parameters takes place simultaneously. All these data suggest that these agents could use the cAMP as second messenger to mediate their effects on CB chemoreception.

Little information is available about the role of PGs in the CB. We have found that indomethacin, a specific inhibitor of the cyclooxygenase, potenciates the low pO_2 evoked release of ^{3}H-CA, demonstrating that the impairment of the PG synthesis facilitates the secretory response of the type I cells. These findings show that endogenous PG participates in the chemotransduction process inhibiting the release of ^{3}H-CA. In fact, in preliminary experiments, we found that PGE_2, at low dosis (0.3 µM), inhibits the release of ^{3}H-CAs induced by hypoxia (7%-O_2 equilibrated solution) in a 55%. As a whole, these results suggests that type I cells may be involved in the inhibitory effect of this PG described in the CB by McQueen and Belmonte (1974), and that the metabolites of the arachidonic acid, more concretely the PGs, constitute an operative second messenger system (Axelrod et al., 1988) in this organ.

The inhibitory action of PGE_2 on the release of ^{3}H-CA contrasts with the increase in CB cAMP induced by this agent (see fig. 1). This discrepancy suggest that cAMP is not the mediator for the effects of PGE_2 on the secretory response of the type I cells because, as already mentioned, there is a potenciation of the release of ^{3}H-CA by increasing the cAMP levels (Pérez-García et al., 1991; Wang et al., 1991). It is possible that the cAMP accumulation induced by PGE_2 takes place in other glomic structure different from type I cells (vessels and/or nervous endings). Further investigations are required to clarify the mechanism by wich PGE_2 inhibits chemoreception and to define the physiological role of prostanoids in the CB.

ACKNOWLEDGEMENTS

Thanks to María Bravo for technical assistance. Work supported by DGICYT grant 89/0358 and Junta de Castilla y León grant 1101/89.

REFERENCES

Almaraz, A., Pérez-García, M.T., and González, C., 1991, Presence of D$_1$ receptors in the rabbit carotid body, Neurosci. Lett., 132:259.

Axelrod, J., Burch, R.M., and Jelsema, C.L., 1988, Receptor-mediated activation of phospholipase A$_2$ via GTP-binding proteins: arachidonic acid and its metabolites as second messengers, TINS, 11:117.

Creese, I., and Fraser, C.M., 1978, "Dopamine receptors". In: Receptor Biochemistry and Methodology, Vol. 8., Alan R. Ligg, Inc., N.Y.

Dinger, B., Almaraz, A., Hirano, T., Yoshizaki, K., González, C., Gomez-Niño, A., and Fidone,S.J., 1991, Muscarinic receptor localization and function in rabbit carotid body, Brain Res., 562:190.

Fidone, S.J., and González, C., 1986, Initiation and control of chemoreceptor activity in the carotid body, in: "Handbook of Physiology". The Respiratory System, sect. 3, Vol II, Part I, A. P. Fishman, ed., American Physiological Society, Bethesda, MD.

Folgering, H., Ponte, J., and Sadig, T., 1982, Adrenergic mechanisms and chemoreception in the carotid body of the cat and rabbit, J. Physiol., 325:1.

Gómez-Niño, A., Dinger, B., González, C., and Fidone, S.J., 1989, Effects of haloperidol on basal and stimulus evoked catecholamine release from rabbit carotid body, Soc. Neurosci. Abstr., 15:753.

Gonsalves, S.F., Smith, E.J., Nolan, W.F., and Dutton, R.E., 1984, b-Adrenoceptor blockade spares chemoreceptor responsiveness to hypoxia, Brain Res., 324:349.

Greengard, P., and Costa, E., 1970, "Role of cyclic AMP in cell function," Raven Press, N.Y.

Hockberger, P.E., and Swandulla, D., 1987, Direct ion channel gating: a new function for intracellular messengers, Cell. Mol. Neurobiol., 7:229.

Joh, T.H., Park, D.H., and Reis, D.J., 1978, Direct phosphorilation of brain tyrosine hydroxylase by cyclic AMP-dependent protein kinase: Mechanism of enzyme activation, Proc. Natl. Acad. Sci. USA., 75:4744.

Kulmacz, R.J., and Lands, W.E.M., 1987, Cyclooxygenase: measurement, purification and properties, In: "Prostaglandins and related substances," C. Benedetto, R.G. McDonald-Gibson, S. Nigam and T.F. Slater, eds., IRL Press, Oxford, UK.

Lahiri, S., Pokorski, M., and Davies, R.O., 1981, Augmentation of carotid body chemoreceptor responses by isoproterenol in the cat. Resp. Physiol., 44:351.

Langer, S.Z., and Arbilla, S., 1990, Presinaptic receptors on peripheral noradrenergic neurons, in: "Presinaptic receptors and the question of autoregulation of neurotransmitter release," Annals of the New York Academy of Sciences, Vol 604, Kalsner and T.C. Westfall, eds. The New York Academy of Sciences, N.Y..

Llados, F., and Zapata, P., 1978,   Effects of dopamine analogues and antagonist on carotid body chemosensors in vitro, J. Physiol., 274:487.

McQueen, D.S., and Belmonte, C., 1974, The effects of prostaglandins E$_2$, A$_2$ ar1 F$_{2a}$ on carotid baroreceptors and chemoreceptors, Quart. J. Experim. Physiol., 59:63.

Milsom, W.K., and Sadig, T., 1983, Interaction between norepinephrine and hypoxia on carotid body chemoreception in rabbits, J. Appl. Physiol. 55:1893.

Mir, A.K., Pallot, D.J., and Nahorski, S.R., 1983, Biogenic amine-stimulated cyclic adenosine-3',5'-monophosphate formation in the rat carotid body, J. Neurochem., 41:663.

Monti-Bloch, L., and Eyzaguirre, C., 1980, A comparative physiological and pharmacological study of cat and rabbit carotid body chemoreceptors, Brain Res., 193:449.

Morita, K., Dohi, T., Kitayama, S., Koyama, Y., and Tsujimoto, A., 1987, Enhancement of stimulation-evoked catecholamine release from cultured bovine adrenal chromaffin cells by forskolin, J. Neurochem.,48:243.

Pérez-García, T., Almaraz, A., and González, C., 1990, Effects of different types of stimulation on cyclic AMP content in the rabbit carotid body: functional significance, J. Neurochem., 55:1287.

Pérez-García, M.T., Almaraz, A., and González, C., 1991, Cyclic AMP modulates differentially the release of dopamine induced by hypoxia and other stimuli and increases dopamine synthesis in the rabbit carotid body, J. Neurochem., 57:1992.

Santiago, M., and Westerink, B.H.C., 1990, Role of adenylate cyclase in the modulation of the release of dopamine: A microdialysis study in the striatum of the rat, J. Neurochem., 55:169.

Sekar, M.C., and Lowell, E.H., 1986, The role of phosphoinositides in signal transduction, J. Membrane Biol., 89:1293.

Starke, K., Gothert, M., and Kilbinger, H., 1989, Modulation of neurotransmitter release by presinaptic autoreceptors, Physiol. Rew., 69:864.

Stjärne, L., 1989, Role of prostaglandins and cyclic adenosine monophosphate in release, in: "The release of catecholamines from adrenergic neurons," D.M. Paton, ed., Pergamon Press, Oxford.

Wang W.-J., Cheng, G.-F., Yoshizaki, K., Dinger, B., and Fidone, S. J., 1991, The role of cyclic AMP in chemoreception in the rabbit carotid body, <u>Brain Res.</u>, 540:96.

Zapata, P., Hess, A., Bliss, E.L., and Eyzaguirre, C., 1969, Chemical, electron microscopic and physiologic observations of the role of catecholamines in the carotid body. <u>Brain Res.</u>, 14:473.

40 DOES ADENOSINE STIMULATE RAT CAROTID BODY CHEMORECEPTORS?

D. S. McQueen

Department of Pharmacology, University of Edinburgh
1 George Square, Edinburgh, EH8 9JZ, Scotland, U.K.

INTRODUCTION

Adenosine can stimulate cat carotid body chemoreceptors (McQueen & Ribeiro, 1981) and its respiratory stimulant action in man has been attributed to excitation of carotid body chemoreceptors (Watt et al, 1987), partly on the evidence that the respiratory stimulant action of adenosine in anaesthetized rat is abolished by cutting both carotid sinus nerves (Monteiro & Ribeiro, 1987).

The present study was undertaken to obtain further evidence concerning the effects of adenosine on rat arterial chemoreceptors by measuring the ventilatory response to adenosine before and after cutting the carotid sinus nerves, and also by recording neural activity from chemoreceptor afferents in the carotid sinus nerve of anaesthetized animals.

METHODS

Experiments were performed on male Wistar rats which were anaesthetized with pentobarbitone (60 mgkg^{-1} i.p., supplemented i.v. as required). After cannulation of the trachea (connected to a pneumograph and electrospirometer), a femoral artery (connected to a B.P. transducer), and a femoral vein (for drug administration), the effects of adenosine on ventilation were studied before and after cutting both carotid sinus nerves. Denervation was confirmed by the abolition of reflex hyperventilation which was seen pre-denervation on intravenous (i.v.) injection of sodium cyanide (2 µmoles).

In other experiments chemosensory discharge was recorded from the peripheral end of a sectioned carotid sinus nerve in the spontaneously breathing anaesthetized rat. The ipsilateral external carotid artery was cannulated in some of these experiments so that drugs could be injected close arterial (i.c.) to the carotid body. Bipolar platinum-iridium wire electrodes were used and the amplified signal was digitized and stored on video tape for subsequent analysis. Action potentials were played back through an oscilloscope and a voltage discriminator, set to count individual action potentials. The discharge was quantified using a microcomputer.

Injections (0.1 ml of drug solution, washed in with 0.2 ml modified Locke solution equilibrated with 5 % CO_2: 95 % air at 37° C) were made over a 2s period into either the venous (i.v.) or carotid (i.c.) catheters. Respiratory responses were also measured during the neural recordings. Arterial blood samples (0.2 ml) were taken at regular intervals and pH, PO_2 and PCO_2 were measured using a Ciba-Corning model 238 blood gas analyzer.

Neurobiology and Cell Physiology of Chemoreception
Edited by P.G. Data *et al.*, Plenum Press, New York, 1993

RESULTS

Ventilation

Adenosine caused a slight dose-related increase in ventilation following i.v. or i.c. injection. The response evoked was slight and belayed in comparison with that caused by sodium cyanide. Cutting both carotid sinus nerves abolished the reflex hyperventilation caused by cyanide, but denervation had little effect on responses to low doses of adenosine, although it did reduce the stimulation seen with high doses injected i.c. (see Fig. 1).

Neural recordings

Chemosensory discharge was recorded from the peripheral end of a sectioned carotid sinus nerve in five experiments. The neural and ventilatory responses to i.c. injections of adenosine was evaluated (i.e. only one carotid nerve was intact). Results from a typical experiment are shown in Fig. 2 from which it can be seen that adenosine caused a slight increase in chemosensory discharge, but this did not correlate well with the concurrently recorded increase in ventilation.

Data from five experiments in which the effects of adenosine on chemoreceptor discharge were studied are shown in Fig. 3. Overall, there was no significant increase in chemosensory discharge when responses in the 20 s period following injection of adenosine were compared with the pre-injection frequency, although there was a trend for i.c. administration to cause a very slight dose-related increase in discharge.

DISCUSSION

The present results confirm that adenosine can stimulate ventilation in the anaesthetized rat, and the responses were similar to those previously reported (Monteiro & Ribeiro, 1987). However, there was only a slightly bigger increase in ventilation when adenosine was injected i.c. as compared with responses to i.v. injection, and the difference was only statistically significant at high doses (see Fig. 1). The same figure shows that cutting the carotid sinus nerves had no significant effect on the ventilatory response to i.v. adenosine, although it abolished the response to intravenous sodium cyanide. Cutting both carotid sinus nerves reduced the hyperventilation evoked by i.c. adenosine, but this reduction was significant only at the highest doses. Perhaps high concentrations of adenosine have some other actions (e.g. in CNS, on vasculature, on peripheral sensors other than the carotid chemoreceptors) that enhances the normal chemosensory input? The finding that low doses of adenosine are associated with respiratory changes, yet cutting the carotid sinus nerves does not significantly alter the response, argues against involvement of peripheral chemoreceptors.

The indirect evidence from the respiratory experiments does not lend support to the hypothesis that adenosine increases ventilation by activating the peripheral chemoreceptors. More direct evidence was obtained by recording chemosensory discharge from the rat carotid nerve. These experiments confirmed that adenosine can cause slight chemoexcitation, although it is more delayed and less intense than the (weak) responses obtained in cats (McQueen & Ribeiro, 1981). Also, there was no clear dose-response

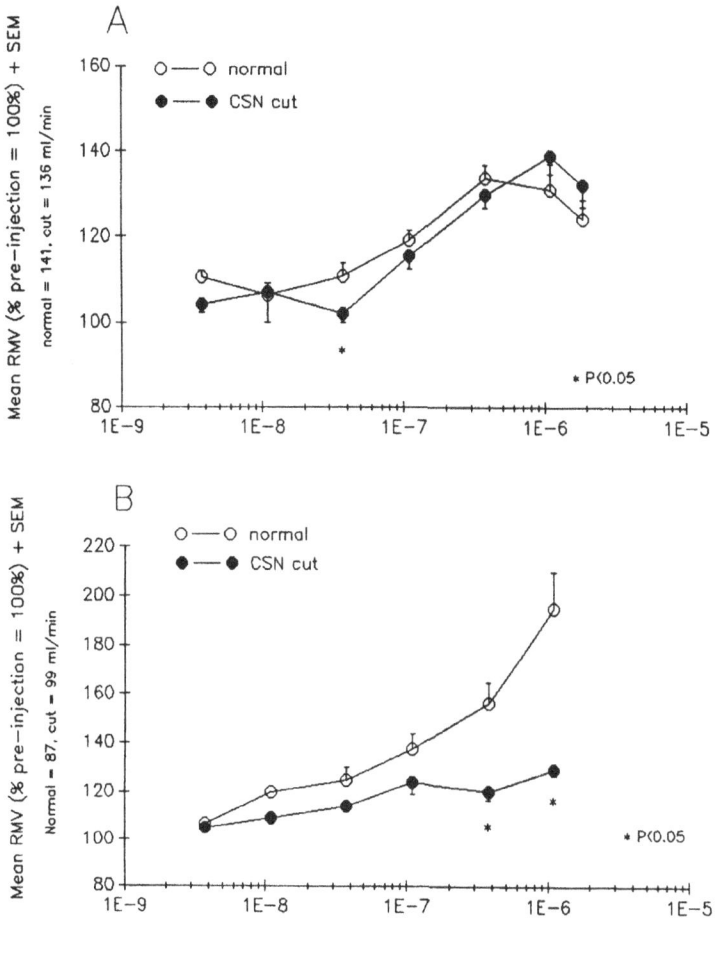

Dose of adenosine injected (moles)

Fig. 1 A. Ventilatory responses to i.v. adenosine before and after cutting both carotid sinus nerves. Pooled
data from six experiments.

B. Ventilatory responses to i.c. adenosine before and after cutting the carotid sinus nerves. Pooled
data from five experiments.

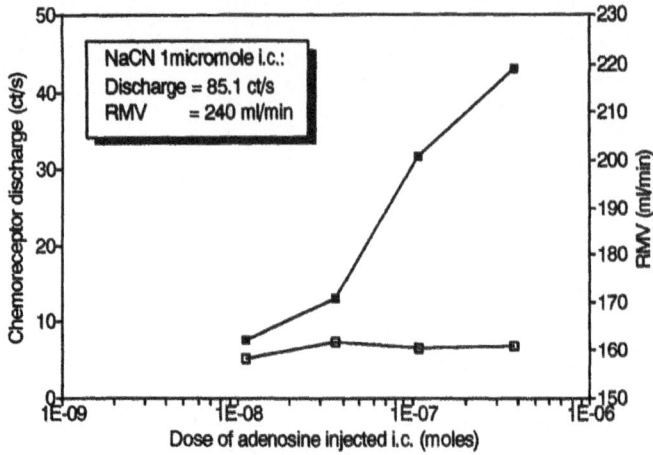

Fig. 2 *Data from a single experiment on a spontaneously breathing anaesthetized rat. Adenosine injected i.c. caused a slight increase in chemoreceptor discharge (open rectangles) at lower, but not higher, doses, whereas ventilation (filled rectangles) increased with high doses. Basal values (mean ± s.e.m.) were: discharge 2.0 ± 0.5 impulses s^{-1}; ventilation R.M.V. was 165 ± 4 ml min^{-1}. Responses to sodium cyanide (1 μmole) are shown in the insert.*

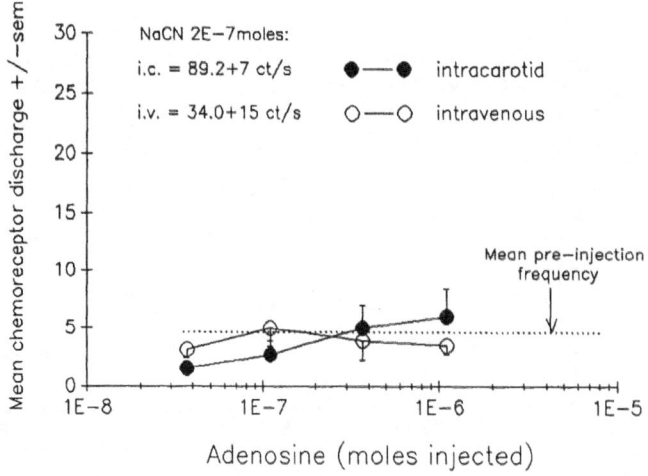

Fig. 3 *Chemosensory discharge measured during the 20 s period following i.v. or i.c. injection of adenosine. Pooled data from five experiments show there was a tendency for discharge to increase with dose on i.c. injection, but not on i.v. injection. In contrast, sodium cyanide caused a substantial excitation of the chemoreceptor units (injections made after adenosine had been administered).*

relationship (see Fig. 3), and there was little or no difference between i.v. and i.c. administration. In some experiments a clear increase in ventilation occurred following adenosine, yet there was only a slight (1-2 impulses per second) increase in chemoreceptor discharge (Fig. 2). Further studies will be needed to establish how adenosine increases ventilation, but it is worth noting that the respiratory response parallel the fall in blood pressure evoked by adenosine.

In conclusion, the evidence from experiments in which respiratory responses to adenosine were measured at the same time as chemosensory discharge was recorded from carotid chemoreceptor afferents in the same animals suggests that adenosine does not, on average, cause excitation of carotid body chemoreceptors. It therefore appears that carotid chemoreceptors do not play a major role in adenosine-induced hyperventilation.

REFERENCES

McQueen, D.S. & Ribeiro, J.A. (1981) Effect of adenosine on carotid chemoreceptor activity in the cat. Br. J. Pharmacol. 74, 129-136.

Monteiro, E. & Ribeiro, J.A. (1987) Ventilatory effects of adenosine mediated by carotid body chemoreceptors in the rat. N.-S. Arch. Pharmacol. 335, 143-148.

Watt, A.H., Reid, P.G., Stephens, M.R. & Rontledge, P.A. (1987) Adenosine-induced respiratory stimulation in man depends on site of infusion - evidence for an action on the carotid body. Br. J. Clin. Pharmacol. 23, 486-490.

41

EFFECTS OF HALOPERIDOL ON CAT CAROTID BODY CHEMORECEPTION IN VITRO

L. Morelli [2], R. Iturriaga [1], D. Spergel [1] and P.G.Data [2]

Department of Physiology [1], University of Pennsylvania,
Philadelphia, PA 19104, USA and Institute of Physiological
Sciences [2],"G. D'Annunzio" University, 66100 Chieti, Italy

INTRODUCTION

A current hypothesis states that carotid body (CB) chemoreception is ultimately mediated by the release of neurotransmitters from the glomus cell[1]. One of the putative neurotransmitters is dopamine which is present in glomus cell dense-core vesicles1 and is released in response to hypoxia[2], low pH and hypercapnia[3]. In the cat CB, however, exogenous dopamine inhibits chemosensory activity in vivo[4,5,6] and in vitro[7]. Elucidation of the role of dopamine in CB chemoreception is further complicated by the finding that haloperidol, a D_2 dopaminergic antagonist, increases the chemosensitivity of the CB to hypoxia[4] and hypercapnia[4,8] in vivo , but decreases the activity and the response to acetylcholine in vitro[9]. Based on their in vitro observations, Nolan et al.[9] proposed an excitatory role for dopamine. However, they did not use other stimuli to test if the effects of haloperidol are specific to O_2 and CO_2 chemoreception. In order to test this possibility we studied the effects of haloperidol on the chemosensory responses in vitro to hypoxia, CO_2, dopamine, sodium cyanide and nicotine.

METHODS

The CB's were excised from cats anaesthetized with sodium pentobarbitone (35 mg/kg). The CB's were perfused and superfused in vitro at 36.5 ± 0.5 °C with a modified Tyrode solution and the chemosensory discharges were recorded from the whole carotid sinus nerve as previously described[10]. The composition of the solution was (in mM) Na^+, 154; K^+, 4.7; Ca^{2+}, 2.2; Mg^{2+}, 1.1; Cl^-, 110; HCO_3^-, 22.4; glutamate, 22.0; glucose, 5.0; HEPES, 5.0 and dextran (5.0 g/l). The perfusate Tyrode was equilibrated at PO_2 of 120 Torr, PCO_2 of 35 Torr and pH of 7.38. The carotid chemosensory responses were studied before, during and after perfusion with Tyrode containing haloperidol (12.5 µM). The tests consisted of responses to interruption of the perfusate flow, hypoxia (PO_2 of 50 Torr), hypercapnia (PCO_2 of 55-65 Torr and pH of 7.20), and bolus injections (0.2 ml) of Tyrode containing dopamine (2-4 µg), sodium cyanide (1-2 µg) and nicotine (2 µg) into the perfusate line.

RESULTS

Effects of haloperidol on chemosensory baseline activity and on responses to dopamine, sodium cyanide and to nicotine

Figure 1 shows the effects of haloperidol on chemosensory responses to dopamine, cyanide and nicotine. During control perfusion with Tyrode, dopamine initially decreased the chemosensory activity followed by an increase presumably due to vascular effects[5] (Fig. 1A). Bolus injections of nicotine and cyanide produced rapid and large increases of chemosensory activity (Fig. 1A). Perfusion of Tyrode containing haloperidol for 20 min reduced the chemosensory baseline activity and nearly abolished the responses to all the pharmacological agents (Fig. 1B). After 20 min of perfusion with Tyrode free of haloperidol, the baseline chemosensory activity was restored to the initial values and the chemosensory responses to cyanide and to nicotine were recovered (Fig. 1C). The inhibitory response to dopamine was not recovered but the excitatory effect was restored.

Effects of haloperidol on chemosensory responses to flow interruption, hypercapnia, and hypoxia

Figure 2 shows the effects of haloperidol on chemosensory responses to perfusate flow interruption and to hypercapnia. Interruption of perfusate flow, as CB tissue PO_2 declined gradually, increased the chemosensory activity. Resumption of perfusate flow

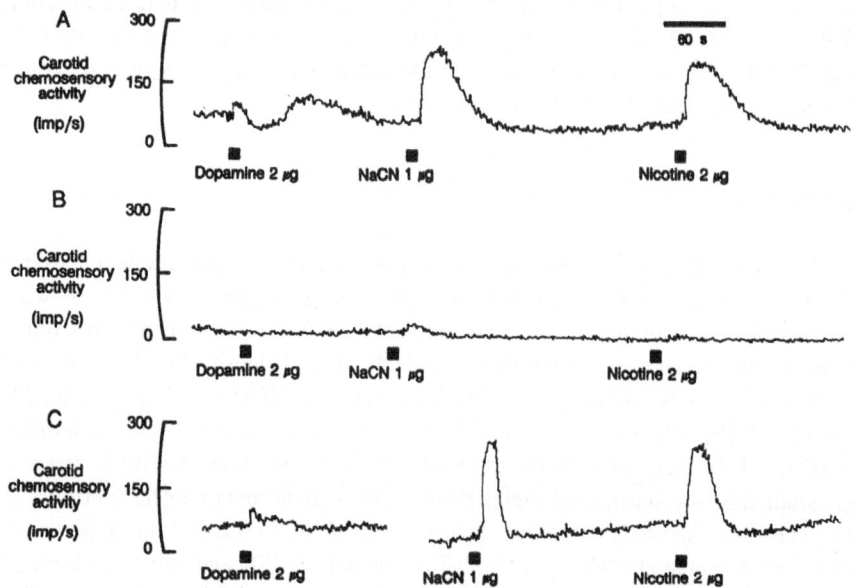

Fig. 1. Effects of haloperidol (12.5 μM) on chemosensory responses to dopamine, sodium cyanide and nicotine. A, control perfusion. B, perfusion with haloperidol. C, recovery. ($PO_2 = 120\,Torr, PCO_2 = 35\,Torr, pH\ 7.38$).

The discrepancy between the augmenting effect of haloperidol in vivo on the chemosensory responses to CO_2 and hypoxia and the non-specific inhibitory effect in vitro needs an explanation. A reasonable explanation for the excitatory effect of dopamine receptor blockade in vivo is that it suppresses the inhibition caused by endogenous dopamine release from the CB. However, Nolan et al[9]. suggested that the excitatory effect induced by dopamine receptor blockade was due to vascular changes. Vascular effects were not present in their in vitro superfused preparation. However, we observed similar effects in our in vitro perfused and superfused CB preparation even though a vascular contribution was present[10]. Furthermore, dopamine initially produced chemosensory inhibition followed by excitation. This dual effect is commonly observed in vivo and is likely to be due to vasoconstriction mediated by adrenoreceptors[4, 5]. Thus, we cannot attribute the observed differences to vascular effects of haloperidol.

A major methodological difference between in vivo and in vitro preparations involves the use of physiological saline solutions in vitro[9, 10] instead of blood which circulates through organs like liver and lungs. These organs can metabolically change many biological activities and inactive agents. Erythrocytes and hepatocytes metabolize haloperidol to reduced-haloperidol in vivo[12]. Reduced-haloperidol binds to D_2 receptors in brain synaptosomal fraction with about an 85-fold lower affinity than haloperidol[13]. However, it is not clear how a decreased affinity for receptors can explain the observed differences. On the other hand, reduced-haloperidol may produce different effects on chemoreceptor activity than haloperidol. Accordingly there are ways to test why haloperidol is excitatory in vivo and inhibitory in vitro.

In summary our results showed that haloperidol in vitro suppresses all the chemosensory responses, and hence do not allow the conclusion that dopamine is an excitatory neurotransmitter for O_2 and CO_2 chemoreception.

ACKNOWLEDGMENTS

We are deeply indebted to Dr. S. Lahiri for advice and encouragement. We also thank Mr. A. Mokashi for his assistance. Supported in part by NIH grants HL-19737, HL-43413 and HL-07027.

REFERENCES

1. *Eyzaguirre, C., Fitzgerald, R. S., Lahiri, S. and Zapata, P.,* 1983, Arterial Chemoreceptors, in: Handbook of Physiology, The Cardiovascular System, Peripheral Circulation and Organ Blood Flow, Am. Physiol. Soc., Williams and Wilkins, Baltimore, p. 557-621.

2. *Rigual, R., Gonzalez, E., Gonzalez, C. and Fidone, S.,* 1986, Synthesis and release of catecholamines by the cat carotid body in vitro effects of hypoxic stimulation, Brain. Res., 374:101-109.

3. *Rigual, R., Lopez-Lopez, J.R. and Gonzalez, C.,* 1991, Release of dopamine and chemoreceptor discharge induced by low pH and high PCO_2 stimulation of the cat carotid body, J. Physiol. Lond., 433:519-531.

4. *Lahiri, S., Nishino, T., Mokashi, A. and Mulligan, E.*, 1980, Interaction of dopamine and haloperidol with O_2 and CO_2 chemoreception in the carotid body, J. Appl. Physiol., 49:45-51.

5. *Llados, F. and Zapata, P.*, 1978, Effects of adrenoreceptor stimulating and blocking agents on carotid body chemosensory inhibition, J. Physiol. Lond., 274: 501-509.

6. *Ponte, J. and Sadler, C. L.*, 1989, Interactions between hypoxia, acetylcholine and dopamine in the carotid body of rabbit and cat, J. Physiol. Lond., 410:395-410.

7. *Zapata, P.*, 1975, Effects of dopamine on carotid chemo- and baroreceptors in vitro, J. Physiol. Lond., 244:235-251.

8. *Donnelly, D.F., Smith, E.J. and Dutton, R.E.*, 1981, Neural response of chemoreceptors following dopamine blockade, J. Appl. Physiol., 50:172-177.

9. *Nolan, W.F., Donnelly, D.F., Smith, E.J. and Dutton, R.E.*, 1985, Haloperidol-induced suppression of carotid chemoreception in vitro, J. Appl. Physiol., 59:814-820.

10. *Iturriaga, R., Rumsey, W. L., Mokashi, A., Spergel, D., Wilson, D.F. and Lahiri, S.*, 1991, In vitro perfused-superfused cat carotid body for physiological and pharmacological studies, J. Appl. Physiol., 70:1393-1400.

11. *Pencek, T. L., Schauff, C.L. and Davis, F. A.*, 1978, The effect of haloperidol on the ionic current in the voltage-clamped node of Ranvier, J. Pharmacol. Exper. Ther., 204:400-405.

12. *Inaba, T., Kalow, W., Someya, T., Takahashi, S., Cheung, S.W. and Tang, S.W.*, 1989, Haloperidol reduction can be assayed in human red blood cells, Can. J. Physiol. Pharmacol., 67:1468-1469.

13. *Bowen, W.D., Moses, E. L., Tolentino, P.J. and Walker, J.M.*, 1990, Metabolites of haloperidol display preferential activity at s receptors compared to dopamine D-2 receptors, Eur. J. Pharmacol., 177:111-118.

42 EFFECT OF ARTERIAL CHEMORECEPTOR STIMULATION: ROLE OF NOREPINEPHRINE IN HYPOXIC CHEMOTRANSMISSION

N.R. Prabhakar, Y.-R. Kou, P. A. Cragg and N.S. Cherniack

Department of Medicine
Case Western Reserve University
Cleveland, Ohio 44106 U.S.A.

INTRODUCTION

Mammalian carotid bodies contain substantial amounts of norepinephrine (NE) (1,2). Despite its abundance in the chemoreceptor tissue, the role of norepinephrine in carotid body function remain uncertain. Several investigators have examined the effects of norepinephrine on carotid body activity (3-6). The reported chemoreceptor responses to NE include inhibition as well as excitation, often inhibition preceeds the excitation.

The physiological effects of norepinephrine are mediated by its action on α- and β-adrenergic receptor subtypes. In the carotid body, the inhibitory actions of NE were attributed to activation α–adrenergic receptors (3); whereas, beta receptors seem to be responsible for the excitation (6). Based on selective ligand binding studies, a-adrenergic receptors have been further classified as α1- and α2- subtypes (7). It is now clear that in many tissues, the inhibitory actions of NE are medited by α2-adrenergic receptor subtype (7). Recently, we reported that the carotid body contains α2-adrenergic receptors and activation of these receptors by α2-agonist inhibits the chemoreceptor response to hypoxia,(8). Whether NE attenuates the hypoxic response and if so, whether α2-adrenergic receptors mediate the effects of NE remain to be investigated.

In the present study we examined the effects of different doses of norepinephrine on baseline carotid body activity and chemoreceptor response to hypoxia both *in* vivo and *in vitro* , the latter to exclude the possible effects of the amine on carotid body blood flow. The contribution of α2-adrenergic receptors was assessed using SKF-86466, a selective α2-adrenergic receptor antagonist blocker (8). Prolonged exposure to hypoxia alters the adrenergic receptor sensitivity in tissue specific manner (7). Therefore, we also examined α2-adrenergic receptor sensitivity in the carotid bodies of cats that were exposed to 24-36 hours of hypoxia.

METHODS

General Preparation of Animals

Adult cats were anaesthetized with an intraperitoneal injection of pentobarbital sodium (35-45 mg/kg). After tracheal intubation, a femoral artery and vein were cannulated for measuring arterial blood pressure and for systemic administration of fluids

Neurobiology and Cell Physiology of Chemoreception
Edited by P.G. Data *et al.*, Plenum Press, New York, 1993

and drugs. To expose the carotid bifurcation the trachea and esophagus were ligated above the site of tracheal cannulation, sectioned, and retracted rostrally. Intracarotid infusion of norepinephrine or α2-agonist was accomplished via a catheter placed in the external carotid artery with its tip close to the carotid body. The dead space of the catheter was kept close to 0.3 ml. After completion of surgical procedures, animals were paralyzed with gallamine triethiodide (3-4 mg/kg; I.V.) and ventilated with a respirator. Endtidal PCO_2 was continuously monitored with an infrared CO_2 analyzer. Arterial blood samples were collected for measurements of PO_2, PCO_2- and pH. The body temperature of the animals was kept at $38 \pm 1°C$ by, means of a servo heating blanket.

Recording of Chemoreceptor Activity In *Vivo*

The technique of recording the activity of chemoreceptors in vivo has been described in detail elsewhere.(9). In brief, either the right or left carotid sinus nerve was isolated, sectioned, desheathed and split into small filaments until clearly identifiable action potentials from 1-3 units were obtained. Chemoreceptor units were identified by their spontaneous sporadic discharge and increased activity during ventilation with hypoxic gas.

Recording of Chemoreceptor Activity *In Vitro*

The carotid body, together with the carotid sinus nerve, was excised from anaesthetized cats. After removing the connective tissue, the carotid body was placed in a perspex chamber containing two compartments, A and B, separated by a partition having a small opening. The carotid sinus nerve was pulled through the small opening and placed on a platform in chamber B, and covered with warm mineral oil. The carotid body in the chamber A was superfused with Krebs solution at a rate of 2-3 ml/min. and at a temperature of $36 \pm 1°C$. In order to prevent the exposure of the carotid body to atmospheric air, the chamber A was closed with a lid. The composition of the superfusing medium was as follows: NaCl 110 mM; KCl 5 mM; $MgCl_2$ 0.5 mM; $CaCl_2$ 2.2 mM; sucrose 54 mM; glucose 5.5 mM; Hepes buffer 5 mM; pH 7.38. The reservoirs containing superfusion medium were bubbled with 100% O_2 and 6% O_2. Hypoxic challenge was initiated by switching the superfusion medium equilibrated with 100% O_2 to one equilibrated with hypoxic gas. The carotid sinus nerve was desheathed, and split into thin filaments until obvious action potentials are obtained. Units responded with an augmented discharge frequency during superfusion with hypoxic medium were considered as chemoreceptors fibers. Drugs to be tested were added to the chamber A.

Drugs Used

Norepinephrine (Levophed; Winthrop Pharmaceuticals) was freshly prepared during each experiment. The stock solutions of guanabenz (α2agonist) and SKF-86466 (an α2-antagonist) were prepared by dissolving each in 0.01 M acetic acid (2 mg/ml). To prevent oxidation the test tubes containing the solutions were covered with an opaque paper. During the experiment, desired doses of the substances were prepared by diluting the stock solutions with Kreb's saline (pH - 7.4) maintained at 37°C.

Data Analysis

The following variables were analyzed: Chemoreceptor activity (impulses/sec), mean systemic arterial blood pressure (mmHg; In Vivo experiments); arterial blood gases and pH. Results are expressed as mean ± standard error of mean statistical analysis was evaluated by a paired t-test or by a one way measures analysis of variance (ANOVA) combined with Tukey's test. P values of less than 0.05 were considered significant.

RESULTS

Carotid Body Responses to Norepinephrine

In vivo studies. The effect of intracarotid administration of norepinephrine on base line chemoreceptor activity was examined during room air ventilation (PaO_2 94±7 mmHg; $PaCO_2$ 33±9 mmHg; n= 6 cats). Different doses of norepinephrine were infused over a period of five minutes. At doses between 0.1 to 5 µg/min., norepinephrine inhibited baseline carotid body activity in a dose dependent manner. Maximum depression of chemoreceptor activity was seen with infusion of 5 µg/min., and it was inhibited by 85 ± 12% (p < 0.01). Increasing the dose of NE to 10 and 20 µg/min., however, augmented the chemoreceptor activity by 22 ± 9 and 27 ± 14%, respectively.

As shown in figure 1, administration of NE as a bolus injection caused an initial inhibition followed by an augmentation of chemoreceptor activity. Instead of a bolus, infusing the same dose of NE, however, produced only an inhibition of the carotid body activity.

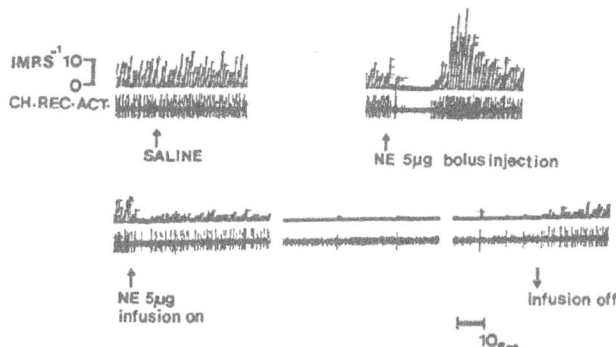

FIGURE 1. Effect of intracarotid infusion of Norepinephrine (NE) on baseline carotid chemoreceptor activity in a cat ventilated with room air. Bolus in;ection of NE caused an inhibition followed by an augmented discharge. Note the infusion of the same dose of NE produced only an inhibition of carotid body activity.

Furthermore, NE (5 µg/min.) had no consistent effects on baseline activity recorded under hyperoxic ventilation (PaO_2 = 360 ±12 mmHg; $PaCO_2$ = 35 ±4 mmHg; Control 2.3 ± 0.6 impulses/sec vs 1.9 ± 0.8 impulses/sec ; P > 0. 5, n = 6) .

Arterial blood pressure was not affected by norepinephrine infusion at doses between 0.1 and 2.5 µg/min. However, increasing the doses to 5 µg elevated arterial blood pressure by 22 ± 8 mmHg.

The effects of intracarotid infusion of norepinephrine (5 µg/min) on carotid body responses to isocapnic hypoxia (PaO_2 36±8 mmHg; $PaCO_2$ 34±10 mmHg) were examined in the same animals as above. In all experiments, NE attenuated the chemoreceptor responses to low pO_2 On average, hypoxic responses were reduced by 44% (p < 0.01).

By contrast, bolus injection of NE had no effect on hypoxic response of the carotid body.

The inhibitory effects of norepinephrine on base line chemoreceptor activity and on hypoxic responses were significantly inhibited after intravenous administration of $\alpha2$-adrenergic receptor antagonist (SKF86466 ; $p < 0.01$) . However, chemoreceptor excitation caused by higher doses of norepinephrine (i.e., 10 and 20 µg/min.) was not affected by SKF- 86466 .

In vitro studies. The effects of different doses of NE on base line carotid body activity and on hypoxic responses were tested on isolated, superfused carotid bodies (n = 6). Norepinephrine had no detectable effect on the base line carotid body activity recorded under hyperoxic conditions (PO_2 386 ± 16 mmHg; PCO_2 4±1 mmHg; $P > 0.05$; n =6). However, superfusing the carotid body with 10, 20 and 40, µg/min. of NE resulted in attenuation of hypoxic response. The inhibitory effects of NE were dose - dependent . On average, at 40 ,µg/min, hypoxic response ($PaO_2 = 40 ± 6$ mmHg) was reduced by 76 ± 12% of the controls ($p < 0.01$) . On the other hand, increasing the dose of NE to 80 and 160 µg/min. increased the hypoxic response by 48% in four experiments, but had no effect in the other two experiments. Superfusing the carotid body with the $\alpha2$-antagonist (700 µg/min.; 5 min.) prevented the inhibitory effects of NE. However, the antagonist had no effect on the excitation observed in four experiments.

Chemoreceptor Responses to $\alpha2$-Adrenergic Receptor Agonist in Cats Exposed to Hypoxia

To test whether prior exposure to hypoxia affects the $\alpha2$-adrenergic receptor sensitivity of the carotid body, unanaesthetized cats were allowed to breathe either room air (21% O_2 ; n = 5 cats), or hypoxic gas (12% O_2 + N_2; n = 5 cats) for 24-36 hours. Subsequently, they were anaesthetized, paralyzed and artificially ventilated with room air. During the course of the acute experiments, arterial blood gases and pH in both groups of animals were kept close to the physiological ranges by adjusting the stroke volume and rate of the respirator, and by injecting saline containing $NaHCO_3$. Chemoreceptor responses to guanabenz were recorded in controls (normoxic animals) and hypoxic animals. In control animals, $\alpha2$-agonist inhibited baseline chemoreceptor activity in a dosedependent manner (n = 8 fibers; 0.5, 1, and 5 µg/min.; $p < 0.01$), and attenuated the hypoxic response by 52% ($p < 0.001$). By contrast, in animals prior exposed to hypoxia, $\alpha2$-agonist produced neither an inhibition of baseline activity, nor a reduction in the hypoxic response with any of the doses tested (n = 9 fibers). It is possible that in hypoxic cats, higher doses of guanabenz are required to produce chemoreceptor inhibition. Therefore, responses of seven of these fibers were tested with higher doses of guanabenz (i.e., 20, 50, 100 µg/min). Activity of these fibers was not affected at doses up to 50 µg/min.; whereas, at 100 µg/ min a small augmentation of chemosensory discharge (+15±3% of controls) was observed.

DISCUSSION

Consistent with the previous reports by others, we observed an inhibition followed by an augmentation of chemosensory discharge with NE. The biphasic pattern of the response, however, was obtained only with bolus administration of NE. On the other hand, intra-carotid infusion of NE at low doses (i.e., up to 5 µg/min.) elicited an inhibition of the base line chemoreceptor activity, and never an excitation. Moreover, the inhibitory effects of NE were discernable under room air ventilation, and were not obvious under hyperoxia. Therefore, besides the dose, the inhibitory actions of NE seem to depend on the mode of administration of the amine as well as on the partial pressure of arterial oxygen. As reported by other investigators (7), we also observed an increase in chemosensory sensory discharge. However, the stimulatory effects were seen only at higher doses of NE, never at low doses.

Norepinephrine at low doses not only affected the base line activity but also significantly attenuated the hypoxic excitation of the carotid body. The inhibitory effects were prevented by systemic administration of SKF-86466, suggesting that α2-adrenergic receptors mediate the inhibitory actions of NE in the carotid body. It is known that activation of vascular α2-adrenergic receptors causes vasoconstriction (10). Therefore, the inhibitory effects of NE on the carotid body activity could be secondary to local blood flow changes. However, the inhibitory actions of NE were seen also in vitro carotid body, where the blood flow changes are effectively absent. Moreover, in the in *vitro* preparation, the effects of NE were obtained under nominal absence of CO_2 in the superfusing medium (PCO_2 = 4 mmHg). Therefore, it seems unlikely that the attenuation of the hypoxic response is secondary to the effect of NE on the CO_2 response of the carotid body. These observations taken together suggest that the α2-adrenergic receptors in the carotid body are coupled to transmission and/or transduction of the hypoxic stimulus.

An interesting observation of the present study is that prior exposure to hypoxia abolished the inhibitory actions of the α2-adrenergic receptors in the carotid body. Lack of α2-adrenergic receptor sensitivity in the hypoxic animals can not be attributed to the acid base and/or blood gas status of the animals because they were comparable with the controls. It seems plausible that the absence of guanabenz response is due to down regulation of α2-adrenergic receptors in the carotid body resulting from prolonged exposure to hypoxia. Further studies however, are necessary to substantiate this notion.

In summary, the present study provides evidence that inhibitory actions of norepinephrine on the carotid body are mediated by α2-adrenergic receptors. Present results also indicate that prolonged exposure to hypoxia down regulate the α2-adrenergic receptors in the carotid body. It is known that chronic hypoxia enhances the carotid body sensitivity to low PO_2 (4) and is associated with alterations in NE dynamics (12). It is possible that lack of α2-adrenergic receptor inhibition contribute, in part, to the enhanced carotid body sensitivity to low PO_2 during chronic hypoxia.

ACKNOWLEDGEMENT

The authors are grateful to Hui-Ping Cao for technical assistance, and to Cheryl Diane Gilliam for assistance with manuscript preparation. This study was supported by grants from the National Heart, Lung and Blood Institute HL-38986, HL-45780, and Research Career Development Award to NRP, HL02599.

REFERENCES

1. *D.M. McDonald*. Peripheral chemoreceptors: Structure - function relationships of the carotid body. In: "Lung Biology in Health and Disease: The Regulation of Breathing," Vol. 17, T.F. Hornbein, ed., Dekker, New York (1981).

2. *S.J. Fidone and C. Gonzalez*. Initiation and control of chemoreceptor activity in the carotid body. In: "Handbook of Physiology Section 3, The Respiratory System," Vol. 2, N.S. Cherniack and T.G. Widdicombe, ed., (1986).

3. *S.R. Sampson, M.J. Aminoff, R.A. Jaffe, and E.H. Vidruk.* A pharmacological analysis of neurally induced inhibition of carotid body chemoreceptor activity in cats. J. Pharmacol. Exp. Ther, 197:119 (1976).

4. *G.E. Bisgard, R.D. Mitchell, and D.A. Herbert.* Effect of dopamine, norepinephrine, and 5-hydroxytryptamine on the carotid body of the dog. Respir. Physiol. 37:61 (1979).

5. *F. Liados and P. Zapata.* Effects of adrenoceptor stimulating and blocking agents on carotid body chemosensory inhibition. J. Physiol (London) 274:501 (1978).

6. *H.Folgering, J. Ponte, and T. Sadig.* Adrenergic mechanisms and chemoreception in the carotid body of the cat and rabbit. J. Physiol (London) 325:1 1982).

7. *D.B. Bylund and V.C. V' Prichard.* Characterization of a1- and a2 adrenergic receptors. Int. Rev. Neurobiol. 24:343 (1983).

8. *Y-R. Kou, P. Ernsberger, P.A. Cragg, N.S. Cherniack, and N.R. Prabhakar.* Role of ~2-adrenergic receptors in carotid body response to hypoxia. Resp. Physiol. 83:353 (1991).

9. *N.R. Prabhakar, S.C. Landis, G.K. Kumar, D. Millikan-Kilpatrick, N.S. Cherniack, and S.E. Leeman.* Substance P and neurokinin A in the cat carotid body: Localization, exogenous effects, and changes in content in response to arterial PO2. Brain Res. 481:205 (1989).

10. *S.Z. Langer and P.E. Hicks.* Alpha-adrenoreceptor subtypes in blood vessels: Physiology and pharmacology. J. Cardio. Pharmacol. 6:S547 (1984).

11. *G.E. Bisgard and M.J. Engwall.* Ventilatory effects of prolonged hypoxia in goats. In: "Hypoxia. The Adaptations,in J.R. Sutton, G. Coats, and J.E. Remmers, eds., Decker Inc., Toronto (1990) .

12. *J.M. Pequignot, J.M. Cottet-Emrad, Y. Dalmaz, and L. Peyrin.* Dompamine and norepinephrine dynamics in rat carotid body during long-term hypoxia. J. Auton. Nerv. Syst, 21:9 (1987).

SECTION IV

Reflex Mechanism

43 CAROTID BODY DENERVATION AND PULMONARY VASCULAR RESISTANCE IN THE RAT

D. Bee and *D. Pallot

Medicine & Pharmacology, Sheffield
University,*Anatomy, Leicester University, UK

INTRODUCTION

Chronic exposure to low oxygen environments causes several adaptive changes in the rat. These include hypertrophy and hyperplasia of the carotid body, right ventricular hypertrophy, pulmonary vascular remodelling and pulmonary hypertension[1,2]. After three weeks in a normobaric chamber at $10\%0_2$ our group has shown that rats have a depressed hypoxic ventilatory response but a greater pulmonary vascular reactivity compared with controls[3,4]. In acute experiments there have been suggestions that the carotid body may have some influence over pulmonary arterial compliance[5,6] and pulmonary hypertension as produced by ventilation hypoxia[7]. The carotid bodies are needed for acclimatisation of the ventilatory reponse[8,9,10] but do they have a role to play in the adaptive processes in the lung? Using bilateral denervation of the carotid bodies we hoped to assess their influence (if any) on these processes.

METHODS

Denervation - 30 day old male wistar rats (Tuck's Labortories, specific pathogen free) were anaesthetised with small mammal imobilon (1:3, subcutaneously). The carotid bodies were exposed via a mid ventral inscission and were denervated by section of the glossophanygeal nerve at the superior point of origin of the carotid sinus nerve. The rats were tested for a hypoxic ventilatory response by giving two breaths of nitrogen. Intact rats showed a rapid increase in breathing frequency which was absent where denervation was successful. The rats were revived with Narcan (subcutaneous). Four days later 4 intact and 4 denervated rats were placed in a normobaric chamber and exposed to $10\%02$ for 3 weeks. They were compared to 2 intact and 2 denervated littermate controls. Food and water wore available *ad libitum*.

Ventilation tests - The rats were anaesthetised with Inactin (BYK, l0Omg/kg, ip). The trachea and femoral artery were cannulated and they were placed supine in a body plethysmograph with the cannulae connected to the outside. Gases were passed over the tracheal port and the hypoxic ventilatory response (HVR) was tested by reducing the FiO_2 from 0.3 to 0.21, 0.15 and 0.1 in three minute stages. Blood pressure, minute ventilation

Neurobiology and Cell Physiology of Chemoreception
Edited by P.G. Data *et al.,* Plenum Press, New York, 1993

(VE) and femoral arterial blood gases were measured at each stage. The percentage change in VE was calculated and plotted against PaO_2.

Physiology and Histology of the Lung

Isolated lung preparation - Rats were anaesthetised with pentobarbitone (60mg/kg, ip) and the trachea cannulated. The ventral thorax was removed and the lungs were isolated in situ. They were ventilated at approx 21/min and perfused with blood of normal temperature, pH and haematocrit at 20ml/min, via a pulmonary artery cannula. Blood was collected from the cannulated left atrium and returned to a heated reservoir for recirculation. When the pulmonary artery pressure (Ppa) had stabilised flow was reduced in steps from 20 to 15, 10 and 5 ml/min. Ppa was measured at each stage and left atrial pressure was zero. The slope of the Ppa - flow line was interpreted as pulmonary vascular resistance (PVR) and the extrapolated intercept is taken as the effective downstream pressure (critical closing pressure of the Starling resistor model).

Heart weights - The atria and large vessels were removed from the hearts collected after the above experiments. The right ventricle (RV) was carefully dissected from the left ventricle and septum (LV+S) and both were weighed. The ratio RV/LV+S was calculated with reference to body weight.

Lung histology - Lungs were fixed with 10% buffered formalin via the trachea at $20cmH_2O$ pressure. They were processed by standard techniques to wax. 5um sections cut transversely from the middle of the left lung were stained with Humberston's for elastin. The number of small (50um external diameter) vessels were counted and the percentage of these showing a double elastic lamina were calculated (%thick-walled-peripheral-vessels, %TWPV).

RESULTS

Ventilation - The initial ventilation /lOOg body weight (mean±sem) were for intact and denervated rats respectively 47.1±2.1 ml/min and 59.4±2.1 in controls and 83.5±9.0 and 60.8±7.6 in chronically hypoxic rats. The individual HVR's for intact (above) and denervated (below) rats are given in figure 1. Clearly the intact chronically hypoxic (CH) rats show a depressed HVR compared with controls. Although the denervated controls had no HVR the CH rats had regained a response. These rats did not respond to the N_2 test when they were denervated.

Pulmonary circulation - The mean values for slopes and intercept are given in Table 1. No difference was found between the normoxic intact and denervated rats. PVR and intercept were raised with hypoxia; the increases being much greater in denervated than control rats.

Right ventricular hypertrophy - The heart weight ratios per lOOg body weight are given in Table 2a. There was little difference between the control intact and denervated rats. Chronic hypoxia caused some RV hypertrophy in intact rats which was greater in denervated rats.

Lung histology - The %TWPV for each group is given in Table 2b. Chronic hypoxia caused a rise in %TWPV. Denervation alone gave a small increase in %TWPV (see controls) but in chronic hypoxia the response was significantly greater with denervation ($p<0,01$).

310

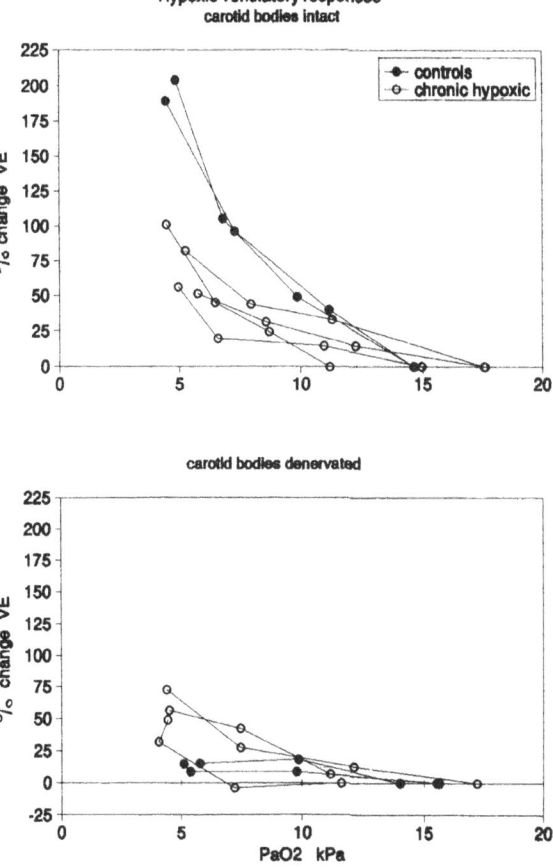

Figure 1 shows the individual hypoxic ventilatory responses for all rats; intact above and denervated below. Chronic hypoxia depresses HVR in intact rats. Denervation abolishes HVR in controls but there is a small response in chronically hypoxic rats.

Table 1 . Pulmonary vascular resistance and closing pressure measurements

	INTACT		DENERVATED	
	C	CH	C	CH
(n)	(2)	(4)	(2)	(4)
Intercept mmHg	5.9±1.4	8.5±0.4	6.5±0.1	13.7±0.6*#
Slope ml/min	0.75±0.01	0.93±0.05	0.68±0.00	1.10±1.13

means±sem, P<0.01, * cf intact; # cf C

Table 2

a)   Right Ventricular Hypertrophy (RV/LV+S/100g BW)

INTACT		DENERVATED	
C	CH	C	CH
0.11	0.21	0.1	0.18
0.18	0.19	0.13	0.25
	0.13		0.22
	0.11		0.16

b)   Pulmonary vascular remodelling (%TWPV)

INTACT		DENERVATED	
C	CH	C	CH
7.61	20.12	10.07	23.05
9.26	18.66	11.16	19.66
	16.84		25.79
	15.89		22.07

DISCUSSION

Pulmonary hypertension is caused by a combination of factors including pulmonary vasoconstriction, raised PVR, arterial thickening and polycythaemia. There is obviously a complex process occurring. As comparisons between the hypoxic responses of the pulmonary vessels and carotid body have often been discussed it seemed reasonable to investigate the possibility of a chronic involvement of the carotid body on PVR.

In our intact rats chronic hypoxia caused a depressed HVR as seen before[3]. Bilateral carotid body denervation completely abolished HVR in control but a small response was achieved in CH rats possibly through recruitment of other chemoreceptors, perhaps the aortic bodies. As yet we do not know the time course for this development. Denervation did not alter the physiology or the heart weights of control rats but raised the amount of peripheral pulmonary vascular thickening. In chronic hypoxia there was a greater degree of RV hypertrophy, pulmonary vascular remodelling and pulmonary hypertension. It is possible that this was caused by a greater degree of hypoxaemia in denervated rats as they had no hypoxic drive to ventilation to relieve their hypoxaemia, at least initially. The recruitment of other chemoreceptors to increase breathing was either too late or not sufficent to alleviate the cardiovascular changes. However we were unable to measure blood gases of rats inside the chamber and can only speculate on differences in arterial oxygenation. While rats were anaesthetised for the respiratory tests denervated rats showed a slightly lower PaO_2 than intact animals while breathing 10%O_2. This again is speculation on the possible cause of the differences and we have yet to show that two levels of chronic hypoxia give significantly different degrees of cardiovascular response and we have no reason to suppose that our meaurements are sensitive to levels of hypoxaemia. This is important as there is considerable variation in individual responses[2].

Acute experiments seem to point to reflex mechanisms that modify HPV and compliance during stimulation of the carotid body[5,6,7] (by hypoxia or almitrine, a specific peripheral chemoreceptor agonist). Levitsky et al[7] showed that unilobar hypoxic vasoconstriction in dogs was attenuated during carotid body stimulation possibly through activation of the para- sympathetic nervous system[11]. These authors imply a dilator action of carotid body stimulation on the pulmonary circulation. If pulmonary vasconstriction was the prime trigger for pulmonary hypertension then our results on chronic changes would support this view. There is however no evidence that RV hypertrophy and pulmonary vascular remodelling are a consequence of high Ppa alone but may be caused directly by hypoxia possibly via mechanisms involving the pulmonary vascular endothelium[12,13.]

In conclusion our preliminary study indicates that some potentiation of the chronic hypoxic adaptive mechanisms in the heart and lung occurs after bilateral carotid body denervation. The reasons for this remain obscure in this study with such small numbers and the need is indicated for physiological monitoring inside the chamber.

ACKNOWLEDGEMENTS

We are grateful to Mrs. Christine Brown for the histological preparations and to Mr. Wayne Sheedy for technical assistance.

REFERENCES

1. D. Bee, D. J. Pallot and G. R. Barer. Division of type 1 and endothelial cells in the hypoxic carotid body. Acta nat. 126:226-229 (1986).

2. C. Hunter, G. R. Barer, J. W. Shaw and E. J. Clegg. Growth of the heart and lungs in hypoxic rodents: a model of human hypoxic disease. Clin. Sci. Mol. Med. 46:375-391 (1974).

3. R. . Wach, D. Bee and G. R. Barer. Dopamine and ventilatory effects of hypoxia and almitrine in chronically hypoxic rats. J. Appl. Physiol. 67:186-192 (1989).

4. D. Bee and R. . Wach. Hypoxic pulmonary vasoconstriction in chronically hypoxic rats. Respir. Physiol. 56:91-103 (1984).

5. J. P. Szidon and J. F. Flint. Significance of sympathetic innervation of pulmonary vessels in response to acute hypoxia. J. Appl. Physiol. 43:65-71 (1977).

6. P. Herve, D. Musset, G. Simmonneau, W. Wagner, P. Duroux. Almitrine decreases the distensibility of the large pulmonary arteries in man. Chest 96:572-577 (1989).

7. M. G. Levitsky, J. C. Newell, J. A. Rrasney and R. E. Dutton. Chemoreceptor influence on pulmonary blood flow during unilateral hypoxia in dogs. Respir. Physiol. 31:345-356 (1977).

8. C. Smith, G. E. Bisgard, . M. Nielsen, L. Daristotle, N. A. Kressin, H. V. Forster and J. A. Dempsey. Carotid bodies are required for ventilatory acclimatisation to chronic hypoxia. J. Appl. Physiol. 60:1003-1010 (1986).

9. G. E. Bisgard, M. . Busch and H. V. Forster. Ventilatory acclimatisation is not dependent on cerebral hypocapnic alkalosis. J. Appl. Physiol. 60:1011-1015 (1985).

10. *H. V. Forster, G. E. Bisgard, B. Rasmussen, J. A. Orr, D. D. Buss and M. Manohar.* Ventilatory control in peripheral chemoreceptor denervated ponis during chronic hypoxaemia. J. Appl. Physiol. 41:878-885 (1976).

11. *L. B. Wilson and M. G. Levitsky.* Chemoreflex blunting of hypoxic pulmonary vasoconstriction is vagally mediated. J. Appl. Physiol. 66:782-791 (1989).

12. *B. E. Robertson, J. B. Warren and P. C. G. Nye.* Inhibition of nitric oxide synthesis potentiates hypoxic vasoconstriction in isolated rat lungs. Exp. Physiol. 75:255-257 (1990).

13. *R. Ross, J. Glomset, B. Kariya and L. Harker.* A platelet dependent serum factor that stimulates the proliferation of arterial smooth muscle cells in vitro. Proc. Natl. Acad. Sci. USA 71:1207-1210.

44 EFFECTS OF CHEMORECEPTOR STIMULATION BY ALMITRINE BISMESYLATE ON RENAL FUNCTION IN CONSCIOUS RATS

R. Behm[*], U. Franz[*], W. H. de Muinck Keizer,
H. Mewes[*], R. Rettig and Th. Unger

Department of Pharmacology and German Institute for High Blood
Pressure Research, Ruprecht-Karls University, D-6900 Heidelberg
[*]Department of Physiology, University of Rostock, D—2500 Rostock,
F.R.G.

INTRODUCTION

A possible link between chemoreceptor activity and kidney function has been demonstrated by several groups (Bardsley and Suggett 1987, Bardsley et al. 1991, Behm et al. 1992, Karim et al. 1987, Honig et al. 1985, Honig 1989, Schmidt et al. 1985). Most of these studies were performed in anaesthetized animals and led to conflicting results with marked differences between hypoxic and pharmacological chemoreceptor stimulation. Hypoxic stimulation of the vascularly isolated carotid bodies resulted in an increase in renal nerve activity (Linden et al. 1981), and in decreases in glomerular filtration rate (GFR), renal blood flow (RBF) and increase in tubular sodium reabsorption (Karim et al. 1987). After sectioning of the renal nerves the same type of chemoreceptor stimulation caused an increase in urinary water and sodium excretion. This shows that the chemoreceptors in the carotid bodies mediate reflex effects on renal function in two ways, one dependent on renal nerves and the other one on the release of one or more hormones. In intact conscious animals it is likely that hypoxic chemoreceptor stimulation activates the neuronal sympathetic pathway thereby masking the humoral axis. As a net result there is antidiuresis and antinatriuresis (Behm 1991).

Long lasting pharmacological stimulation of the carotid chemoreceptors can be achieved by the use of almitrine bismesylate. This substance acts as a respiratory stimulant by directly inducing a specific and long lasting excitation of the peripheral arterial chemoreceptors (Laubie and Schmitt 1980). Almitrine does not, however, simulate the effects of arterial hypoxia. Thus, the substance can modify the metabolism of dopamine in rat carotid bodies by producing decreases in dopamine content and utilization rate, whereas hypoxia elicits an increase in dopamine content (Pequignot et al. 1987). Furthermore, pharmacological antagonists of dopamine receptors potentiate the sensitivity of arterial chemoreceptors to hypoxic stimulation. In contrast, they do not alter the chemoreceptor response to almitrine (Bisgard 1981, O'Regan et al. 1987). Differences between the effects of pharmacologic and hypoxic stimulation of the carotid bodies on renal function in conscious animals may, therefore, be expected in conscious animals.

Indeed, several groups have demonstrated an almitrine-induced and chemoreceptor specific natriuresis in the anaesthetized (Bardsley and Suggett 1987, Bardsley et al. 1991, Honig et al. 1985) and conscious (Behm et al. 1987) animal.

The aim of the present study was to investigate the effect of almitrine on renal function in conscious intact and carotid body denervated rats using a new and more precise method of urine sampling.

METHODS

Animals and surgery

The right kidney was removed under light ether anaesthesia from 40 young rats weighing approximately 150 g. Animals were allowed 6 to 8 weeks for recovery from surgery. Then, the animals were anaesthetized again (chloral hydrate, 40mg/kg ip), and surgery for carotid body denervation (CBD) and catheter placement was performed. CBD was achieved after a ventral midline incision by cutting the left and right carotid sinus nerves, at the glossopharyngeal sinus nerve junctions. A sham operation (SO) consisting of superficial cervical muscle dissection was performed in all control rats. Carotid bodies were considered denervated if arterial pCO_2 was elevated above 43 mmHg 3 days after surgery.

Following closure of the neck incision, all rats were instrumented with catheters placed in the abdominal aorta via the right femoral artery and placed in the inferior vena cava via the right femoral vein. The catheters were tunnelled under the skin and exteriorized at the nape of the neck. The instrumentation with a ureter catheter was performed according to the method of Horst et al. (1988) and modified as follows: The abdomen was opened via a midline incision extending from the urinary bladder to the sternum. A special catheter was tunnelled under the skin from the abdomen up to the nape of the neck perforating the lateral abdominal wall. The ureter was carefully mobilized and gently freed from surrounding tissue at a length of 10 mm cranially from the level of the aortic bifurcation. A transverse incision was made in the ureter. The catheter was filled with heparinized saline and advanced 5 to 10 mm into the ureter and tied into place. The free end of the catheter was sewed to the skin in a way that the dripping urine would not stain the animal. Experiments were performed two days after recovery from surgery. Rats were considered for the protocol only, if they showed normal postsurgical food and water intake.

Experimental protocol

All experiments were performed between 08.00 and 12.00 a.m. Animals were placed in a Plexiglas chamber of sufficient size to allow free movement. The arterial catheter was connected to a Statham P23Db pressure transducer and a Gould Brush recorder to continuously measure mean arterial blood pressure (MABP) and heart rate (HR). The venous catheter was connected to an infusion pump (Unita, Braun-Melsungen, FRG) and, following a prime injection (0.3ml/100g body wt), an infusion of 4% polyfructosan (Inutest, Laevosan, Linz, Austria) in sterile saline was started at a rate of 17 ml/min. The exposed end of the ureter catheter was extended with a PE tubing to allow for urine collection outside of the chamber. After an equilibration period of 1 h, the following experimental protocol was employed. After 1 h of control measurement, almitrine or vehicle were injected intravenously by a bolus injection of 0.2 ml. Consecutive urine collections in 15 min periods were obtained throughout the protocol. Arterial blood samples were taken 30 min prior to as well as 15 min and 60 min after the bolus injection to assess arterial blood gases and plasma inulin concentrations. MABP and HR were monitored throughout the protocol. Urine volume was measured by weight.

Data were analyzed by Student's t-test for paired and unpaired samples as appropriate. Statistical significance was accepted if $P < 0.05$.

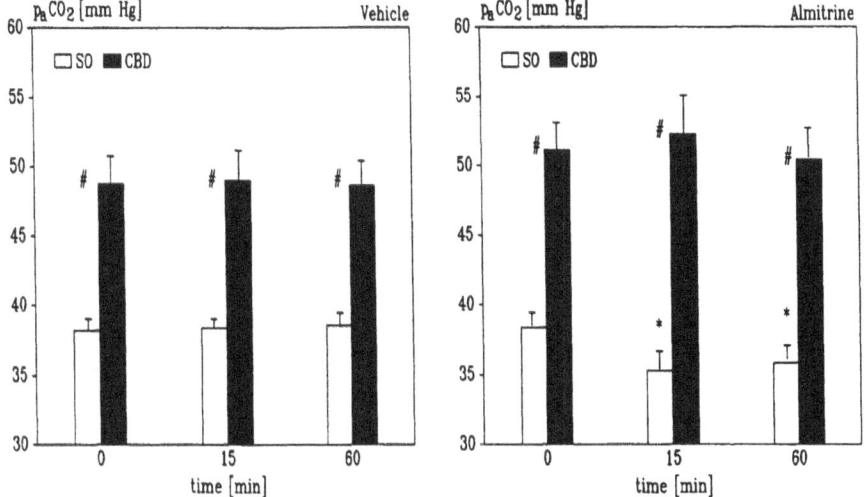

Fig.1. Effects of intravenous injections of vehicle (solvent) and almitrine bismesylate on arterial carbon dioxide tension ($p_a CO_2$) in sham operated (SO) and carotid body denervated (CBD) rats. Values are means ± S.E.M.

  * $P < 0.05$ compared to baseline, # $P < 0.05$ between groups.

RESULTS

The effectiveness of chemoreceptor denervation is documented by significant differences between SO and CBD rats in arterial PCO_2 48 8 ± 2.1 vs 38.2 ± 0 8 mmHg, $p < 0.05$, Fig 1) and PO_2 (71.7 ± 2.2 vs. 83.0 ± 1.6 mmHg, $p < 0.05$, not shown) at control conditions Bolus injections of almitrine induced hyperventilation associated with a significant decrease in arterial PCO_2 in SO but not in CBD rats.

Table 1. Effects of intravenous injections of vehicle (V) and almitrine bismesylate (A) on glomerular filtration rate (GFR), urine flow rate (V) and urinary sodium (U_{Na}-V) and potassium excretion (U_K·V) in sham operated (SO) and carotid body denervated (CBD) conscious rats.
Values are means ± S.E.M.
* P < 0.05 compared to baseline.

time [min]		0	15	30	45	60
GFR [μl/min]	SO-V	535±30	640±110	431±120	430±70	582±65
	SO-A	499±60	615±40	435±62	555±101	510±92
	CBD-V	415±57	454±76	371±47	319±48	401±33
	CBD-A	470±59	552±80	468±63	470±61	455±63
V̇ [μl/min]	SO-V	24.5±3.1	32.4±3.5 *	30.8±6.1	25.2±3.7	23.4±4.0
	SO-A	26.4±2.4	50.5±4.8 *	46.9±7.3 *	41.0±5.1 *	38.4±4.3 *
	CBD-V	30.3±5.8	40.8±8.0 *	37.9±6.7 *	29.3±6.1	29.7±4.4
	CBD-A	32.3±3.6	43.8±5.7 *	40.9±5.6 *	30.7±2.8	29.2±3.4
U_{Na}·V [μmol/min]	SO-V	1.33±.40	1.53±.50	1.40±.60	1.09±.40	0.86±.30 *
	SO-A	0.90±.21	1.63±.35 *	2.14±.51 *	2.19±.49 *	2.52±.55 *
	CBD-V	1.62±.61	2.04±.66 *	2.29±.75 *	1.88±.61	1.78±.60
	CBD-A	1.60±.39	2.56±.70 *	2.43±.64 *	1.82±.41	1.66±.40
U_K·V [μmol/min]	SO-V	2.50±.38	2.60±.30	2.57±.63	2.22±.42	2.23±.38
	SO-A	2.17±.32	2.69±.37 *	2.65±.41	2.52±.40	2.58±.38
	CBD-V	2.61±.41	2.94±.49	2.76±.54	2.53±.39	2.58±.29
	CBD-A	2.56±.27	2.31±.27	2.44±.28	2.52±.30	2.44±.20

Table 1 summarizes the effects of iv bolus injections of almitrine on GFR, urine flow as well as on urinary sodium and potassium excretion in SO and CBD rats. At baseline there was no significant difference in these parameters between SO and CBD rats. GFR was not affected by almitrine administration in either group. In SO rats almitrine elicited a sustained and significant increase in urine excretion, whereas in CBD rats only a transient increase was observed, which was not significantly different from the effects of vehicle alone. A striking and sustained increase in sodium excretion was observed after almitrine injection in SO rats. This effect was much less pronounced in CBD rats. Almitrine had no significant longer-lasting effects on urinary potassium excretion in SO and CBD rats There were no significant effects of almitrine on MABP and HR in either group.

DISCUSSION

The major findings of the present study are: 1) Intravenous bolus injections of almitrine were associated with a sustained diuresis and natriuresis while MABP remained constant. 2) GFR was unaffected by almitrine injection. 3) chemoreceptor denervation abolished the effects of almitrine on urinary flow rate indicating that carotid chemoreceptors essentially mediated this response. These results suggest that almitrine has effects on carotid bodies and renal function which are qualitatively different from those caused by hypoxia.

Both, arterial hypoxia and hypoxic perfusion of the carotid bodies cause renal vasoconstriction with a decrease in GFR and RBF in intact kidneys of anaesthetized and conscious rats. It has been demonstrated that these effects are mediated through the renal nerves (Karim et al. 1987, Schmidt et al. 1985). In contrast, in denervated kidneys chemoreceptor stimulation induced by perfusion of the vascularly isolated carotid sinus region causes a striking diuresis and natriuresis. This response was also abolished by crushing the carotid bodies (Karim et al. 1987).

The question arises as to whether this apparently humorally mediated effect is functionally comparable with the almitrine-induced natriuresis. Almitrine acts directly on the arterial chemoreceptors. Recently, Bardsley, et al. (1991) demonstrated that iv bolus injections of almitrine do not affect GFR and RBF in intact kidneys of anaesthetized rats, although a significant diuresis and natriuresis was observed. The same result were seen in denervated kidneys. In addition, acute and chronic administration of almitrine did not alter norepinephrine levels or turnover in the kidney indicating that sympathetic renal nerve activity remained unchanged (Dalmaz et al. 1987).

The findings of the present study performed in conscious rats confirm the results obtained in anaesthetized animals. Unaffected MABP and GFR suggest that the sympathetic system was not activated by almitrine-induced chemoreceptor stimulation. We assume that almitrine exerts its action on the kidney via a humoral pathway. At present, it remains unknown which hormone (s) is/are activated by chemoreceptor stimulation. Several hormones have already been excluded: 1) Keeping right atrial pressure constant, Karim et al. (1987) could demonstrate that the atrial natriuretic factor is not involved in the chemoreceptor-mediated natriuresis. 2) Almitrine induced a natriuresis in Brattleboro rats which lack vasopressin (Bardsley et al. (1991)). 3) It has been shown (Bardsley and Suggett 1987) that the adrenal gland is not essential for the almitrine-induced natriuresis.

It is conceivable that a hypothalamic natriuretic hormone may play a role in almitrine-induced natriuresis. In the rat, electrophysiological data demonstrate that carotid chemoreceptor afferents project to hypothalamic magnocellular neurons via the preoptic area (Harris et al. 1984). Further work is needed to elucidate whether this hypothalamic hormone or any other hormones are released by carotid chemoreceptor stimulation.

ACKNOWLEDGMENT

This investigation was supported in part by a grant from the Deutsche Forschungsgemeinschaft (DFG) (nn 47/2-4).

REFERENCES

Bardsley, P. A., and Suggett, A. J., 1987, The carotid body and natriuresis effect of almitrine bismesylate, Biomed. Biochim. Acta, 46 : 1017.

Bardsley, P. A. , Johnson, B. F. , Stewart, A. G. , and Barer, G. R., 1991, Natriuresis secondary to carotid chemoreceptor stimulation with almitrine bismesylate in the rat: The effect on kidney function and the response to renal denervation and deficiency of antidiuretic hormone, Biomed. Biochim. Acta, 50 :175.

Behm, R , 1991, Role of the peripheral arterial chemoreceptors in electrolyte homeostasis and cardiovascular regulation, Proceedings of the 7th International Hypoxia Symposium, Lake Louise, Canada.

Behm, R., Gerber, B., Griffel, D., Spee, B., and Zingler, C., 1987, Comparison of the response of renal sodium excretion to almitrine and hypoxia in conscious normotensive and spontaneously hypertensive rats, Biomed Biochim Acta, 46: 999.

Behm, R., Mewes, H., de Muinck Keizer, W. H., Rettig, R., and Unger, Th., 1992, Cardiovascular and renal response to hypoxia in chemoreceptor denervated rats, (submitted).

Bisgard, G. E., 1981, The response of few-fiber carotid chemoreceptor preparation to almitrine in the dog, Can. J. Physiol. Pharmacol, 59: 396.

Dalmaz, Y., Tavitian, E., Pequignot, J.M., Durra, A. , and Peyrin, L. , 1987, Acute and chronic effects of almitrine on sympathetic renal activity, systemic arterial blood pressure, hydromineral balance in normotensive rats, Biomed. Biochim. Acta, 46: 1023.

Harris, M. C., Ferguson, A. V., and Banks, D., 1984, The afferents pathway for carotid body chemoreceptor input to the hypothalamic supraoptic nucleus in the rat, Pflügers Arch., 400: 80.

Honig, A., Wedler, B., Zingler, C., Ledderhos, C., and Schmidt, M., 1985, Kidney function during arterial chemoreceptor stimulation, III. Long-lasting inhibition of renal tubular sodium reabsorption due to pharmacologic stimulation of the peripheral arterial chemoreceptors with almitrine bismesylate, Biomed. Biochim. Acta, 44: 1659.

Horst, P. -J., Bauer, M., Veelken, R. and Unger, Th., 1988, A new method for collecting urine directly from the ureter in conscious unrestrained rats, Renal. Physiol. Biochem. , 11: 325.

Karim, F., Poucher, S. M., and Summerill, R. A., 1987, The effects of stimulating carotid chemoreceptors on renal haemodynamics and function in dogs, J. Physiol. , 392: 451.

Laubie, M., and Schmitt, H., 1980, Long-lasting hyperventilation induced by almitrine: evidence for a specific effect on carotid and thoracic chemoreceptors, Eur. J. Pharmacol. , 61: 125.

Linden, R. J., Mary, D. A. S. G., Weatherill, D., 1981, The response in renal nerves to stimulation of atrial receptors, carotid sinus baroreceptors and carotid chemoreceptors, Q. J. Exp. Physiol., 66: 179.

O'Reagan, R.G., Kennedy, M., and Przybyszewski, A.W., 1987, Carotid body responses to administration of almitrine bismesylate, in:"Chemoreceptors in Respiratory Control", J.A. Ribeiro, and D. J. Pallot, eds., Croom Helm, Beckenham, U.K., p. 386.

Pequignot, J.M., Tavitian, E., Boudet, C., Evrard, Y., Claustrie, J., and Peyrin, L., 1987, Inhibitory effect of almitrine on dopaminergic activity of rat carotid body, J. Appl. Physiol., 63:746.

Schmidt, M., Ledderhos, C., and Honig, A., 1985, Kidney function during arterial chemoreceptor stimulation. I. Influence of unilateral renal nerve section, bilateral cervical vagotomy, constant artificial ventilation, and carotid body chemoreceptor inactivation, Biomed. Biochim. Acta, 44:695.

45

CARBON DIOXIDE-SENSITIVE LARYNGEAL RECEPTORS AND THEIR REFLEX EFFECTS

A. J. Bradford*, R.G. O'Regan, P. Nolan and D. McKeogh

*Department of Physiology, Royal College of Surgeons in Ireland, Dublin 2 and Department of Physiology and Histology, University College, Dublin 2, Ireland

INTRODUCTION

It has been knowm for more than a century that upper airway (UA) CO_2 reflexly inhibits breathing through the superior laryngeal nerve (SLN)[1,2] and it has been shown that SLN afferents are sensitive to CO_2[3]. However, in these studies, the larynx was opened by a ventral midline incision and/or was not in receipt of the pressures, airflows, temperatures and humidities associated with breathing.

We have developed an isolated, artificially-ventilated laryngeal preparation in anaesthetized cats which reproduces physiological conditions in order to study laryngeal receptor and reflex responses to CO_2.

METHODS

Adult cats of either sex were anaesthetized with pentobarbitone sodium (induction 40-48 mg. kg^{-1} i.p., rnaintenance 6-12 mg. i.v. as required), paralysed (pancuronium, 0.8 mg. i.v. as required) and artificially ventilated through a low cervical tracheostomy. The larynx was isolated and artificially ventilated as described previously. In nine cats, recordings were made of electroneurographic activity from single or few-fibre units of the peripheral end of the cut internal branch of the SLN using conventional techniques. Electroneurographic data was quantified as described previously[4].

For the reflex studies, nine additional, anaesthetized (pentobarbitone sodium, induction 30 mg. kg^{-1} i.p., maintenance 6 mg. i.v. as required) spontaneously breathing cats were used. The laryngeal innervation was carefully preserved. Laryngeal ventilation was synchronised with the animals spontaneous breathing using electronically controlled solenoid valves triggered by the spontaneous tracheal airflow signal recorded using a pneumotachograph attached to the low cervical tracheostomy. During inspiration, the valve to a vacuum source was opened drawing gas mixtures caudally over the larynx and during expiration, the valve to a positive pressure source was opened, pushing warmed, humidified gas mixtures cranially over the larynx. Raw and integrated electromyographic (EMG) activity was recorded from the genioglossus (GG) and diaphragm muscles. In both preparations, UA temperature, pressure, airflow and CO_2 concentration and arterial blood pressure were continually monitored. The larynx was ventilated with room air during

Neurobiology and Cell Physiology of Chemoreception
Edited by P.G. Data *et al.*, Plenum Press, New York, 1993

control periods and with mixtures of 5 and 9% CO_2 with 21% O_2 and the balance N_2 for 1-2 minutes during trials to determine responses to CO_2.

RESULTS

A total of 93 sensory units were recorded from the SLN and categorized in accordance with the protocol previously described[4]. There were 32 quiescent mechanoreceptors sensitive to negative pressure, 24 quiescent mechanoreceptors sensitive to positive pressure, 17 tonically active mechanoreceptors sensitive to positive pressure, 7 tonically active mechanoreceptors sensitive to negative pressure, 8 cold airflow receptors and 5 fibres with no consistent response to occlusion of the oral cannula.

The effects of 5 and 9% CO_2 are summarized in Fig. 1. The discharge of a substantial fraction of all categories of fibres was significantly altered by 5 and 9% CO_2 in a dose-dependent manner.

Tonically active positive and negative pressure receptors and quiescent positive pressure receptors were inhibited whereas quiescent negative pressure receptors, cold airflow receptors and fibres with no consistent response to occlusion were excited. Receptor discharge was unaffected by UA hypoxia and by systemic hypercapnia and asphyxia.

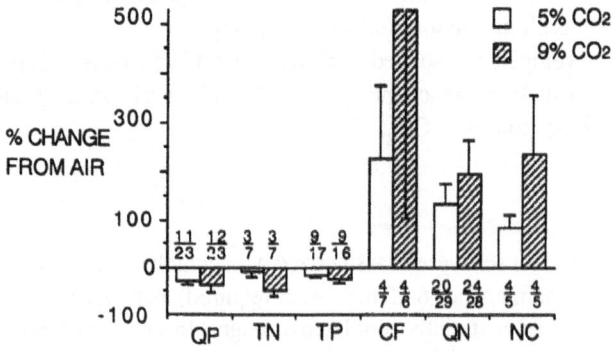

Fig. 1. The effect of 5% CO_2 (open bars) and 9% CO_2 (striped bars) on the 6 categories of SLN afferents. The bars represent the mean % change ($\pm$ SE) in activity per respiratory cycle compared to air. The fractions for each bar indicate the fraction of fibres tested which showed a significant change in activity (P < 0.05, Student's paired t test). QP = quiescent fibres responsive to positive pressure. TN = tonic fibres responsive to negative pressure, TP = tonic fibres responsive to positive pressure, CF = cold airflow receptors. QN = quiescent fibres responsive to negative pressure. NC = fibres with no consistent response to occlusion.

For the reflex studies, phasic GG EMG activity was abolished and respiratory frequency was increased by cessation of laryngeal ventilation or by cutting both SLNs whilst the larynx was being ventilated. Introducing 5 or 9% CO_2 into the gas ventilating the larynx caused marked, dose-dependent excitation of the GG muscle (40 out of 46 trials) and a decrease in respiratory frequency (36 out of 46 trials) as shown in Fig. 2. These responses were unaffected by peripheral arterial chemoreceptor denervation, were greatly reduced by cutting one SLN and were abolished by cutting both SLNs.

DISCUSSION

The present results show that under physiological conditions, laryngeal receptors are either excited or inhibited by 5 and 9% CO_2. The significance of the excitatory effects of CO_2 on cold airflow receptors and on fibres with no consistent response to occlusion is unclear, especially since the adequate stimuli of the latter category of fibres were not determined.

However, since CO_2 excited quiescent negative pressure receptors and inhibited quiescent positive pressure receptors, we hypothesized that these reciprocal effects may contribute to reflex control of UA patency. It is known that negative UA pressure reflexly excites and positive UA pressure reflexly inhibits the GG muscle, a major pharyngeal dilator[5,6]. This reflex is mediated through the SLN but the receptors have not been identified. We reasoned that laryngeal CO_2, by exciting and inhibiting negative and positive receptors respectively, would also reflexly excite the GG muscle.

The present reflex studies show that phasic GG EMG activity, under the conditions of the present experiments, was dependent on SLN afferent activity and that intralaryngeal CO_2 does indeed reflexly excite the GG muscle and inhibit breathing. The GG excitation is not secondary to the depression of ventilation

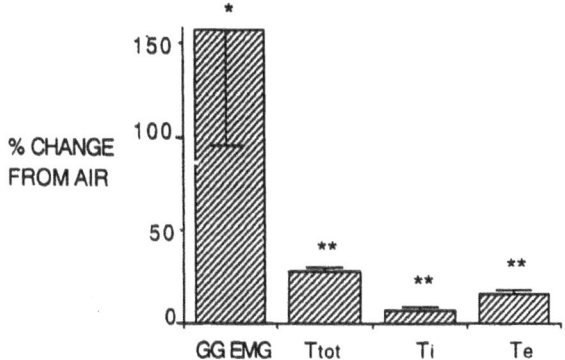

Fig. 2. *Effect of 9% CO_2 on mean peak integrated genioglossus EMG acvity (GG EMG), total respiratory cycle duration (Ttot), inspiratory time (Ti) and expiratory time (Te). Values are mean % changes ($\pm$SE) for all trials in which a significant effect occurred. * P 0.05 (n=40) and ** P < 0.01 (n=36) vs. air control. Student's t test for Daired data.*

since it was also observed in trials where breathing was unaffected. Taken together, our receptor and reflex studies suggest that quiescent pressure responsive receptors may be of importance in mediating the reflex thoracic and UA respiratory muscle responses to respiratory-related UA stimuli.

We now propose that CO_2-sensitive SLN afferents may be important in ventilatory control and in promoting UA patency, especially during obstructive apnea when, in addition to greater negative pressures being developed below pharyngeal obstruction, there may also be raised concentrations of CO_2. The resultant augmentation of UA muscle activity and inhibiton of diaphragmatic activity would counteract phayngeal collapse by dilating the pharynx and by reducing the negative pressure generated by the thoracic muscles[7].

REFERENCES

1. *F. Kratchmer*, Uber reflexe von der Nasenschleimhaut auf Athmung und Kreislauf, Sber. Akad. Wiss. Wien. 62:147 (1870).

2. *H. A. Boushey, and S. Richardson*, The reflex effect of intralaryngeal carbon dioxide on the pattern of breathing, J. Physiol. 228:181 (1973).

3. *H. A. Boushey, P.S. Richardson, J. G. Widdicombe, and J. C. M. Wise*, The response of laryngeal afferent fibres to mechanical and chemical stimuli, J. Physiol. 240:153 (1974).

4. *A. Bradford, C. Bannon, P. Nolan, and R. G. O'Regan*, A study of the effects of airway carbon dioxide ($P_{aw}CO_2$) on superior laryngeal nerve afferents using an isolated, artificially ventilated closed laryngeal preparation in the anaesthetized cat, in: "Chemoreceptors and Chemoreceptor Reflexes, H. Acker, A. Trzebski, and R. G. O'Regan, ed., Plenum Press, New York (1990).

5. *O. P. Mathew, Y. K. Abu-Osba, and B. T. Thach*, Influences of upper airway pressure changes on genioglossus muscle activity, J. Appl. Physiol. 52:438 (1982).

6. *O. P. Mathew, Y. K. Abu-Osba, and B. T. Thach*, Genioglossus muscle responses to upper airway pressure changes: afferent pathways, J. Appl. Physiol. 52:445.

7. *J. E. Remmers, W. J. deGroot, E. K. Sauerland, and A. M. Anch*, Pathogenesis of upper airway occlusion during sleep, J. Appl. Physiol. 44:931 (1978).

46 VENTILATORY RESPONSE AT THE ONSET OF EXERCISE: AN UPDATE OF THE NEUROHUMORAL THEORY

P. Cerretelli [1-2], L. Xi [3], F. Schena [1], C. Marconi [2], B. Grassi [2], G. Ferretti [1] and M. Meyer [3]

[1]Dept. of Physiology, Univ. of Geneva, CH; [2]I.T.B.A. C.N.R., Milano, I; [3]Max-Planck-Institute for Exp. Med., Göttingen, FRG

INTRODUCTION

There appears to be agreement on the fact that the ventilatory response at the onset of a constant-load exercise is characterized in man by two components (Dejours, 1963): a) an abrupt reflex response ("phase I", ph1 ventilatory response, according to the definition of Wasserman, 1978), most probably originating from the contracting muscles, whose afferent pathway is presumably represented by the group III spinal afferent fibers (Mitchell et al., 1977; Mitchell, 1990); b) a delayed response ("phase II and III", ph2 and ph3), mainly controlled by the endogenous production of CO_2. The way these components interact has been for a long time, and still is, a matter for discussion.

The repetition with more sophisticated equipement of a classic experiments earlier carried out by Lefrançois and Dejours (1964), by Asmussen (1973) and by Ward et al. (1983) allowed us to gain a better insight into the problem.

SUBJECTS, METHODS AND EXPERIMENTAL PROCEDURE

Six untrained healthy male subjects (age 33.7±7.6 yr; body weight 68.2±4.6 kg) volunteered for the study. They were familiar with the laboratory procedures but unaware, as were the laboratory operators, of the aim of the investigation.

Pulmonary ventilation ($\dot{V}_E$), end tidal PCO_2 ($PETCO_2$) and gas exchange ($\dot{V}O_2$ and $\dot{V}CO_2$) were determined breath-by-breath (BbB) by means of a MMC 4400tc SensorMedics cart. The respiratory variables were collected during three rectangular, 5 min duration work loads (50, 100, 150 W), carried out on an electrically braked bicycle ergometer (Bosch ERG 551). Together with ph1, also ph2 and ph3 and the kinetics of gas exchange (t1/2 $\dot{V}O_{2on-}$, t1/2 $\dot{V}CO_{2on-}$) were obtained.

Cardiac output ($\dot{Q}$) was determined on a beat-by-beat basis (bbb) by impedance cardiography (details of the method in Meyer et al., 1989). $\dot{Q}$ readjustment kinetics was determined bbb from the onset of the 50 W load on 4 of the subjects in normocapnia.

The experimental procedure was as follows. The subject sat for 5 min on the bicycle ergometer either breathing room air (normocapnic mode, N), or hyperventilating ($\dot{V}_E =$ x 3 the resting level) for the first 4 min and resuming spontaneous $\dot{V}_E$ over the 5th minute (hypocapnic mode, H). He was then asked to carry out the imposed load without warning, starting from a standardized position with the left pedal arm kept in a frontal horizontal plane. The load was attained within 3 s. The pedalling rate was set at 60 ± 2 r.p.m. On the subjects for whom both BbB gas exchange and bbb $\dot{Q}$ were assessed, the CO_2 flow to the lungs ($\dot{Q}CO_2$) from the onset of exercise could be calculated as follows:

$$\dot{Q}CO_2 = \dot{Q} \cdot C\bar{v}CO_2 \tag{1}$$

where:

$$C\bar{v}CO_2 = CaCO_2 + \dot{V}CO_2 \ (BbB)/\dot{Q} \ (\text{average over 1 breath period}) \tag{2}$$

Thus, all variables on the right side of eq. (2) were measured, except $CaCO_2$ which was calculated from $PETCO_2$ ($\sim PaCO_2$) from Dill's nomogram (Dill et al., 1937).

RESULTS

The ventilatory response at 50 W is shown for a typical subject in Fig. 1 for both normocapnic (N) and hypocapnic (H) conditions. In the N mode, $\dot{V}_E$ appears to increase abruptly at the onset of work (ph1). This increase is followed by a short transient drop, likely a consequence of the decrease of $PETCO_2$ elicited by the preceding hyperventilatory reaction. When $PETCO_2$ resumes a threshold level of ~39 mmHg, $\dot{V}_E$ starts increasing (ph2) until a steady state is attained (ph3). The hypocapnic tracing (H) shows, immediately after ph1, a prolonged drop of $\dot{V}_E$ even below pre-exercise values. At the same $PETCO_2$ threshold, $\dot{V}_E$ increases as for N. The duration and amplitude of ph1 are approximately the same in N and H. The consequences of increasing work load from 50 to 150 W is shown in Fig. 2. The $\dot{V}_E$ time course is rather similar to that observed in Fig. 1. The 6 investigated subjects behaved substantially in the same manner.

The BbB $\dot{V}_E$, $\dot{V}O_2$ and $\dot{V}CO_2$ transients in normocapnia are shown in Fig. 3 for one of the subjects, along with the kinetics of readjustment of $\dot{Q}$. From the above data, the calculated $\dot{Q}CO_2$ (see eq. 1 and 2) during the first minute of exercise is plotted in Fig. 4 together with the corresponding ventilatory response. In contrast to $\dot{V}_E$, which increases rather suddenly at work onset (ph1), $\dot{Q}CO_2$ does not appear to change significantly over the first 10-15 sec after exercise onset. Similar results were obtained for all investigated subjects (n=4).

DISCUSSION

The amplitude of ph1 was approximately the same both in N and in H. This finding is in contrast with the results of Asmussen (1973) who did not observe ph1 in two subjects after volitional hyperventilation in O_2, and of Ward et al. (1983) who found that ph1, after volitional hyperventilation, was drastically reduced, but is in agreement with the original results of Lefrançois and Dejours (1964), who showed that after hyperventilation

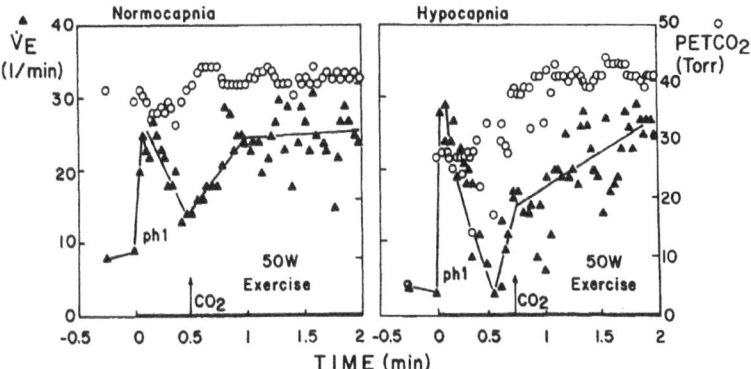

Fig. 1. Ventilatory response to exercise (50W) in normocapnia (left panel) and in hypocapnia (right panel) in a typical subject. $\dot{V}_E$ = pulmonary ventilation; $PETCO_2$ = end tidal CO_2 partial pressure.

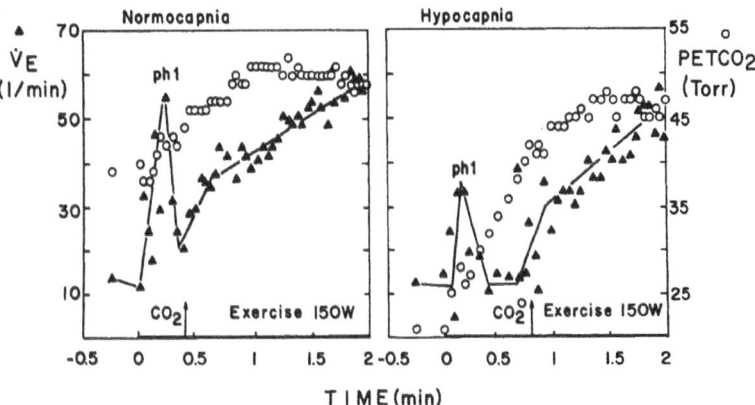

Fig 2 Ventilatory response to exercise (150 W) in normocapnia (left panel) and in hypocapnia (right panel) in a typical subject. $\dot{V}_E$ = pulmonary ventilation; $PETCO_2$ = end tidal CO_2 partial pressure.

ph1 is independent of the PACO₂ level. The present results cannot be attributed to the different work load imposed on the subjects. The latter, indeed, contrary to the finding of Asmussen and Nielsen (1948) does not affect the amplitude of the $\dot{V}_E$ response (personal observation, 1991). According to Lefrançois and Dejours (1964), ph1 is of neural origin and lasts only for one or two breaths. The $\dot{V}_E$ response is then followed by the chemical response elicited by the work-dependent increase in blood PCO₂. Thus, Dejours' neurohumoral theory (Dejours, 1963) is confirmed, only needs to be updated. Indeed, there appears to be a total dichotomy between the neurogenic and the humoral determinants of $\dot{V}_E$, which is clearly shown by the present results. ph1, independent of its origin, after a few seconds following work onset tends to subside. In normocapnic subjects this tendency is almost masked by the incoming chemical drive. By contrast, it becomes striking when work is preceded by volitional hyperventilation. Since the activity of the cortical and/or subcortical (hypothalamic) motor centers is likely constant during constant-load exercise, the present results are only compatible with a reflex drive of the type "exercise reflex" as hypothesized by Zuntz and Geppert (1886) rather than with the feed-forward mechanism described by Eldridge et al. (1981). The observed transient ph1 drive could indeed depend on the input from group III spinal afferent fibers in response to muscle contraction. In fact, this response is characterized by a strong initial firing activity, followed by rapid adaptation (Mitchell et

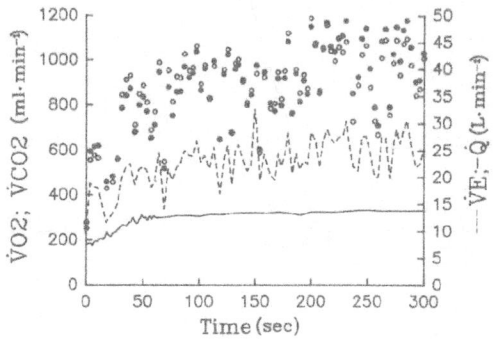

Figure 3 - Ventilatory ($\dot{V}_E$), gas exchange ($\dot{V}O_2$ and $\dot{V}CO_2$) and cardiac output ($\dot{Q}$) kinetics of readjustment to a 50 W exercise in normocapnia (typical subject).

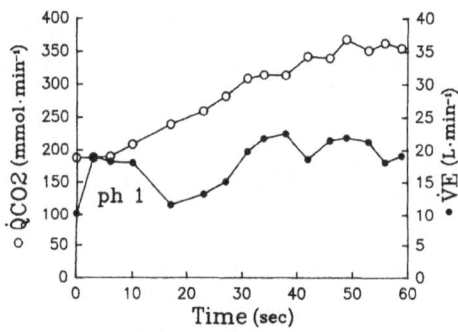

Figure 4. Pulmonary ventilation ($\dot{V}_E$) and calculated CO_2 flow to the lungs ($\dot{Q}CO_2$) during the first minute of a 50 W exercise in normocapnia (same subject as for Fig. 3).

al. 1977; Mitchell, 1990) whose time course (t1/2 ~10-20 sec) appears to match surprisingly well that of ph1 in the absence of the CO_2 stimulus.

An exercise reflex from the contracting limbs, however, is not the only potential determinant of ph1. Alternatively, chemical and/or mechanical inputs from the heart and/or the lungs have been postulated (Wasserman et al., 1974; Jones et al., 1982). These could potentially act on the breathing centers and sustain ph1. Recent preliminary studies by our group, particularly experiments on heart and lung transplant recipients (Cerretelli et al., 1989) appear strongly against such mechanisms. In addition, the possibility that ph1 be driven by an increased CO_2 load in the presence of CO_2 receptors in the lungs and/or in the right heart can be ruled out by the results of the present experiments, showing that in N, but particularly in H conditions, ph1 is dissociated from the hypothesized increase of $\dot{Q}CO_2$, which, in fact, could not be shown.

CONCLUSIONS

From the present preliminary experiments it appears that ph1:

1) Is likely generated by a phasic peripheral input since it subsides shortly after the onset of exercise, in spite of the continuing discharge of the motor cortex. Its time course is similar to that of the firing pattern of group III afferent fibers described by Mitchell et al. (1977) an by Mitchell (1990).

2) Is independent of the pre-exercise $PaCO_2$ and of the work load.

3) Is also independent of changes of the CO_2 load to the lungs.

Thus the original hypothesis of Zuntz and Geppert (1886), who attributed ph1 to a muscle "reflex" is essentially confirmed, except for the allegation that the latter reflex is of chemical nature. In fact, it could be elicited exclusively by motion and, besides, its amplitude is essentially not influenced by the muscle mass recruited or by changes of important variables such as tissue PO_2 (Miyamura et al. 1990) and PCO_2.

ACKNOWLEDGEMENTS

This work was partially supported by Grant n. 3228719.90 of the Fond National Suisse for the Scientific Research.

REFERENCES

Asmussen, E., 1973. Ventilation in the transition from rest to exercise. Acta Physiol. Scand. 89:68.

Asmussen, E. and M. Nielsen, 1948. Studies on the initial changes in respiration in the transition from rest to work and from work to rest. Acta Physiol. Scand. 16: 270.

Cerretelli, P., B. Grassi, G. Ferretti, A. Colombini, M. Rieu, L. Xi, M. Meyer and C. Marconi, 1989. Regulation of ventilation at exercise after lung and heart denervation in humans. FASEB J. 3:A855 (Abstract).

Dejours, P., 1963. The regulation of breathing during muscular exercise in man. A neuro-humoral theory. In: "The regulation of Human Respiration", D.J.C. Cunningham and B.B. Lloyd, ed. Blackwell, Oxford (pp. 535-547).

Dill, D.B., H.T. Edwards and W.V. Consolazio, 1937. Blood as a physicochemical system. XI. Man at rest. J. Biol. Chem. 118:635.

Eldridge, F.L., D.E. Millhorn and T.G. Waldrop, 1981. Exercise hyperpnea and locomotion: parallel activation from the hypothalamus. Science 211: 844.

Jones, P.W., A. Huszczuk and K. Wasserman, 1982. Cardiac output as a controller of ventilation through changes in right ventricular load. J. Appl. Physiol. 53: 218.

Lefrançois, R. and P. Dejours, 1964. Etude des relations entre stimulus ventilatoire gaz carbonique et stimulus ventilatoire neurogeniques de l'exercice musculaire chez l'homme. Rev. Franc. Etudes Clin. Biol. 9:498.

Meyer, M., P. Cerretelli, C. Marconi, M. Rieu and C. Cabrol, 1989. Cardiorespiratory adjustment to exercise after cardiac transplantation. In: "Clinical Aspects of O_2 Transoprt and Tissue Oxygenation", K. Reinhart and K. Eyrich, ed. Springer Verlag, Berlin, (pp. 477-499).

Mitchell, J.H., 1990. Neural control of the circulation during exercise. Med. Sci. Sports Exercise 22:141.

Mitchell, J.H., W.C. Reardon and D.J. McCloskey, 1977. Reflex effects on circulation and respiration from contracting skeletal muscle. Am. J. Physiol. 233: H374.

Miyamura, M., L. Xi, K. Ishida, F. Schena and P. Cerretelli, 1990. Effects of acute hypoxia on ventilatory response at the onset of submaximal exercise. Jpn. J. Physiol. 40:417.

Ward, S.A., B.J. Whipp, S. Koyal and K. Wasserman, 1983. Influence of body CO_2 stores on ventilatory dynamics during exercise. J. Appl. Physiol. 55:742.

Wasserman, K., 1978. Breathing during exercise. N. Engl. J. Med. 298:780.

Wasserman, K., B.J. Whipp, R. Casaburi and W.L. Beaver, 1977. Carbon dioxide flow and exercise hyperpnea. Cause and effect. Am. Rev. Respir. Dis. 115:225.

Wasserman, K., B.J. Whipp and J. Castagna, 1974. Cardiodynamic hyperpnea: hyperpnea secondary to cardiac output increase. J. Appl. Physiol. 36:457.

Zuntz, N. and J. Geppert, 1886. Uber die Natur der normalen Atemreize und den Ort ihrer Wirkung. Pflügers Arch. ges. Physiol. 38:337.

47 CAROTID CHEMORECEPTOR REFLEX CARDIOINHIBITORY RESPONSES: COMPARISON OF THEIR MODULATION BY CENTRAL INSPIRATORY NEURONAL ACTIVITY AND ACTIVITY OF PULMONARY STRETCH AFFERENTS

M. de Burgh Daly

Department of Physiology, Royal Free Hospital School of Medicine
London NW3 2PF, U.K.

The effectiveness of some excitatory inputs to cardiac vagal motoneurones (CVMs) is modified by phasic changes in the respiratory cycle. There are two mechanisms by which this modulation is brought about: changes in central inspiratory neuronal activity and changes in activity of pulmonary stretch afferents, both acting contemporaneously. Accordingly, during the inspiratory phase of the respiratory cycle the CVMs are partly or wholly refractory to excitatory inputs from carotid chemoreceptors and baroreceptors (Davidson, Goldner & McCloskey, 1976; Gandevia, McCloskey & Potter, 1978; McAllen & Spyer, 1978a, b; Potter, 1981) and from cardiac ventricular C fibres (Daly, Kirkman & Wood, 1988).

Daly & Kirkman (1988) studied another excitatory input to the CVMs and found in contrast to the above results that the bradycardia evoked by stimulation of pulmonary C fibre afferents was not subject to respiratory modulation, or only weakly so. This implied that the pulmonary C fibre input to the CVMs is unaffected by changes in either the central inspiratory neuronal activity or the activity of pulmonary stretch afferents. Further experiments were therefore carried out to test this hypothesis which in the event was confirmed.

The respiratory modulation by the two mechanisms was studied in separate series of experiments. It was necessary, however, to establish the viability of the preparations by demonstrating the known respiratory modulation of another excitatory input. In fact the opportunity was taken of making a quantitative comparison of the effects of changes in central inspiratory neuronal activity and activity of pulmonary stretch afferents on the cardioinhibitory responses evoked by stimulation of three other groups of receptors, the arterial baroreceptors, carotid chemoreceptors and cardiac C fibre endings.

In the first series of experiments the effects of single inflations of the lungs were studied on the cardioinhibitory responses elicited by stimulation of individual groups of cardiovascular receptors during a period of cessation of inspiratory neuronal activity evoked by electrical stimulation of the central cut ends of both the superior laryngeal nerves. In the second, phasic changes in central inspiratory neuronal activity were measured quantitatively as changes in the volume of the pneumothorax during temporary

Neurobiology and Cell Physiology of Chemoreception
Edited by P.G. Data *et al.*, Plenum Press, New York, 1993

interruption of artificial respiration, the volume of the lungs being held constant at their end-expiratory level. In this way the activity of slowly adapting pulmonary stretch receptors was maintained constant.

METHODS

Cats of either sex were anaesthetized with a mixture of α-chloralose (52 mg kg^{-1}) and urethane (520 mg kg^{-1}) administered intraperitoneally. An open pneumothorax was created and respiration was maintained artificially by a Starling 'Ideal' pump, the lungs collapsing passively against a resistance of 0.5-3.0 cm H_2O pressure.

In the first series of experiments tests of controlled single inflations of the lungs from a syringe were carried out during temporary interruption of artificial respiration and during central apnoea induced by electrical stimulation of the central cut ends of the superior laryngeal nerves with pulses of 2-4V, 2ms duration at a frequency of 30-50Hz. A stimulus strength was used just sufficient to stop respiratory movements for a period of 20s, but without having any chronotrophic effect on the heart (Daly & Kirkman, 1989).

In the second series of experiments, movements of the lungs were temporarily stopped by disconnecting the lungs from the pump, the lungs being held at constant volume at their end-expiratory level (expriratory resistance 1.5-3.0 cm H_2O pressure). Changes in volume of the pneumothorax (Vpn) provided a quantitative measure of the phasic output of the respiratory centres through movements of the diaphragm and rib cage (Daly, 1991). For this purpose, the two tubes that had been passed through the chest wall were connected to a pneumotachograph (Godart).

Reflex cardioinhibitory responses were elicited as follows: the carotid body chemoreceptors on one side or both sides simultaneously were stimulated by sodium cyanide (0.01-0.1% w/v) injected into the common carotid arteries; pulmonary C fibres by injections of phenylbiguanide (200 µg ml^{-1}) into the right atrium via a catheter inserted into the right external vein; cardiac C fibre endings by veratridine (50 µg ml^{-1}) injected into the catheterized left atrium; and the arterial baroreceptors by a sustained rise in blood pressure in the aortic arch and carotid circulation, thereby eliciting an arterial 'baroreflex'. For this purpose a Swan-Ganz balloon-tipped monitoring double-lumen catheter was inserted into the descending thoracic aorta via a femoral artery. The pressure was raised by controlled partial inflation of the balloon.

RESULTS

Effects of lung inflation on cardioinhibitory reflexes

The four cardioinhibitory reflexes were elicited from the carotid chemoreceptors, arterial baroreceptors, cardiac C fibre receptors and pulmonary C fibre endings under conditions in which the central inspiratory neuronal activity was inhibited by electrical stimulation of the central cut ends of the superior laryngeal nerves and the lungs were held static in their expiratory position. The test was then repeated, the lungs being inflated during the period of maximum response. The results are summarized in Fig.1 in which the lungs were inflated with 22.7 ± 1.2 (mean ± s.e.m.)ml kg^{-1} of room air, corresponding to a maintained inflation pressure of 8.1± 1.3 mmHg.

It will be noted that lung inflation had the greatest effect on the bradycardia which was elicited by stimulation of the carotid bodies. On the other hand there was no effect on the bradycardia elicited by excitation of pulmonary C fibre endings.

Figure 2 shows a comparison of the effectiveness of the pulmonary input driven by

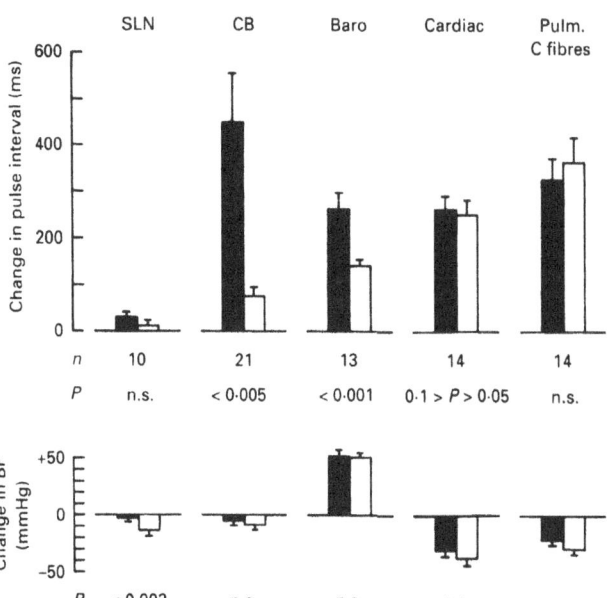

Fig. 1. The effects of inflation of the lungs on the cardioinhibitory responses elicited by electrical stimulation of the superior laryngeal nerves (SLN), stimulation of the carotid body chemoreceptors (CB, intracarotid sodium cyanide, 6.6 ± 0.4 mg kg⁻¹), elicitation of the arterial baroreflex (Baro, increase in blood pressure, 51.0 ± 4.4 mmHg), and stimulation of the cardiac receptors (Cardiac, left atrial veratridine, 3.3 ± 0.3 μg kg⁻¹) and pulmonary C fibres (Pulm. C fibres, right atrial phenylbiguanide, 15.1±0.9 mg kg⁻¹). All tests carried out during electrical stimulation of the SLNs to inhibit central inspiratory neuronal activity. Filled rectangles, excitation of cardioinhibitory responses, the lungs being held at their expiratory volume; Open rectangles, excitation of cardioinhibitory responses with superimposition of inflation of the lungs (From Daly and Kirkman, 1989).

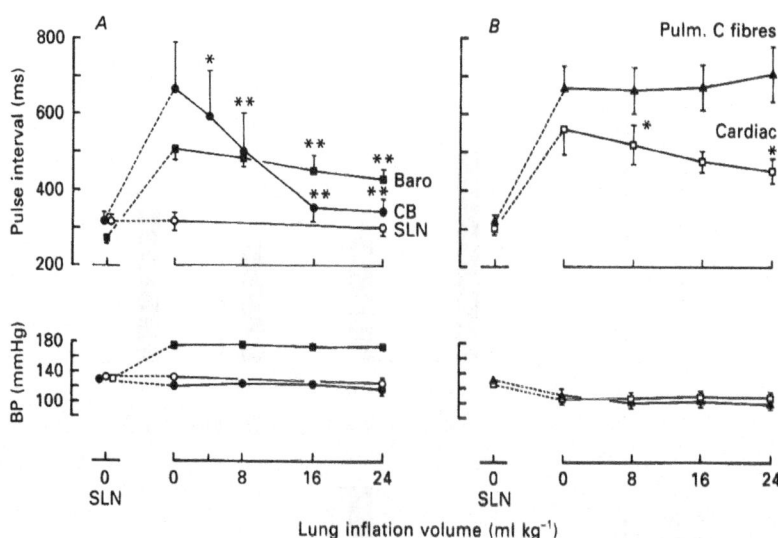

Fig. 2. The effects of inflation of the lungs with different volumes of gas on the cardioinhibitory responses to stimulation of various cardiovascular receptor groups. Positive pressure ventilation. Open pneumothorax. Each value for pulse interval and mean arterial blood pressure was obtained under conditions of cessation of the central inspiratory neuronal activity produced by electrical stimulation of a superior laryngeal nerve. *Left hand* section of each panel, control observations during stimulation of a superior laryngeal nerve with lungs held at the end-expiratory pressure. *Right hand* section of each panel shows the effects of stimulation of the arterial baroreflex (*Baro*; arterial blood pressure increase of 43.7 ± 5.0 mmHg; n=7), the carotid body chemoreceptors (*CB*; intracarotid sodium cyanide, 5.9 ± 0.6 μg kg^{-1}; n=7), pulmonary C fibres (*Pulm. C fibres*; right atrial phenylbiguanide, 17.9 ± 1.9 μg kg^{-1}; n=8), and cardiac receptors (*Cardiac*; left atrial veratridine, 4.5 ± 0.7 μg kg^{-1}; n=5). The cardioinhibitory reflexes were elicited at zero lung volume (lungs at expiratory pressure), and combined with inflation of the lungs with 4, 8, 16 and 24 ml. kg^{-1} room air. Similar control observations were made on the effects of inflation of the lungs during electrical stimulation of the superior laryngeal nerve alone (SLN;n=9). Values are the means ± SEM (From Daly and Kirkman, 1989).

single inflations of the lungs with increasing volumes of 4, 8, 16 and 24 ml kg^{-1}, selected in random order. Increasing the volume of lung inflation caused a progressive reduction in the size of the cardioinhibitory responses elicited by stimulation of the carotid chemoreceptors, the arterial baroreflex and the cardiac receptors. The suppression of the bradycardia with increasing lung volume was greatest in the case of the carotid chemoreceptor-induced bradycardia, the size of the response at 24 ml kg^{-1} lung inflation being only 6.8 $\pm$ 2.4% of that occurring without inflation of the lungs. The corresponding values for the arterial baroreflex and cardiac receptor responses were 66.4 $\pm$ 4.6 and 69.7 $\pm$ 8.7% respectively.

On the other hand, the cardiac response to pulmonary C fibre stimulation was unaffected by inflation of the lungs (Fig. 2). Of the eight series of tests carried out, however, lung inflation increased the pulmonary C fibre - evoked bradycardia in four test, reduced it slightly in two, and had no effect in the remaining two tests.

Selective surgical denervation of the lungs (Daly & Scott, 1958) abolished the Hering-Breuer respiratory reflex evoked by lung inflation and the effects of lung inflation on the cardioinhibitory responses elicited by stimulation of the carotid chemoreceptors, arterial baroreceptors and cardiac receptors. The bradycardia of pulmonary C fibre stimulation was reduced in size by 90% in half the tests or abolished. Control tests of electrical stimulation of the superior laryngeal nerves alone which caused cessation of respiratory movements had no effect on pulse interval.

Effects of central respiratory activity on cardioinhibitory reflexes

When artificial respiration was temporarily interrupted, rhythmic respiratory movements, measured as the changes in volume of the pneumothorax (Vpn), continued. The pulse interval measured at the peak of inspiration was compared with that at the end-expiratory phase of respiration. Without stimulation of any of the cardiovascular receptor groups, no significant change in pulse interval occurred (Fig. 3A). Typical responses to stimulation of the four groups of cardiovascular receptor groups are shown in Fig. 3B-E.

The averaged results are summarized in Fig.4. Stimulation of the carotid body receptors caused bradycardia, the pulse interval, measured during the expiratory phase of the respiratory cycle, increasing by 380.5 $\pm$ 52.5 ms from a control value of 308.3 $\pm$ 10.5 ms. During the inspiratory phase of the cycle there was an acceleration of the heart corresponding to a reduction in pulse interval of 338.8 $\pm$ 47.4 ms and this response was associated with a Vpn of 25.8 $\pm$ 1.3 ml kg^{-1}. The pulse interval was, therefore, reduced to a value of only 57.5 $\pm$ 13.6 ms greater than the control value without stimulation of the carotid bodies. Central inspiratory neuronal activity thus reduced pulse interval to 15.1% of the expiratory value without chemoreceptor stimulation.

The central inspiratory drive was less effective in altering the reflex cardioinhibitory responses from the arterial baroreceptors and cardiac receptors, the corresponding values being 42 and 51% respectively. In contrast, the bradycardia evoked by pulmonary C fibre stimulation was not significantly affected by the central inspiratory drive (Fig.4).

To see whether the degree of modulation of an excitatory input from each cardiovascular receptor group was affected by the size of the central inspiratory drive, the data were replotted so as to relate the inspiratory and expiratory values of pulse interval to the level of Vpn. The results for the carotid chemoreceptor input are shown in Fig.5 and indicate that the expiratory bradycardia was considerably suppressed by central inspiratory neuronal activity over the observed limits of range of spontaneous changes in Vpn. Control observations carried out during temporary interruption of artificial respiration alone indicate there were no changes in pulse interval or blood pressure at any level of Vpn.

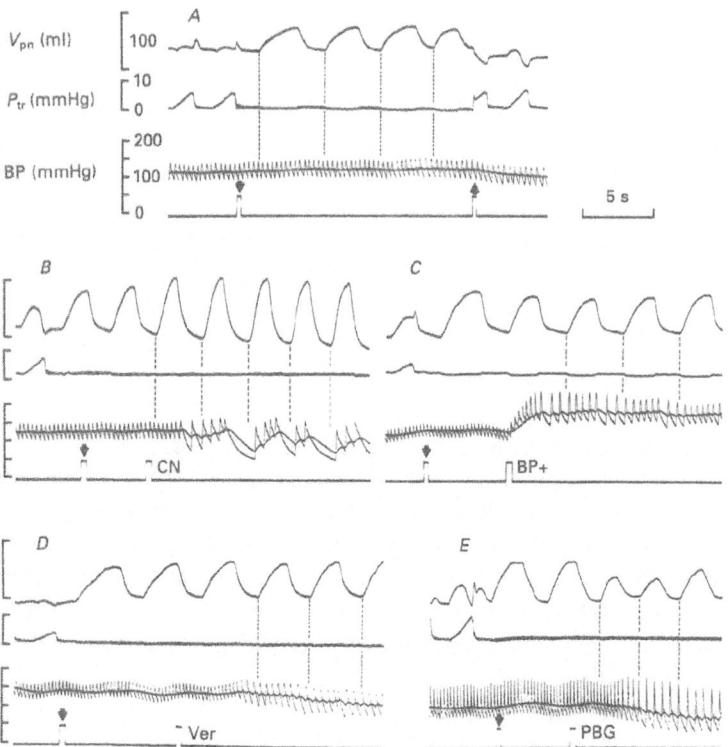

Fig. 3   The effects of phasic central inspiratory neuronal activity on the cardioinhibitory responses elicited by stimulation of various cardiovascular receptor groups. Bilateral upper thoracic sympathectomy. Positive pressure ventilation. Open pneumothorax. Each reflex cardioinhibitory response was studied under conditions in which artificial respiration was temporarily interrupted, with the lungs held at the expiratory pressure, 1.5 mmHg, as indicated by arrow in each panel. A, cessation of artificial respiration alone between arrows. B, stimulation of the carotid body chemoreceptors by intracarotid cyanide (CN; 4.8 µg kg⁻¹); C, elicitation of the arterial baroreflex alone by raising the arterial pressure by 40 mmHg; D, stimulation of the cardiac receptors by left atrial injection of veratridine (Ver; 4.1 µg kg⁻¹); E, stimulation of pulmonary C fibres by right atrial injection of phenylbiguanide (PBG; 16.2 µg kg⁻¹). Vertical interrupted lines indicate beginning of each inspiratory phase of the respiratory cycle. Records from above downwards: Vpn, changes in volume of the pneumothorax (inspiration upwards); Ptr, tracheal pressure; BP, phasic and mean arterial blood pressure. Time calibration, 5 s (From Daly, 1991).

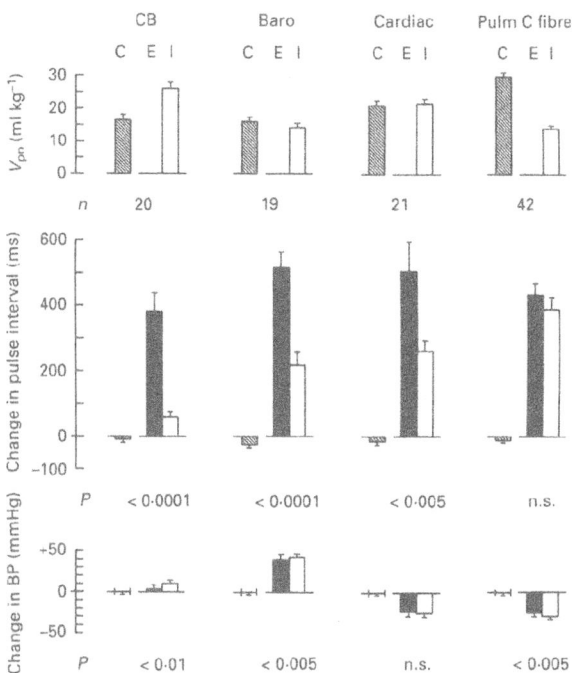

Fig. 4. *Summary of the effects of changes in central respiratory neuronal activity (E, expiration; filled blocks. I, inspiration; open blocks) on the cardioinhibitory responses to stimulation of the carotid body chemoreceptors (CB; intracarotid sodium cyanide, 4.9 ± 0.3 µg kg^{-1}), the arterial baroreflex (Baro; increase in blood pressure, 39.5 ± 2.9 mmHg), cardiac receptors (Cardiac; left atrial veratridine, 4.0 ± 0.2 µg kg^{-1}), and pulmonary C fibre endings (Pulm. C fibres; right atrial phenylbiguanide, 18.6 ± 1.1 µg kg^{-1}). C (cross-hatched blocks), control changes in physiological variables without stimulation of cardiovascular receptors. All tests carried out during temporary interruption of artificial respiration with the lungs held at their expiratory volume. From above downwards: changes in Vpn, volume of pneumothorax; changes in PI, pulse interval; and changes in BP mean arterial blood pressure. Statistical values refer to comparisons of paired data in filled (E) and open (I) blocks. n, number of tests (From Daly, 1991).*

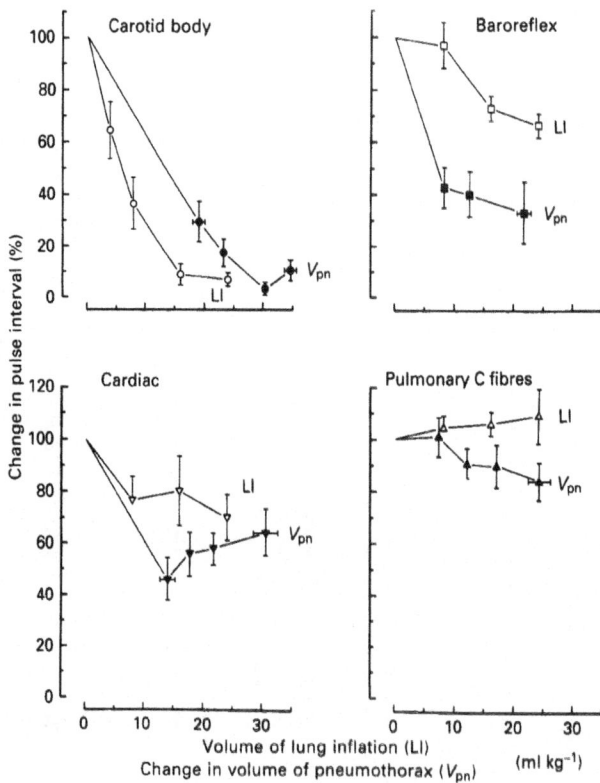

Fig.5. Normalized data showing the effects of phasic cental inspiratory neuronal activity, measured as changes in volume of the pneumothorax (Vpn), on the cardioinhibitory responses to stimulation of the carotid body chemoreceptors, the arterial baroreflex, cardiac receptors and pulmonary C fibre endings. The increase in pulse interval evoked by stimulation of each of the four inputs is taken as 100%, measured during the expiratory phase of the respiratory cycle, the expiratory volume of the pneumothorax being taken as zero. The inspiratory values observed at increasing levels of Vpn are expressed as a percentage of that occurring at zero volume. For comparison, the figure shows the normalized data for the effects of inflation of the lungs with different volumes of room air on the cardioinhibitory responses elicited by stimulation of the same four groups of cardiovascular receptors in the absence of activity of the central inspiratory neurones. The data is taken from Fig. 5 of Daly & Kirkman (1989). The increase in pulse interval evoked by stimulation of each of the four inputs is taken as 100%, measured during cessation of central inspiratory neuronal activity evoked reflexly by electrical stimulation of the superior laryngeal nerves. Filled symbols, changes in volume of the pneumothorax (Vpn), representing the changes in central inspiratory neuronal activity; Open symbols, lung inflation volume (From Daly, 1991).

In Fig.5, the data has been normalized for comparison of the effects of changes in Vpn on the bradycardia evoked by all four inputs from cardiovascular receptors. The degree of suppression was greater in the case of the carotid chemoreceptor input than with the arterial baroreceptor or cardiac receptor inputs. On the other hand, the cardiac response to pulmonary C fibre stimulation was not significantly affected by the central inspiratory activity at any level of Vpn up to 24 ml kg^{-1} (Fig.5).

Effects of interrution of the cardiac sympathetic pathway

The differential modulation of the cardiac responses by lung inflation and by the central inspiratory drive occurred independently of the integrity of the sympathetic supply to the heart which indicates that the cardiac efferents involved were largely fibres running in the vagus nerves. In these experiments the stellate ganglia and sympathetic chains down to the level of T3 were surgically excised (Fig.3).

DISCUSSION

These experiments demonstrate that activity of pulmonary stretch afferents driven by lung inflation and the central inspiratory neuronal drive can independently modulate the vagally-induced cardioinhibitory response elicited by stimulation of the carotid chemoreceptors and arterial baroreceptors (Koepchen et al. 1961a, b; Davidson et al. 1976; McAllen & Spyer, 1978a, b; Potter, 1981; Gilbey et al. 1984) and of cardiac C fibre afferents. It is clear, however, that the three inputs are not equally affected by the two respiratory mechanisms. In this connexion, one of the advantages of the choice of method for measuring the phasic changes in the central inspiratory neuronal activity, as changes in Vpn, enable a quantitative comparison to be made of the results obtained in experiments in which the effects of inflation of the lungs were studied. Comparison of the normalized data is shown in Fig.5 from which it may be seen that the input from the carotid bodies is the most affected cardioinhibitory reflex, being almost completely suppressed at levels of lung inflation and Vpn of 20-30 ml kg^{-1} (Daly & Kirkman, 1989; Daly, 1991). This range of volumes is within the normal physiological range, taking the normal resting tidal volume of the cat to be about 15 ml kg^{-1}. The unexpected finding was that, in contrast, the bradycardia evoked by stimulation of pulmonary C fibre endings was hardly affected by either respiratory mechanism. The differential nature of the respiratory modulations of the effectiveness of the four cardiovascular inputs occurred independently of the integrity of the sympathetic innervation of the heart and lungs, so that the vagus nerves respresented the sole pathway.

The mechanisms by which the cardioinhibitory reflex responses are respiratory modulated have been discussed by Daly & Kirkman (1989) and Daly (1991) together with the pertinent literature. The CVMs in the cat are situated largely in the nucleus ambiguus but also in the dorsal vagal motor nucleus. The former group can be activated synaptically by excitation of all four groups of cardiovascular receptors and are the main site for respiratory 'gating' of excitatory inputs, at least from chemoreceptors and baroreceptors. If the CVMs are a homogenous group, the differential modulation by lung inflation, of the cardioinhibitory responses to stimulation of chemoreceptors, baroreceptors and cardiac receptors could be explained, as proposed by Potter (1981), by the fact that pulmonary stretch afferents have an independent effect earlier in each vagal excitatory pathway, although not directly on afferent terminals. Daly & Kirkman (1989) concluded that the failure of pulmonary stretch afferents to affect appreciably the cardioinhibitory response evoked by excitation of pulmonary C fibre endings was due to the absence of a neural integration between the two inputs, which led them to postulate there are two populations of CVMs, one which can be inhibited by lung inflation, the other cannot. The results obtained on the modulation of the excitatory inputs by central inspiratory neuronal activity are not incompatible with this hypothesis. The possibility that the differential modulation described here is determined, at least to some extent, by the differing patterns of evoked activity in various functional groups of inspiratory and expiratory neurones has been considered. It is known that reciprocal and nonreciprocal effects on inspiratory and expiratory medullary neurones and in activity of inspiratory and expiratory nerves occur with respect to respiration, to the input from pulmonary stretch afferents and as a result of excitation of arterial chemoreceptors, baroreceptors and pulmonary C fibres. This appears, however, to be an unlikely explanation (Daly, Jordan & Spyer, 1992).

SUMMARY

The differential nature of the modulation of excitatory inputs from various cardiovascular receptor groups by pulmonary stretch afferents, driven by lung inflation, and the central inspiratory neuronal activity, has been demonstrated. At levels of activity of pulmonary stretch afferents and of central inspiratory drive that almost completely suppress the bradycardia of stimulation of the carotid bodies, the response to stimulation of the arterial baroreceptors and cardiac C fibre endings was reduced by 30-58%. In contrast, the bradycardia evoked by pulmonary C fibre stimulation was not significantly affected by either respiratory mechanism.

ACKNOWLEDGEMENT

This work was supported by the British Heart Foundation.

REFERENCES

Daly, M. de B. 1991, Some reflex cardioinhibitory responses in the cat and their modulation by central inspiratory neuronal activity, J. Physiol., 439: 559-577.

Daly, M. de B., Jordan, D., and Spyer, K.M., 1992, Modification of respiratory activities during stimulation of carotid chemoreceptors, arterial baroreceptors and pulmonary C fibre afferents in the anaesthetized cat, J. Physiol., 446: 466 P.

Daly, M. de B., and Kirkman, E.,1988, Cardiovascular responses to stimulation of pulmonary C fibres in the cat: their modulation by changes in respiration, J. Physiol., 402: 43-63.

Daly, M. de B., and Kirkman, E., 1989, Differential modulation by pulmonary stretch afferents of some reflex cardioinhibitory responses in the cat, J. Physiol., 417: 323-341.

Daly, M. de B., Kirkman, E., and Wood, L.M., 1988, Cardiovascular responses to stimulation of cardiac receptors in the cat and their modification by changes in respiration, J. Physiol., 407: 349-362.

Daly, M. de B., Scott, M.J., 1958, The effects of stimulation of the carotid body chemoreceptors on heart rate in the dog, J.Physiol., 144: 148-166.

Davidson, N.S., Goldner, S., and McCloskey, D.l., 1976, Respiratory modulation of baroreceptor and chemoreceptor reflexes affecting heart rate and cardiac vagal efferent nerve activity, J. Physiol., 259: 523-530.

Gandevia, S.C., McCloskey, D.l. & Potter, E.K., 1978, Inhibition of baroreceptor and chemoreceptor reflexes on heart rate by afferents from the lungs, J. Physiol., 276: 369-382.

Gilbey, M.P., Jordan, D., Richter, D.W., and Spyer, K.M., 1984, Synaptic mechanisms involved in the inspiratory modulation of vagal cardio-inhibitory neurones in the cat, J. Physiol., 356: 65-78.

Koepchen, H.P., Lux, H.D., and Wagner, P. H., 1961a, Untersuchungen uber Zeitbedarf und zentrale Verarbeitung des pressoreceptorischen Herzreflexes, Pflugers Arch., 273: 413-430.

Koepchen, H.P., Wagner, P. H., and Lux, H.D., 1961b, Über die Zusammenhänge zwischen zentralen Erregbarkeit, reflektorischem Tonus und Atemrhythmus bei der Nervösen Steuerung der Herzfrequenz, Pflugers Arch., 273: 443-465.

McAllen, R.M., and Spyer, K.M., 1978a, Two types of vagal preganglionic motoneurones projecting to the heart and lungs, J. Physiol., 282: 353-364.

McAllen, R.M., and Spyer, K.M., 1978b, The baroreceptor input to cardiac vagal motoneurones, J. Physiol., 282: 365-374.

Potter, E.K., 1981, Inspiratory inhibition of vagal responses to baroreceptor and chemoreceptor stimuli in the dog, J. Physiol., 316: 177-190.

48 THE MODULATION OF PERIPHERAL CHEMORECEPTOR INPUT BY CENTRAL NERVOUS SYSTEM HYPOXIA

N. H. Edelman, J. E. Melton and J. A. Neubauer

Division of Pulmonary & Critical Care Medicine
Department of Medicine
University of Medicine & Dentistry of New Jersey
Robert Wood Johnson Medical School
New Brunswick, New Jersey (USA)

Production of progressive isocapnic brain hypoxia by carbon monoxide (CO) inhalation in anesthetized, vagotomized, peripherally chemodenervated cats results in an initial reduction of phrenic neurogram amplitude followed by a secondary decrease in phrenic burst frequency as the level of hypoxia increases (Melton et al., 1988; Melton et al., 1990; Wasicko et al., 1990). After arterial O_2 content (CaO_2) is reduced by approximately 50%, the phrenic neurogram becomes silent. Severe depression of the phrenic neurogram during hypoxia does not inhibit the respiratory response to CO_2, however, suggesting that processing of central chemoreceptor afferent information is unaffected by hypoxia (Van Beek et al., 1984; Melton et al., 1988).

Although hypoxia does not appear to attenuate the response of the respiratory control system to central respiratory stimuli, there is evidence that the respiratory response to peripheral chemoreceptor stimulation is blunted during brain hypoxia. Prolonged (20-25 min.) isocapnic hypoxia in both humans (Easton et al., 1986) and anesthetized cats (Vizek et al., 1987) resulted in an initial brisk increase in respiratory output followed by a secondary reduction to a level intermediate to normoxic and peak hypoxic levels. In anesthetized cats, carotid sinus nerve (CSN) activity increased during the initial exposure to hypoxia and associated increase in ventilatory output but remained high during the secondary decrease in ventilation. Based on this observation, Vizek et al. (1987) concluded that the reduction in ventilation seen during prolonged hypoxia is central rather than peripheral in origin.

We have assessed the effect of brain hypoxia on peripheral chemoreceptor activity by monitoring the response of the phrenic neurogram to 30 sec. supramaximal CSN stimulations during isocapnic CO hypoxia in peripherally chemodenervated, vagotomized cats (Melton et al., 1992). Stimulations were performed at 3 min. intervals during progressive hypoxic depression to phrenic silence. The degree of phrenic stimulation was calculated as the difference between the peak phrenic amplitude of the last five pre-stimulation phrenic bursts and the last five bursts during CSN stimulation. The effect of hypoxia on the peak phrenic amplitude response to CSN stimulation in 12 cats is shown in Figure 1. In this figure, all of the CSN stimulations performed during hypoxia are grouped according to degree of hypoxia at the time of stimulation and the peak phrenic amplitude response is expressed as a percentage of the response under control hyperoxic conditions. In spite of a progressive decrease in unstimulated peak phrenic amplitude during hypoxia, the mean difference between stimulated and unstimulated phrenic amplitude was unchanged from values measured under control conditions indicating that peripheral chemoreception is unimpaired. As the hypoxia becomes more severe, however, the intact peripheral stimulatory response is imposed upon a progressive central depression of phrenic amplitude with the net result that the absolute value of phrenic amplitude during CSN stimulation is decreased. This finding is completely consistent with the previous observation of Vizek et al. (1987) that phrenic activity in the anesthetized cat falls secondarily during prolonged hypoxia in spite of the fact that CSN activity is undiminished.

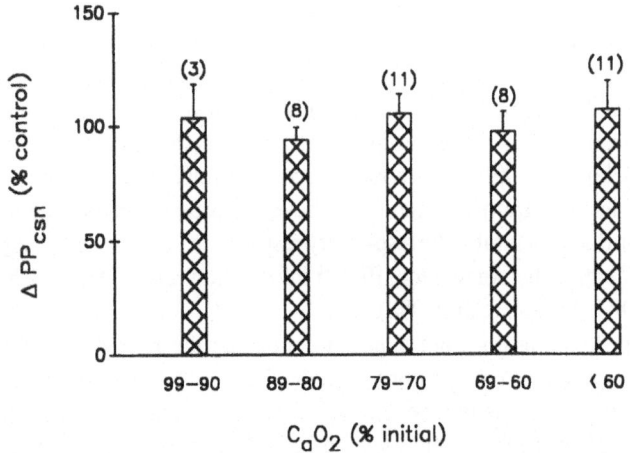

Fig. 1. *The difference between peak phrenic amplitude during supramaximal carotid sinus nerve stimulation and unstimulated peak phrenic amplitude (DPPcsn) as a function of arterial O_2 content. Values of DPPcsn are normalized to hyperoxic control values. All values are $x \pm SEM$, sample size in parentheses. There was no significant effect of hypoxia on DPPcsn.*

A secondary depression of ventilation during sustained isocapnic hypoxia is also seen in conscious humans (Easton et al., 1986; Easton et al., 1988; Bacon et al., 1990; Bascom et al., 1990) but in this case the response to peripheral chemoreceptor stimulation appears to be depressed. The cause of this depression is unresolved. In a study of conscious humans, Bascom et al. (1990) observed that the increase in ventilation in response to successive exposures to transient (1 min.) severe hypoxia during sustained moderate hypoxia was progressively attenuated while the ventilatory response to transient CO_2 was unaffected. Based on this observation, a failure of peripheral chemoreceptor input rather than central afferent processing during sustained hypoxia was inferred. Following sustained (25 minutes) hypoxia and 7 minutes of room air recovery, Easton et al. (1988) noted a decrease in hypoxic ventilatory response in humans during a subsequent hypoxic exposure. Full recovery of hypoxic ventilatory response was only seen if the room air recovery period was extended to 60 min. or if recovery took place under hyperoxic conditions for shorter periods of time. This hypoxic attenuation of peripheral chemoreceptor stimulation was attributed to accumulation of inhibitory neurotransmitters in the vicinity of the central respiratory neuron pool. The findings in both of these human studies contrast with the observation in the cat that the increase in phrenic neurogram activity during CSN stimulation was unchanged during hypoxic respiratory depression (Melton et al., 1992).

Although the increase in phrenic neurogram amplitude seen in anesthetized cats in response to supramaximal peripheral chemoreceptor stimulation was not attenuated by hypoxia, the absolute value of peak phrenic amplitude during stimulation decreased as hypoxia became more severe (Melton et al., 1992). This observation contrasts sharply with the response of the phrenic neurogram to central chemoreceptor stimulation during this stress. There was no significant difference in the maximum value of peak phrenic amplitude achieved during CO_2 rebreathing under hyperoxic and hypoxic conditions despite the fact that phrenic amplitude was near zero at the time rebreathing was initiated during hypoxia (Melton et al., 1988). This difference in peripheral and central chemoreceptor responses during hypoxia may represent a fundamental difference in the central processing of these stimuli. Recordings of single unit respiratory-related activity from both the dorsal and ventral respiratory groups of the cat during hypercapnia and hypoxia indicate differences in the respiratory neuronal response to these stimuli (St. John, 1981). Hypercapnia resulted in augmentation of single unit activity in both brain areas while peripheral chemoreceptor stimulation with hypoxia increased single unit activity primarily in the dorsal respiratory group. St. John (1981) concluded that central chemoreceptor afferents are widely distributed among respiratory neurons while peripheral chemoreceptor afferents are directed to a more discrete population of neurons. Thus, CO_2 presumably stimulates the entire pool of premotor neurons which contribute to the phrenic neurogram resulting in maximal phrenic stimulation at high levels of CO_2. This universal stimulation of the phrenic premotor neuron pool is sufficient to overcome the inhibitory effect of central hypoxia. Conversely, peripheral chemoreceptor stimulation is directed to a much more limited pool of premotor neurons whose output is then added to the larger population of unexcited premotor neurons.

During normoxia, cessation of peripheral chemoreceptor stimulation does not result in an immediate return of respiration to unstimulated levels. Rather, there is a slow decay of respiratory output (time constant of approximately 1 min.) after the peripheral respiratory stimulus is withdrawn (Eldridge, 1974; Eldridge and Gill-Kumar, 1978). This phenomenon, which was originally termed respiratory afterdischarge (Eldridge, 1974) and more recently short-term potentiation of respiration (Wagner and Eldridge, 1991), appears to originate in the brainstem (Eldridge, 1974; Eldridge, 1976; Eldridge and Gill-Kumar, 1978) and may be an important factor stabilizing ventilation during periods of varying respiratory drive (Younes, 1989). Although the cellular mechanism of respiratory afterdischarge is unknown, Wagner and Eldridge (1991) have recently suggested that it is analogous to short term use-dependent potentiation described in other neuronal systems and, as such, is a synaptic phenomenon. Since hypoxia could potentially interfere with synaptic events, particularly those which are energy requiring, we have examined the effects of this stress on respiratory afterdischarge in the anesthetized cat model (Melton et al., 1992).

To produce respiratory afterdischarge, the cut CSN of an anesthetized, paralyzed, vagotomized cat was supramaximally stimulated for 30 sec. while respiratory output was monitored from the phrenic neurogram. The time constant of respiratory afterdischarge

(t) was determined by plotting the peak amplitude of each post-stimulation phrenic burst against time after cessation of stimulus. The best fits of the resultant curves were exponential and t was calculated as -l/b, where b is the shape constant of the exponential. After hyperoxic control CSN stimulations, cats were made hypoxic with 0.5% CO. CSN stimulations were performed at 3-5 min. intervals during hypoxia until phrenic output was reduced to zero.

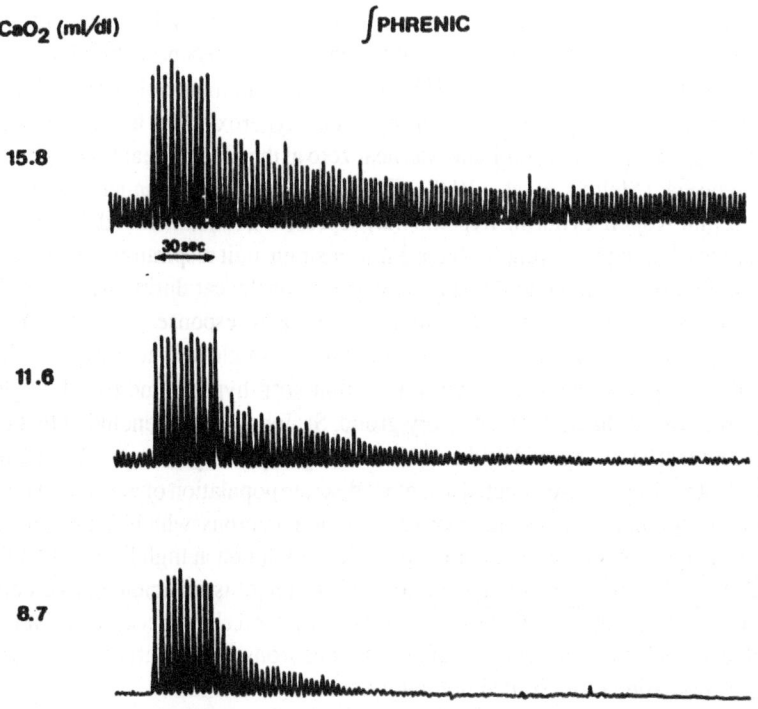

Fig. 2. The effect of supramaximal carotid sinus nerve stimulation (between arrows) on the integrated phrenic neurogram of a representative cat during hyperoxia (top panel) and subsequent carbon monoxide hypoxia. Note that the time required for phrenic amplitude to return to pre-stimulus levels decreases as hypoxia progresses.

A series of CSN stimulations in a representative cat during progressive hypoxia are shown in Figure 2. In this figure it can be clearly seen that the length of time over which afterdischarge can be sustained is decreased as the degree of hypoxia increases. This effect can be seen more clearly in Figure 3 which shows the calculated values of t from all stimulations in 12 cats as a function of degree of hypoxia. t decreased as a linear function of the reduction of arterial O_2 content over the range of hypoxia studied. This effect of hypoxia to blunt afterdischarge has been previously inferred from studies in conscious humans who demonstrate a ventilatory undershoot during the transition from prolonged (25 min.) isocapnic hypoxia to hyperoxia (Georgopoulos et al., 1990).

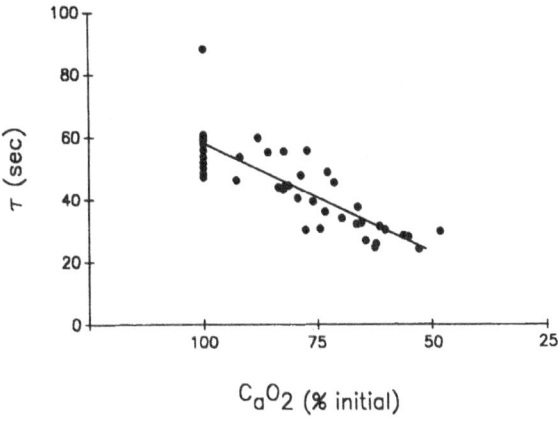

Fig. 3. The time constant of respiratory afterdischarge (t) as a function of arterial O_2 content (CaO_2) normalized to initial hyperoxic values. Control values of t $(CaO_2=100\%)$ are mean values of at least 3 control stimulations in each cat. All other points are single stimulations. t is a linear function of CaO_2 over the range of hypoxia used in these studies.

Although these studies did not directly address the mechanism of the hypoxic attenuation of afterdischarge, if one assumes that afterdischarge is an example of a use-dependent alteration of synaptic excitability as has been suggested by Wagner and Eldridge (1991), specific means by which hypoxia might affect this process can be proposed. Several different forms of use-dependent alteration of synaptic excitability with widely varying time constants have been described in various regions of the brain (Kennedy, 1989). Those forms with the shortest time constants (msec-sec) appear to result entirely from stimulation-dependent presynaptic uptake of Ca^{++} with resultant increased release of neurotransmitter. Augmented release continues following cessation of the stimulus until intracellular Ca^{++} is restored to basal level (Zucker, 1989). Synaptic alterations with time constants in the minutes to days range appear to involve a mixture of both pre- and postsynaptic events (Racine et al., 1983). Long-term potentiation (LTP), the most commonly studied of these phenomena, appears to be initiated by simultaneous depolarization of the postsynaptic membrane and binding of glutamate to postsynaptic N-methyl-D aspartate (NMDA) receptor sites resulting in activation of an inward Ca^{++} current (Kennedy, 1989). Maintenance of LTP may depend upon subsequent activation of pre-synaptic terminals by a retrograde messenger, perhaps nitric oxide, released from the postsynaptic terminals as part of the cascade of events resulting from the increase in postsynaptic intracellular Ca^{++} (O'Dell et al., 1991; Schuman and Madison, 1991) .

If respiratory afterdischarge is analogous to LTP, hypoxia may have some very specific effects on this process. Because initiation of LTP requires postsynaptic depolarization (Brown et al., 1988; Kennedy, 1989), any inhibitory neuromodulator which hyperpolarizes the postsynaptic membrane could prevent or reduce potentiation. g-aminobutryic acid (GABA), a major inhibitory neurotransmitter in the brain, may be such an agent. The brain extracellular concentration of GABA increases during hypoxia (Weyne et al., 1977; Hagberg et al., 1985) and direct application of GABA has been shown to hyperpolarize neurons of the nucleus tractus solitarius (NTS) in the rat (Nakagawa et al., 1991), the region of the brainstem to which CSN afferents are known to project monosynaptically (Davies and Kalia, 1981). Thus, GABA hyperpolarizes some subpopulation of the NTS neurons which presumably represent the postsynaptic terminal of the first carotid sinus nerve synapse. Taken together, these observations suggest that hypoxia may reduce afterdischarge by resulting in increased GABA-mediated inhibition at CSN synaptic sites in the NTS.

The effect of hypoxia to reduce afterdischarge may have important physiological implications. Afterdischarge is probably the most important factor stabilizing ventilation during periods of varying respiratory drive and protecting against periodic breathing (Younes, 1989). In the total absence of afterdischarge, hypocapnia resulting from a hyperventilatory stimulus could result in ventilatory undershoot, manifested as hypoventilation and/or apnea, upon cessation of hyperventilation. The resultant increase in $PaCO_2$ may then initiate another cycle of respiration. Thus, by reducing afterdischarge, hypoxia may predispose to periodic breathing. Afterdischarge also appears to be decreased during sleep (Daristotle et al., 1991) suggesting that imposition of hypoxia during sleep may cause more instability of the respiratory control system than either of these two conditions alone.

REFERENCES

Bacon, D. S., Afifi, M. S., Griebel, J. A., and Camporesi, E. M., 1990, Cerebrocortical oxygenation and ventilatory response during sustained hypoxia, Respir. Physiol., 80:245.

Bascom, D. A., Clement, I. D., Cunningham, D. A., Painter, R., and Robbins, P. A., 1990, Changes in peripheral chemoreflex sensitivity during sustained, isocapnic hypoxia, Respir. Physiol., 82:161.

Brown, T. H., Chapman, P. F., Kairiss, E. W., and Keenan, C.L., 1988, Long term synaptic potentiation, Science, 242:724.

Daristotle, L., Petrozzino, J., and Santiago, T. V., 1991, Effect of sleep-wake state on ventilatory afterdischarge (Abstract), FASEB J. 5:A1478.

Davies, R. 0., and Kalia, M., 1981, Carotid sinus nerve projections to the brain stem in the cat, Brain Res. Bull., 6:531.

Easton, P. A., Slykerman, L. J., and Anthonisen, N. R., 1986, Ventilatory response to sustained hypoxia in normal adults, J. Appl. Physiol. 61:906.

Easton, P. A., Slykerman, L. J., and Anthonisen, N. R., 1988, Recovery of the ventilatory response to hypoxia in normal adults, J. Appl. Physiol. 64:521.

Eldridge, F. L., 1974, Central neural respiratory stimulatory effect of active respiration, J. Appl. Physiol., 37:723.

Eldridge, F. L., 1976, Central neural stimulation of respiration in unanesthetized decerebrate cats, J. Appl. Physiol., 40:23.

Eldridge, F. L., and Gill-Kumar, P., 1978, Lack of effect of vagal afferent input on central neural respiratory after discharge. J. Appl. Physiol., 45:339.

Georgopoulos, D., Bshouty, Z., Younes, M., and Anthonisen, N.R., 1990, Hypoxic exposure and activation of the afterdischarge mechanism in conscious adults, J. Appl. Physiol., 69:1159.

Hagberg, H. A., Lehmann, A., Sandberg, M., Nystrom, B., Jacobson, I., and Hamberger, A., 1985, Ischemia-induced shift of inhibitory and excitatory amino acids from intra- to extracellular compartments, J. Cereb. Blood Flow Metab , 5:413.

Kennedy, M. B., 1989, Regulation of synaptic transmission in the central nervous system: long-term potentiation, Cell, 59:777.

Melton, J. E., Neubauer, J. A., and Edelman, N. H., 1985, CO_2 sensitivity of the cat phrenic neurogram during hypoxic respiratory depression, J. Appl. Physiol., 65:736.

Melton, J. E., Neubauer, J. A., and Edelman, N. H., 1990, GABA antagonism reverses hypoxic respiratory depression in the cat, J. Appl. Physiol., 69:1296.

Melton, J. E., Yu, Q. P., Neubauer, J. A., and Edelman, N. H., 1992, The effect of brain hypoxia on peripheral chemoreception and respiratory afterdischarge in the cat, J. Appl. Physiol. 73:2166.

Nakagawa, T., Wakamori, M., Shirasaki, T., Nakaye, T., and Akaike, N., 1991, Gamma-aminobutyric acid-induced response in acutely isolated nucleus solitarii neurons of the rat, Am. J. Physiol., 260: (Cell Physiol. 29): C745.

'O' Dell, T.J., Hawkins, R.D., Kandel, E.R., and Arancio, O., 1991, Tests of the roles of two diffusible substances in long-term potentiation: evidence for nitric oxide as a possible early retrograde messenger, Proc. Nat. Acad. Sci. USA, 88:11285.

Racine, R. J., Milgram, N. W., and Hafner, S., 1983, Long-term potentiation phenomena in the rat limbic forebrain, Brain Res., 260:217.

Schuman, E. M., and Madison, D. V., 1991, A requirement for the intracellular messenger nitric oxide in long-term potentiation, Science, 254:1503.

St. John, W. M., 1981, Respiratory neuron responses to hypercapnia and carotid chemoreceptor stimulation, J. Appl. Physiol., 51:816.

Van Beek, J. H., Berkenbosch, A., DeGoede, J., and Olievier, C. N., 1984, Effects of brainstem hypoxemia on the regulation of breathing, Respir. Physiol., 57:171.

Vizek, M., Pickett, C. K., and Weil, J. V., 1987, Biphasic ventilatory response of adult cats to sustained hypoxia has central origin, J. Appl. Physiol., 63:1884.

Wagner, P. G., and Eldridge, F. L., 1991, Development of short-term potentiation of respiration, Respir. Physiol., 83:129.

Wasicko, M. J., Melton, J. E., Neubauer, J. A., Krawciw, N., and Edelman, N. H., 1990, Cervical sympathetic and phrenic nerve responses to progressive brain hypoxia, <u>J. Appl. Physiol.</u>, 68:53.

Weyne, J., VanLeuven, F., and Leusen, I., 1977, Brain amino acids in conscious rats in chronic normocapnic and hypocapnic hypoxemia, <u>Respir. Physiol.</u>, 31:231.

Younes, M., 1989, The physiological basis of central apnea and periodic breathing, <u>Curr. Pulmonol.</u>, 10:265.

Zucker, R. S., 1989, Short term synaptic plasticity, <u>Ann. Rev. Neurosci.</u>, 12:13.

49 CHEMOREFLEXOGENIC VENTILATORY DRIVE IN HUMANS ADAPTED TO UNUSUAL ENVIRONMENTS

B. Grassi[1], G. Ferretti[2], M. Costa[1], C. Marconi[1] and P. Cerretelli[1,2]

[1]I.T.B.A. of C.N.R., Milano (I); [2]Dept. of Physiol. CMU, Univ. of Geneva (CH)

INTRODUCTION

Pulmonary ventilation ($\dot{V}E$) increases following a decrease in the inspired O_2 partial pressure or an increase in the inspired CO_2 partial pressure. These two stimuli in combination act synergistically on $\dot{V}E$ (Lloyd et al., 1958; Loeschke and Gertz, 1958). The hypoxic and hypercapnic ventilatory responses show a substantial interindividual variability (Weil et al., 1970; Lambertsen, 1960). The hypoxic ventilatory response is known to be lower than normal in various groups of subjects, such as endurance athletes (Byrne-Quinn et al., 1971) or swimmers (Bjurstrom and Schoene, 1986). Lambertsen (1960) observed a remarkable range of variation in the hypercapnic ventilatory response, and defined subgroups of people characterized by "low", "normal" and "high" ventilatory response to inhaled CO_2.

Over the last 15 years several climbers have reached the top of mountains higher than 8500 m without supplemental O_2. Before these extraordinary performances, most exercise physiologists believed that an arterial partial pressure of O_2 lower than 40 Torr (as expected at those great heights) would be incompatible with the effort of climbing, particularly in those extreme conditions. It was therefore hypothesized that those individuals could be characterized by an extremely high aerobic power and/or by an enhanced hypoxic ventilatory response, allowing them to mantain a relatively high arterial O_2 saturation (SaO_2) at the expenses of an extreme hyperventilation.

On the other hand, over the last few years several divers were able to perform breath-hold dives deeper than 100 m. Some of these divers were found (Ferretti et al., 1991) to be able to prolong a maximal dry breath-hold at rest well beyond the end-tidal partial pressures of O_2 and CO_2 ($PETO_2$ and $PETCO_2$) expected at the breath-hold breaking point (Otis et al., 1948). To explain these results a blunted hypoxic and/or hypercapnic ventilatory response were hypothesized as adaptative or genetically induced phenomena.

VENTILATORY ADAPTATIONS IN ELITE CLIMBERS

In 1986 Oelz et al. measured the maximal aerobic power of 6 climbers who had reached at least one of the four highest peaks in the world without supplemental O_2. Maximal O_2 consumption ($\dot{V}O_2max$) values of 59.5 ± 6.2 ($\bar{x} \pm SD$) ml.min^{-1}.kg^{-1} were obtained, which are higher than those normally found in untrained subjects, but remarkably lower than those observed in elite endurance athletes, and substantially equivalent to those of well-trained nonathletic subjects.

Neurobiology and Cell Physiology of Chemoreception
Edited by P.G. Data *et al.*, Plenum Press, New York, 1993

As far as the hypoxic ventilatory response is concerned, a brisker response could have allowed the climbers to mantain a relatively high SaO_2 even at extreme altitude. The hypoxic ventilatory response of elite climbers has been evaluated by several authors, with conflicting results. Some authors (Schoene, 1982; Schoene et al., 1984; Masuyama et al., 1986) have indeed described a positive relationship between a brisk hypoxic ventilatory response and altitude performance, but such relationship was not confirmed by others (Milledge et al., 1983; Oelz et al., 1986; Schoene et al., 1987). It must be pointed out that Oelz et al. (1986) examined the hypoxic ventilatory response on the same climbers on whom the $\dot{V}O_2$max data mentioned above were obtained. At least in part such conflicting results could be attributable to the different methods employed to determine the hypoxic ventilatory response (Ward et al., 1989). However, Andean Indians and the Sherpas of the Himalayas, populations that are characterized by an extraordinary working capacity at altitude, even at great heights, show a reduced hypoxic ventilatory response compared to sea level residents (Severinghaus et al., 1966; Lahiri and Milledge, 1965; Milledge and Lahiri, 1967).

From the above contradictory observations it may be concluded that a brisk hypoxic ventilatory response does not seem to be necessary for extreme physical performance at great heights. This conclusion, hovewer, should be sustained by further investigations.

VENTILATORY ADAPTATIONS IN ELITE BREATH-HOLD DIVERS

Recent studies performed on three extreme breath-hold divers (Ferretti et al., 1991; Ferrigno et al., 1991) led to hypothesize a series of physiological adaptations possibly taking place in such divers.

One of these divers was the first man to dive below 100 m in open sea while holding his breath. The divers (a 57 years-old male, and his two daughters, 28 and 30 years-old) were able to prolong a resting dry breath-hold in the supine position up to a maximum of 5 min, to be compared with a maximum of 2.5 min in a group of non-divers controls. $PETO_2$ and $PETCO_2$ values obtained in the divers and in controls before and at the end of maximal dry breath-holds are plotted in the figure on a O_2-CO_2 diagram (Rahn and Fenn, 1955). On the diagram are also drawn: the SaO_2 isopleths calculated for humans at sea level; the normal alveolar air curve; the breath-hold breaking point curve (Otis et al., 1948); the lines defining regions of normal and impaired contrast discrimination performance (Otis et al., 1946).

The controls, as well as one of the divers interrupted their maximal dry breath-holds essentially on the breaking point curve, at the boundary of the severe hypoxia region. Such an alveolar gas composition in normal subjects would exert, at rest, a respiratory drive from combined hypoxia and hypercapnia leading to an about 7 times greater VE than during normoxic normocapnia (Lloyd et al., 1958; Loeschke and Gertz, 1958). At the end of the maximal dry breath-holds SaO_2 values in the diver (0.78) as well as in the controls (0.83) approached those of the corresponding calculated isopleths. By contrast, the other two divers could prolong their maximal dry breath-holds to reach $PETO_2$ and $PETCO_2$ values well beyond the breaking point line, to the boundary of the expected anoxic unconsciousness. Such values of alveolar air would generate in resting normal subjects a combined central and peripheral respiratory stimulation leading to VE values about 10 times greater than during normoxic normocapnia (Lloyd et al., 1958; Loeschke and Gertz, 1958). From these results it may be hypothesized that blunted ventilatory responses to hypoxia and/or to hypercapnia may represent adaptive phenomena taking place in these divers.

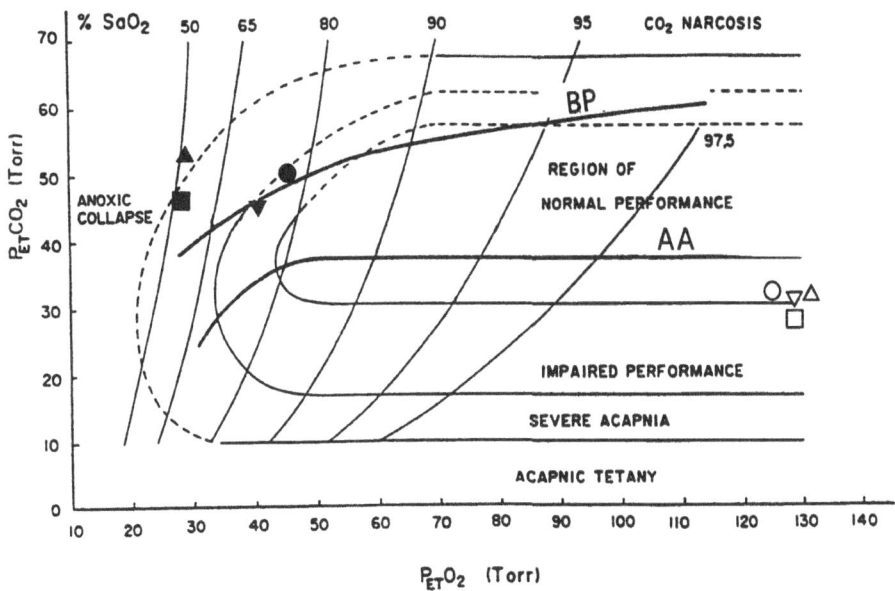

Figure 1. End tidal O_2 and CO_2 partial pressures($PETO_2$ and $PETCO_2$) obtained in the divers (triangles and squares) and in the controls (circles) before (open Symbols) and at the end (filled symbols) of maximal dry breath-holds, plotted on the O_2-CO_2 diagram (see text). AA = alveolar air curve; BP = breaking point curve.

Preliminary experiments on ventilatory control performed on the same divers seem to only partially confirm such hypothesis. Indeed, whereas the hypercapnic ventilatory response seems reduced in the divers compared to that observed in controls, the hypoxic ventilatory response (evaluated according to the same method utilized by Oelz et al., 1986) does not seem different in the two groups.

With respect to these observations, it is noteworthy that during a dive the external pressure generated on the thorax of the diver is transmitted to the alveolar gases. Therefore, even during a 100 m-deep, 2 min-long dive, the alveolar O_2 partial pressure should remain quite high for most of the dive, reaching values significantly lower than the pre-dive level only during the last few seconds of the ascent (Olszowka and Rahn, 1987). Whereas those last few seconds are undoubtely important for the risk of the so-called "ascent black-out", they may not lead to a physiological adaptation in the diver. On the other hand, the external pressure on the thorax would mantain for most of the dive a significantly high alveolar CO_2 partial pressure (Olszkowka and Rahn, 1987), that should even generate an "inverse passage" of CO_2 from the alveoli to the capillary blood, forcing the diver to "resist" the drive to breathe coming from central chemoreceptors, and possibly leading to some kind of physiological adaptation. In this context it is noteworthy that the same divers showed during breath-hold dives other adaptative phenomena, aimed at preserving as much O_2 as possible for vital organs. Such phenomena comprehend a particularly marked bradycardic response (Ferrigno et al., 1991), and a shift to anaerobic energy metabolism even in the presence of a low metabolic output, presumably as a consequence of peripheral vasoconstriction (Ferretti et al., 1991).

Being the three divers members of the same family, however, these results should be interpreted with caution, since they could be representative of a genetic factor rather than of a physiological adaptation.

CONCLUSIONS

Over the last few years some extraordinary sport performances in extreme environments led physiologists to hypothesize that altered hypoxic and/or hypercapnic ventilatory responses could represent physiological adaptations in some subjects. The hypothesis that extreme altitude climbers are characterized by an enhanced hypoxic ventilatory response is supported by some authors, but denied by others. The fact that populations living at high altitude show a lower hypoxic ventilatory response compared to lowlanders seems to indicate that a brisk response is not necessary for extreme physical performance at great heights. On the other hand, extreme breath-hold divers seem characterized by a blunted hypercapnic ventilatory response, allowing them to resist the drive to breathe coming from the presumably elevated arterial CO_2 partial pressure during breath-hold dives.

REFERENCES

Bjurstrom, R.L. and R.B. Schoene (1986). Ventilatory control in elite synchronized swimmers. Am. Rev. Respir. Dis. 133 (Suppl.): A135.

Byrne-Quinn, E., J.V. Weil, I.E. Sodal, G.F. Filley and R.F. Grover (1971). Ventilatory control in the athlete. J. Appl. Physiol. 30: 91-98.

Ferretti, G., M. Costa, M. Ferrigno, B. Grassi, C. Marconi, C.E.G. Lundgren and P. Cerretelli (1991). Alveolar gas composition and exchange during breath-hold diving and dry breath-holds in elite divers. J. Appl. Physiol. 70: 794-802.

Ferrigno, M., B. Grassi, G. Ferretti, M. Costa, C. Marconi, P. Cerretelli and C.E.G. Lundgren (1991). Electrocardiogram during deep breath-hold dives by elite divers. Undersea Biomed. Res. 18: 81-91.

Lahiri, S. and J.S. Milledge (1965). Sherpa Physiology. Nature 207: 10-12.

Lambertsen, C.J. (1960). Carbon dioxide and respiration in acid base homeostasis. Anesthesiology 21: 642-651.

Lloyd, B.B., M.G.M. Jukes and D.J.C. Cunningham (1958). The relation between alveolar oxygen pressure and the respiratory response to carbon dioxide. Quart. J. Exp. Physiol. 43: 214-227.

Loeschke, H.H. and K.H. Gertz (1958). Einfluss des 02-Druckes in der Einatmungszeit auf die Atemtätigkeit des Menschen, geprüft unter Konstanthaltung des alveolaren CO_2-Druckes. Pflüegers Arch. 267: 460-477.

Matsuyama, S., H. Kimura, T. Sugita, T. Kuriyama, K. Tatsumi, F. Kunitomo, S. Okita, H. Tojima, Y. Yuguchi, S. Watanabe and Y. Honda (1986). Control of ventilation in extreme-altitude climbers. J. Appl. Physiol. 61: 500-506.

Milledge, J.S. and S. Lahiri (1967). Respiratory control in lowlanders and sherpa highlanders at altitude. Respir. Physiol. 2: 310-322.

Milledge, J.S., M.P. Ward, E.S. Williams and C.R.A. Clarke (1983). Cardiorespiratory response to exercise in men repeatedly exposed to extreme altitude. J. Appl. Physiol. 55: 1379-1385.

Oelz, O., H. Howald, P.E. di Prampero, H. Hoppeler, H. Claassen, R. Jenni, A. Buhlman, G. Ferretti, J.-C. Brückner, A. Veicsteinas, M. Gussoni and P. Cerretelli (1986). Physiological profile of world-class high-altitude climbers. J. Appl. Physiol. 60: 1734-1742.

Olszowka, A.J. and H. Rahn (1987). Breath-hold diving. In:"Hypoxia and Cold", J.R. Sutton, C.S. Houston and G. Coates, Eds. Praeger, New York (USA) (pp. 417-428) .

Otis, A.B., H. Rahn, M.A. Epstein and W.O. Fenn (1946). Performance as related to composition of alveolar air. Am. J. Physiol. 146: 207-221.

Otis, A.B., H. Rahn and W.O. Fenn (1948) . Alveolar gas changes during breath-holding. Am. J. Physiol. 152: 674-686.

Rahn, H. and W.O. Fenn (1955). "A Graphical Analysis of the Respiratory Gas Exchange. The O_2-CO_2 Diagram". American Physiological Society, Washington, DC (USA).

Schoene, R.B. (1982). Control of ventilation in climbers to extreme altitude. J. Appl. Physiol. 53: 886-890.

Schoene, R.B., P.H. Hackett and R.C. Roach (1987) . Blunted hypoxic chemosensitivity at altitude and sea level in an elite high altitude climber. In: "Hypoxia and Cold", J.R. Sutton, C.S. Houston and G. Coates, Eds. Praeger, New York (USA) (pp. 532-535) .

Schoene, R.B., S. Lahiri, P.H. Hackett, R.M. Peters, J.S. Milledge, C.J. Pizzo, C.J. Sarnquist, S.J. Boyer, D.J. Graber, K.H. Maret and J.B. West (1984). Relationship of hypoxic ventilatory response to exercise performance on Mount Everest. J. Appl. Physiol. 56: 1478-1483.

Severinghaus, J.W., C.R. Bainton and A. Carcelan (1966). Respiratory insensitivity to hypoxia in chronically hypoxic man. Respir. Physiol. 1: 308-334.

Ward, M.P., J.S. Milledge and J.B. West (1989) . "High Altitude Medicine and Physiology". Chapman and Hall Medical, London (UK) (pp. 88-89).

Weil, J.V., E. Byrne-Quinn, I.E. Sodal, W.O. Friesen, B. Underhill, G.F. Filley and R.F. Grover (1970). Hypoxic ventilatory drive in normal man. J. Clin. Invest. 49: 1061-1072.

50 RESPIRATORY AND CARDIOVASCULAR ACTIVITIES IN CAROTID BODY RESECTED HUMANS

Y. Honda and M. Tanaka

Department of Physiology, School of Medicine
Chiba University, Chiba, 280 Japan

INTRODUCTION

It is known that the carotid bodies is an important chemoreceptor to mediate respiratory and cardiovascular responses to hypoxia. Therefore, the knowledge of these response activities during hypoxia in humans without these receptors may contribute to clarify the role of the carotid bodies in man. This study is important because considerable species differences have been reported in respiratory and cardiovascular responses to hypoxia (Comroe, 1939; Fitzgerald and Lahiri, 1986; Daly, 1986).

In the present study, we have undertaken to examine this problem during sustained mild hypoxia.

METHODS

Two male subjects, whose carotid bodies were bilaterally resected ca. 40 yrs ago (BR), were examined. Their pulmonary functions were in mild obstructive pattern and exhibited moderate hypoxemia with hypercapnia (Table 1). As a control, 8 normal

Table 1. Physical characteristics, pulmonary function and blood gas of the patients with bilateral carotid body resection (BR).

Subject	age	yr of surgery	%VC	$FEV_{1.0\%}$	pH	Pao_2	$Paco_2$
N.K.	65	1952	82.5	37.1	7.42	62.9	51.0
M.K.	83	1948	94.7	57.6	7.38	80.0	45.7

healthy adults were also examined. Both subject groups were rebreathed in a closed circuit with a $CO2$ absorber and maintained mild isocapnic-hypoxia (Sao_2 being approximately 80%) for approximately 20 min. Ventilatory parameters were measured by a hot-wire flowmeter (Minato RF-2), O_2 and CO_2 concentrations by a rapid-response gas analyzer (Sanei 1H21), Sao_2 by a pulse oximeter (Ohmeda, Biox III), and cardiac parameters by an impedance cardiograph (Nihon Kohden).

Neurobiology and Cell Physiology of Chemoreception
Edited by P.G. Data *et al.*, Plenum Press, New York, 1993

RESULT

Figs. 1 and 2 illustrate respiratory and cardiovascular responses to sustained mild hypoxia in 2 BR patients. Hypoxia induced practically no change in respiratory variables while heart rate (HR) progressively increased, blood pressure (BP) and cardiac output slightly increased and stroke volume (SV) and systemic total peripheral vascular resistance (TPVR) exhibited more or less the same level.

Table 2 shows that the control subjects increased ventilation mainly by increasing tidal volume (V_T) in hypoxia whereas this response was nearly zero in the BR subjects. The main differences in circulatory variables in hypoxia between control and BR patients were that marked elevation in HR and depression in SV in the former while small rise in HR and no appreciate change in SV in the latter.

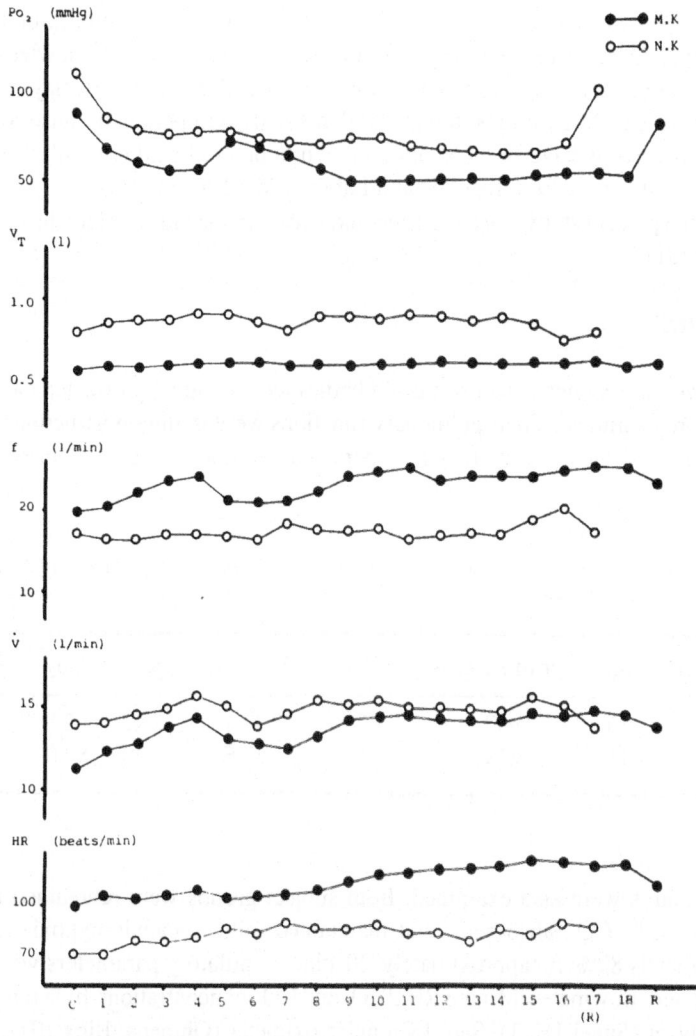

Fig. 1. *Effects of sustained isocapnic hypoxia on the respiratory variables in two patients with bilateral carotid body resection.*

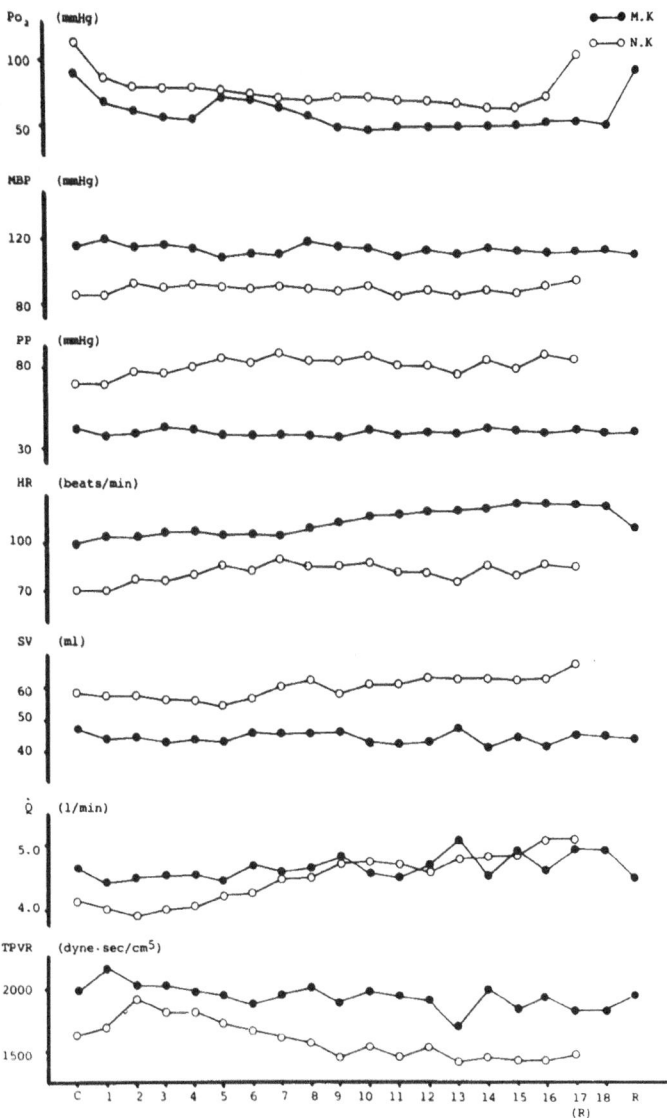

Fig. 2. *Effects of sustained isocapnic hypoxia on circulatory variables in the same patients with Fig. 1.*

Table 2 *Comparison of relative ventilatory response to mild sustained hypoxia between control and two BR subjects.*

Subject	$V_T(\%)$	$f(\%)$	$\dot{V}(\%)$
Control(n=8)	+ 20±7	+3±3.6	+22±7.5
N.K.	+ 1.7	+3.1	+ 4.6
M.K.	+ 0.4	+4.0	+ 5.4

Values in the control are mean±S.E.

DISCUSSION

Fig. 3 indicates the possible factors affecting cardiovascular activities in hypoxia. The primary effect of carotid body activities defined as CB are bradycardia, negative left ventricular inotropic response, decreased cardiac output and increased systemic and pulmonary vascular resistance (Daly, 1986). These changes are most clearly demonstrated in diving mammals. This primary effect is substantially revered by the pulmonary inflation reflex defined as IR. Moreover, final feature of the cardiovascular activities are more or less modified by catecholamines, autonomic nervous tone, P_{ao_2} and P_{aco_2} levels, anesthesia, and local vascular and myocardiac condition (defined OM).

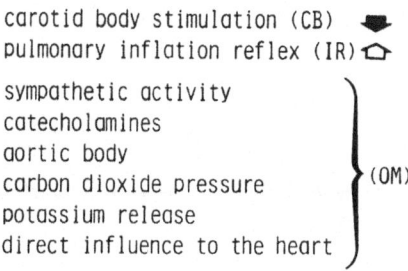

Fig. 3. Possible factors contributing cardiovascular activity in hypoxia.

The overall circulatory response to hypoxia in normal subjects (defined OR) may be determined by the sum of CB, IR and OM as explained by Fig. 3, i. e., OR=CB+IR+OM. Due to the lack of both carotid body and hypoxic hyperventilation, the overall response in the carotid body-resected patient (O' R') may be solely determined by other modifying influences on the circulation (O' M'), namely O' R' = O' M'. If we further postulate in relative magnitude that OM equals O'M', then CB+IR can be calculated as OR-O'R'. The upper section of Fig. 4 illustrates the result of this calculation.

When we take into consideration Daly ' s statement (1986) that the primary effects of carotid body stimulation are the elevation of MBP, $\dot{Q}$ and TPR and the depression of HR and SV as depicted in the lower section of Fig. 4, comparison of the upper and lower figures leads us to assume that MBP, SV and TPR are predominantly determined by carotid body stimulation whereas HR is effectively reversed from the depressed state by carotid body stimulation to augmentation by IR.

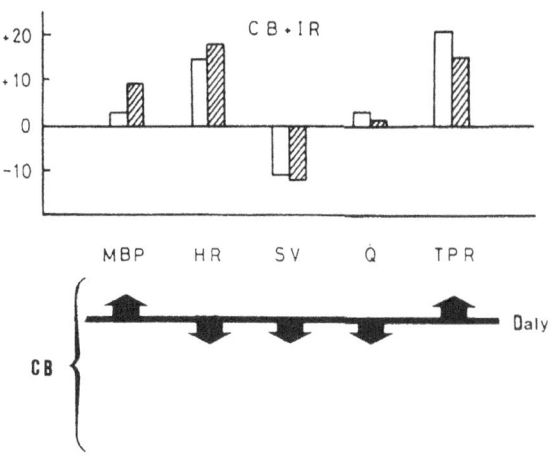

Fig. 4. Comparison of the calculated magnitude of carotid body stimulation (CB) plus pulmonary inflation reflex (IR) to the direction of primary CB effect proposed by Daly.

Although we postulated that effects of CB, IR and OM are simply additive, Hilton and Marshall (1982) found a significant augmentation of sympathetic activity and adrenaline release by carotid body stimulation in cats. If this is the case in man, the magnitude of the CB's effect may have been underestimated. However, Cherniack et al. (1991) recently demonstrated marked sympathetic activity by hypoxic stimulation in anesthetized and peripheral chemodenervated cats. Therefore, substantial amount of sympathetic contribution to circulatory activity in hypoxia seems to be expected without carotid body activation.

CONCLUSION

Respiratory and circulatory activities to moderate hypoxia in two carotid body resected patients were examined, and influence of carotid body stimulation and pulmonary inflation reflex was analyzed.

REFERENCES

Cherniack, N. S., Mitra, J., Prabhakar, N., Lust, D. and Eransberger, P., 1991, Effect of CNS hypoxia on respiratory and sympathetic activity, 1991 Oxford Conference: 5th Meeting on Control of Breathing and Its Modelling Perspective (Abstract) p.19.

Comroe, J. H., Jr., 1939, The location and function of the chemoreceptors of aorta, Am. J. Physiol. 127 : 176-191.

Daly, M. De B., 1986, Interaction between respiration and circulation. In : Handbook of Physiology. The respiratory system. Control of Breathing, edited by A.P. Fishman, N.S. Cherniack, J.G. Widdicombe and S.R. Geiger, Bethesda; MD, Am.Physiol.Soc., Sect. 3, Vol. II, Pt 2, pp. 529-594.

Fitzgerald, R. S. and Lahiri, S., 1986, Reflex responses to chemoreceptor stimulation. In : Handbook of Physiology. The Respiratory System. Control of Breathing, edited by A. P. Fishman, N. S. Cherniack, J. G. Widdicombe and S. R. Geiger, Bethesda; MD, Am. Physiol. Soc., Sect. 3, Vol. II, Pt. 1, pp. 313-362.

Hilton, S. M. and Marshall, J. M., 1982, The pattern of cardiovascular response to carotid chemoreceptor stimulation in the cat. J. Physiol. 326: 365-378.

51 EFFECTS OF OXYGEN TESTS ON THE VENTILATORY RESPOSES OF THE CAT AND RHESUS MONKEY TO CHANGES IN ARTERIAL POTASSIUM

D. J. Paterson and P.C.G. Nye

University Laboratory of Physiology
Parks Road, oxford OX1 3PT, UK

INTRODUCTION

Raising the concentration of arterial plasma potassium ($[K^+]_a$) to levels seen in heavy exercise in man (7-8mM)[1-2] excites the arterial chemoreceptors of the anaesthetised cat[3,4] and causes an increase in ventilation[5]. The excitation of chemoreceptor discharge by hyperkalaemia is further enhanced by hypoxia and virtually abolished with short latency by abruptly replacing the inspiratory gas with oxygen[6]. Furthermore the stimulatory effects of potassium are abolished by peripheral chemoreceptor denervation[5]. We show here that ventilation in the decerebrate cat and sedated rhesus monkey is increased during hyperkalaemia and that this effect is also virtually abolished by abrupt replacement of inspired gas with 100% oxygen.

METHOD

Six cats (2.2-4.8 kg) were anaesthetised with halothane (2.5%) in oxygen. Cannulae were placed in the (1) trachea (2) left femoral vein for infusing KCl (3) left femoral artery for recording blood pressure (4) right femoral artery for withdrawal of blood for measurement of $[K^+]$, pH, PCO_2 and PO_2. The cats were then decerebrated as previously described by Paterson & Nye[7]. In addition, two adult rhesus monkeys (6 and 9kg) were premedicated with ketamine (22mg/kg i.m.) and anaesthetised with halothane (2%) in 100% oxygen during tracheal cannulation and catheterisation of the right jugular vein, right femoral vein and both femoral arteries. All wounds were sutured and sprayed at regular intervals with lignocaine. After surgery was completed the halothane was stopped and the monkeys were sedated and given analgesia with a continuous infusion of ketamine (2-5mg/kg/hr i.v.). Arterial blood pressure was measured continuously. All animals were made hypoxic F_1O_2 ca. 10%) and ventilation was measured with a pneumotachometer. Tracheal gases were continuously measured by mass spectrometry. The ventilatory drive from the arterial chemoreceptors was estimated by recording the reduction in ventilation caused by abrupt switches of inspired gas to 100% O_2 before, during and after a KCl infusion. End-tidal PCO_2 was held constant by manually adjusting inspired PCO_2 throughout the experimental run. KCl (150mM) was infused into the central venous line in the monkey and into the femoral vein in the cat for 1 min at 1ml/kg/min and thereafter reduced to 0.5 ml/kg/min by a variable speed peristaltic pump. In one monkey

Neurobiology and Cell Physiology of Chemoreception
Edited by P.G. Data *et al.*, Plenum Press, New York, 1993

this was repeated in euoxia. Arterial blood (1ml) was sampled at regular intervals and analysed for pH, blood gases, and for K^+ by flame photometry.

RESULTS

Raised $[K^+]$a (mean$\pm$SEM 7.4+0.3mM in the cat n=18 tests and 7.0$\pm$0.4mM in the monkey n=6 tests) always increased ventilation (fig 1,2a). In the hypoxic cats (PaO_2 48.2$\pm$1.4 Torr) ventilation was increased by 42$\pm$8% (p<0.01) at a $PaCO_2$ of 25.7$\pm$1.6 Torr (fig 3). In the monkey euoxic ventilation was increased by ca. 40% and by as much as 250% in hypoxia (PaO_2 39.6$\pm$1.4) at a $PaCO_2$ of 29.3$\pm$1.4 Torr (fig1). This effect in hypoxia was virtually abolished in both cat and monkey within 4s of an abrupt switch to 100% oxygen (fig 1,2a) and was markedly reduced during euoxia in the one monkey tested. Arterial pH (pHa) and arterial blood gases were not significantly changed by the KCl infusion in either the cat or the monkey. In two cats and both monkeys oxygen tests resulted in an apnoea for 30-75s both before and during the infusion of KCl.

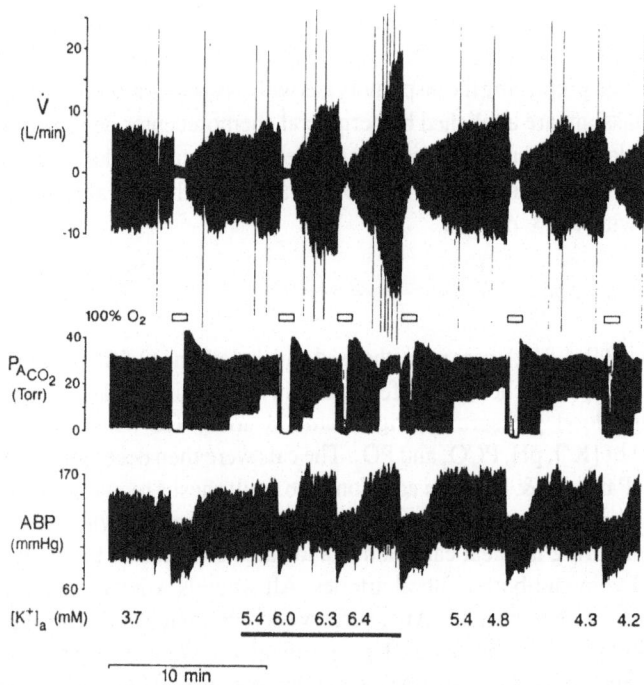

Figure 1. Ventilatory response of a hypoxic (PaO₂ 39 Torr) monkey to abrupt switches of 100% oxygen (hollow boxes) before, during and after a KCl infusion. Traces from the top down: Tracheal airflow, alveolar PCO₂ and arterial blood pressure (ABP). The dark bar indicates the duration of the KCl infusion. Note that raising plasma K⁺ increased ventilation and that to keep PACO₂ constant CO₂ was added to the inspirate. A switch to 100% O₂ virtually abolished the excitation of ventilation by hyperkalaemia. When the KCl was stopped ventilation decreased with the fall in [K⁺]ₐ. (from Paterson et al. [8])

1
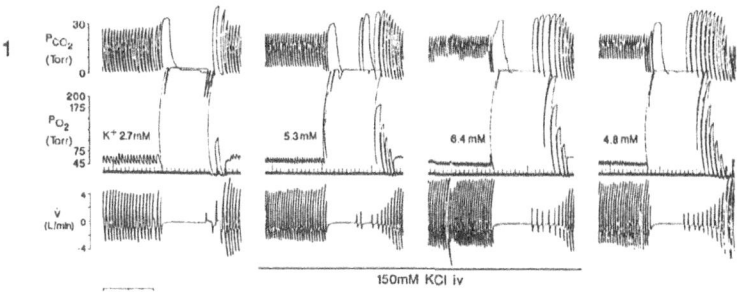

P_{CO_2}
(Torr)

30
0

P_{O_2}
(Torr)

200
175
75
45

K⁺ 2.7mM   5.3mM   6.4 mM   4.8 mM

$\dot{V}$
(L/min)

4
0
-4

150mM KCl iv

1 min

2
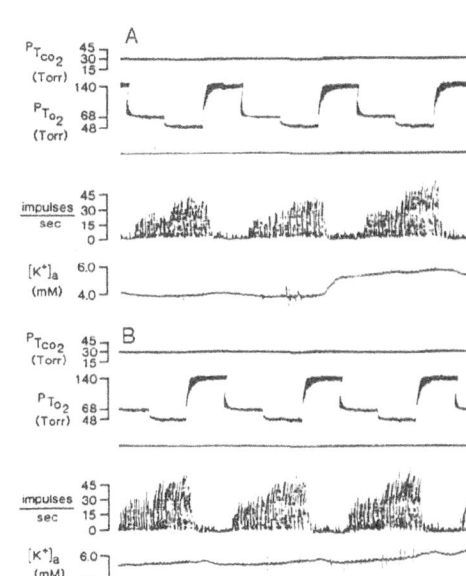

$P_{T_{CO_2}}$
(Torr)

45
30
15

A

$P_{T_{O_2}}$
(Torr)

140
68
48

impulses / sec

45
30
15
0

$[K^+]_a$
(mM)

6.0
4.0

$P_{T_{CO_2}}$
(Torr)

45
30
15

B

$P_{T_{O_2}}$
(Torr)

140
68
48

impulses / sec

45
30
15
0

$[K^+]_a$
(mM)

6.0
4.0

results reflect well the previous reported response of the arterial chemoreceptors to K^+ during hypoxia and hyperoxia (fig 2b).

ACKNOWLEDGEMENTS

This work was supported by the Wellcome Trust and Medical Research Council. DJP is BHF Lecturer and PCGN was a Wellcome Senior Lecturer.

REFERENCES

1. *J.L. Medbo and O.M. Sejersted*, Plasma potassium and high intensity exercise. J.Physiol. 421:105-122 (1990).

2. *D.J.Paterson, J.S.Friedland, D.A.Bascom, I.D.Clement, D.A.Cunningham, R.Painter and P.A.Robbins*, Changes in arterial K^+ and ventilation during exercise in normal subjects and subjects with McArdle's syndrome J.Physiol. 429:339-348 (1990).

3. *R.A.F. Linton and D.M. Band*, The effect of potassium on carotid chemoreceptor activity and ventilation in the cat. Respir. Physiol.59:65-70 (1985).

4. *D.J. Paterson and P.C.G. Nye*, The effect of beta adrenergic blockade on the carotid body response to hyperkalaemia in the cat. Respir. Physiol.74:229-238 (1988).

5. *D.M.Band, R.A.F.Linton, R.Kent and F.L.Kurer*, The effect of peripheral chemoreceptor denervation on the ventilatory response to potassium. Respir.Physiol. 60:217-225 (1985).

6. *R.E. Burger, J.A. Estavillo, P. Kumar, P.C.G. Nye and D.J. Paterson*, Effects of potassium, oxygen and carbon dioxide on the steady-state discharge of cat carotid body chemoreceptors. J. Physiol.401:519-531 (1988).

7. *D.J. Paterson and P.C.G. Nye*, Effect of oxygen on potassium-excited ventilation in the decerebrate cat. Respir. Physiol. 84:223-230 (1991).

8. *D.J.Paterson, K.L.Dorrington, D.H.Bergel, G.Kerr, R.C.Miall, J.F.Stein and P.C.G. Nye*, Effect of potassium on ventilation in the rhesus monkey. Expt Physiol. 77:217-220 (1992).

9. *D.M. Band and R.A.F. Linton*, The effect of potassium on carotid body chemoreceptor discharge in the anaesthetised cat. J. Physiol. 381:39-47 (1986).

10. *J.R. Sneyd, R.A.F. Linton and D.M. Band*, Ventilatory effects of potassium during hyperoxia, normoxia and hypoxia in anaesthetised cats. Respir. Physiol. 72:59-64 (1988).

52 THERMAL EFFECTS UPON THE CHEMO-SENSORY DRIVE OF VENTILATION

P. Zapata, C. Larrain, R. Fadic, B. Ramirez and H. Loyola

Laboratory of Neurobiology
Catholic University of Chile
Santiago 1, Chile

THE CAROTID BODIES AS THERMOSENSORS

It has been shown that the frequency of chemosensory discharges recorded from carotid bodies superfused in vitro is highly dependent on temperature (Gallego et al., 1979), presenting both dynamic and static components (Eyzaguirre and Zapata, 1984). Such thermal influence had also been observed in carotid bodies in situ, in which the rate of chemosensory discharges increased in response to local warming of the arterial blood circulating through the carotid bifurcation (McQueen and Eyzaguirre, 1974). The high energies of apparent activation (μ) and high thermal coefficients (Q_{10}) exhibited by the rate of discharges of these preparations indicate that the carotid body chemoreceptors fulfil the criteria for being considered as potential thermosensors.

Hyperthermia, acting on all body tissues, increases oxygen consumption and carbon dioxide production, and the resulting changes in the levels of these gases in the blood may augment the chemoreceptor activity of the carotid body. In the blood, as temperature rises, the pK' of Henderson-Hasselbach equation decreases, the pH of arterial blood plasma falls and acidity stimulates arterial chemoreceptors; this effect may be summated to that of hypoxaemia, as the O_2 transport capability of the erythrocytes is reduced by the shift to the right of the oxyhemoglobin dissociation curve, under combined thermal and Bohr effects. On glomus tissue itself, the raised temperature increases the rate of sensory discharges generated from this organ, while that of the baroreceptors of the nearby carotid sinus is minimally affected. Studies with carotid bodies or tissue slices in vitro indicate that the resting membrane potential of glomus cells is rapidly modified by local thermal changes (see Eyzaguirre and Zapata, 1984).

The resting levels of carotid chemosensory discharges maintain a certain degree of tonic reflex excitation of ventilatory centres ("chemosensory drive"), demonstrated by the immediate, transient decreases in tidal volume in response to breathing 100% O_2 for a few seconds (Dejours, 1963). It is known that the carotid chemosensory discharges are briefly silenced by this test (see Eyzaguirre and Zapata, 1984). Therefore, the increased chemosensory discharge evoked by a rise in blood temperature should enhance the chemosensory drive upon resting ventilation in a normoxic environment, unless temperature would evoke a larger increase in ventilation through its action on non-chemosensory controlling mechanisms (e.g., central thermoreceptors).

For the above reasons, we have studied the possible contribution of carotid afferents to the ventilatory changes induced by raising the body temperature (T_B) of pentobarbitone anesthetized cats by external heat. It is known that pentobarbitone renders T_B more dependent on environmental temperature (Refinetti and Carlisle. 1989). Furthermore, the transient ventilatory depression evoked by hyperoxia is well manifested under pentobarbitone anesthesia (Gautier et al., 1986).

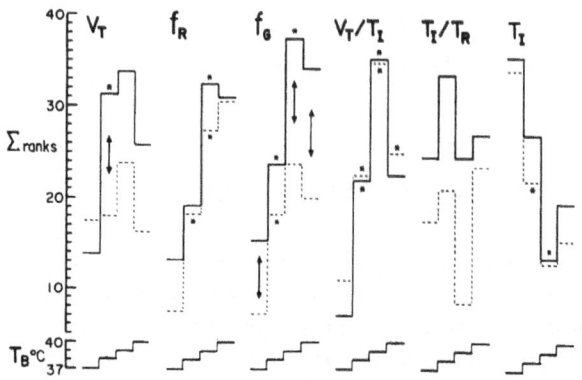

Fig. 1. Summary of changes in some ventilatory variables and indexes evoked by unitary increases in body temperature (T_B). Values obtained in steady-state conditions before (continuous lines) and after (interrupted lines) bilateral carotid neurotomy (BCN), and expressed as sums of ranks (minimum 5, maximum 40 for 8 conditions in 5 cats). Statistical significance of differences ascertained by Quade's tests. Asterisks, significantly different ($p < 0.05$) from preceding value, recorded at 1°C lower T_B. Double headed arrows, significant differences between conditions before and after BCN, but at same T_B.

VENTILATORY EFFECTS OF HYPERTHERMIA

We observed that raising T_B from 37 to 40°C increased respiratory frequency (f_R), tidal volume (V_T) and frequency of spontaneous gasps (f_G), leading to a consecutive reduction in $P_{ET}CO_2$ (Fadic et al., 1991). After bilateral carotid neurotomy (BCN), raising T_B still induced hyperventilation, and the increase in f_R was maintained, but the increase in f_G was less pronounced and no significant change in V_T was observed.

Figure 1 summarizes the changes in the main ventilatory variables recorded from 5 cats in steady-state condition when raising T_B in unitary steps from 37 to 40°C, before and after BCN. Results are expressed as sums of ranks, to discard interindividual variations and to emphasize significant changes. In comparison to the steady ventilatory values recorded at different temperatures in the intact condition, significantly lower values in V_T at 38°C, and in f_G at 37, 39 and 40°C were observed after BCN. These differences may be of consequence since f_G is the most important ventilatory variable for the control of alveolar surfactant in resting conditions (Oyarzun et al., 1991).

It must be noted that the mean inspiratory flow (V_T/T_I index) was significantly increased, without significant changes in breath timing (T_I/T_R index), when T_B rose from 37 to 39°C. Increases in V_T/T_I without changes in T_I/T_R had been observed when rewarming cats from 34 to 38°C (Gautier and Gaudy, 1986). We observed shortenings in both T_I and T_E along the 37-39°C range, but no further changes were observed with the following raise in T_B to 40°C. The fact that similar increases in f_R and decreases in T_I and T_E in response to step increases in T_B from 37 to 40°C occurred before and after BCN indicates that the carotid chemoreceptors have little influence on the hyperthermic tachypnea observed along this range.

As a whole, the above results indicate that the carotid chemosensors contribute to the hyperventilation induced by hyperthermia, but they are not the only source for such effect.

We also observed that hyperoxic tests induced transient decreases in V_T and increases in $P_{ET}CO_2$, which were significantly larger at 40°C than at 37°C (Fadic et al., 1991). This is illustrated in Fig. 2. The transient ventilatory depression evoked by hyperoxia disappeared after BCN. These results suggest that the peripheral chemosensory drive on ventilation during air breathing was enhanced at higher temperatures.

CHEMOSENSORY DISCHARGES IN HYPERTHERMIC CONDITIONS

We observed that raising T_B from 35 to 40°C increased the frequency of chemosensory discharges (f_x) recorded from one carotid (sinus) nerve in some spontaneously breathing cats. This effect was more commonly seen after BCN (Fig. 3A). In other animals, the simultaneously increased alveolar ventilation (assessed by the fall in $P_{ET}CO_2$) counteracted the above effect. The curves describing the chemosensory responses to increasing levels of CO_2 at different temperatures showed that raising T_B displaced the sensitivity to the left without affecting the slope (Fig. 3B). A multiple correlation analysis of global data showed predicted increases in f_x in response to raising T_B, at different CO_2 levels (Loyola et al., 1991).

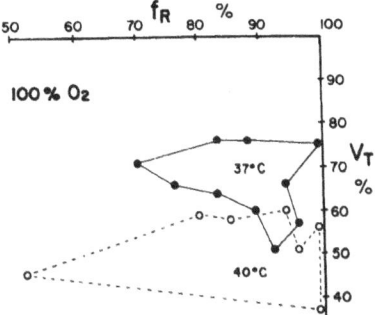

Fig. 2. Plotting of minimal tidal volume (V_T; y-axis) against minimal instantaneous respiratory frequency (f_R; x-axis) in response to 1-min periods of 100% O_2 breathing, performed at 37 and 40°C, in 10 cats with intact carotid nerves (in 3 of them the manoeuvre was only performed at 37°C).

With regard to the association between pulmonary ventilation and chemosensory discharges as affected by temperature, Fig. 4 plots mean inspiratory minute volume ($\dot{V}_I$) against mean f_x and T_B, in 5 cats with unilateral carotid neurotomy (UCN) and 11 cats with BCN. $\dot{V}_I$ was calculated by multiplying mean V_T by mean f_R, obtained along 1-min steady-states. An ANOVA justified the use of the regression model for these analyses ($p<0.001$). Consistent and smooth associations were observed in the UCN condition, but this picture was disrupted after BCN. In cats with UCN, V_I increases were observed in response to T_B rises without variation in f_x but they were more pronounced in response to the combined increases in f_x and T_B. Sequential sum of squares analysis confirms that the residual contribution of f_x to $\dot{V}_I$ control after consideration of T_B effect was small but significant in cats with UCN ($p<0.005$), but not significant after BCN ($p<0.1$). The dependence of $\dot{V}_I$ on the low range of f_x observed within isothermal conditions in cats with UCN probably rests on the ventilatory drive provided by the still intact contralateral carotid nerve. Otherwise, $\dot{V}_I$ was still enhanced in response to T_B rises after BCN, and the residual isothermal variations in $\dot{V}_I$ did not appear to depend on the f_x originated from the aortic bodies.

The possibility that the hypoxic responsiveness of arterial chemoreceptors could be modified by temperature was checked by giving the cat with BCN 100% N_2 to breath for periods of 5, 10 and 15 s, at different degrees of T_B. The magnitudes of the responses were related to the duration of the tests. The maximal f_xs attained, especially in response to 5 s tests, were higher when elicited at higher T_B, but the change in frequency (Δfx) must be calculated to correct for the thermally-induced changes in basal f_x. When chemosensory responses were measured in terms of Δfx evoked by tests of a given duration, no significant correlations with T_B were observed. The time courses of the responses obtained at different T_Bs were also similar.

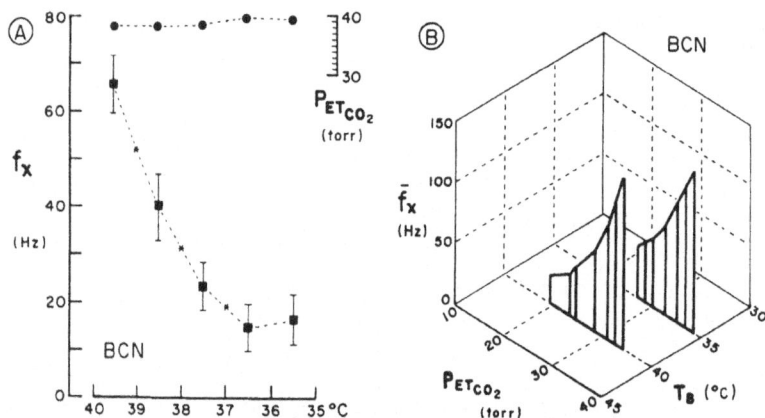

Fig. 3. A. Relationships between frequency of chemosensory discharges (left ordinate), averaged $P_{ET}CO_2$ (upper right ordinate) and T_B (abscissa). Data obtained at different stages of cooling in one cat in which BCN was performed at 40°C. Means ± SDs of 30 periods of 1 s at each thermal step. Asterisks, significant differences ($p<0.005$) between contiguous thermal steps, ascertained by Student's t-tests after Bonferroni correction. The experiment here illustrated is one of the few in which $P_{ET}CO_2$ was not modified along this thermal range. B. Mean chemosensory frequencies (y-axis) at different levels of $P_{ET}CO_2$ (x-axis) attained by additive increases of ventilatory dead space at two different degrees of T_B (z-axis). Data obtained from another cat with BCN.

Hyperoxic tests (breathing 100% O_2 for 15 s) silenced chemosensory activity when administered at any given T_D. Three parameters of these responses were studied in cats with BCN: the slope of the decrease in f_x, the time for maximal inhibition and the half-recovery time (period between start of O_2 test and recovery to half of basal f_x). However, no significant differences in these parameters at different degrees of T_B were revealed by Quade's tests. The fact that chemosensory silencing following removal of normoxic stimulation was not significantly modified within the 35-40°C range suggests t hat the possibly enhanced O_2 consumption of glomus tissue at higher T_B was well satisfied by concurrent P_aO_2 levels.

RESPIRATORY MUSCLES CONTRIBUTION TO HYPERTHERMIC HYPERVENTILATION

Recently, we have recorded the EMG activity of the diaphragmatic and interchondral muscles in cats subjected to increases in T_B from 35 to 40°C (Ramirez, Larrain and Zapata, unpublished observations). Paired disk electrodes were placed under the right costal and sternal portions of the diaphragmatic vault, through a midline laparotomy and after retraction of the liver and other abdominal viscera, while paired needle electrodes were inserted parasternally on the interchondral muscle of the right 5th intercostal space, after retraction of the pectoralis major and minor and rectus abdominis (which in the cat extends up to the first costal cartilage). These differential recordings provided adequate signal-to-noise ratios and common mode rejection of EKG and other unwanted artifacts. We analyzed the instantaneous frequencies of muscle action potentials, the duration of the discharge, and the rectified and integrated electromyograms (IEMEGs) on a breath by breath basis.

The maximal rates of discharges of the diaphragm and interchondral muscles occurred during inspiratory flows. However, the diaphragm showed prominent preinspiratory bursts, while the interchondral muscle action potentials were prolonged beyond the end of inspiratory flow. On a breath-by-breath basis, the maximal instantaneous frequencies of these muscles were directly correlated with V_Ts. Their increases in response to 100% N_2 tests and NaCN injections were also correlated with the simultaneous increases in V_T, at all T_Bs between 35 and 40°C.

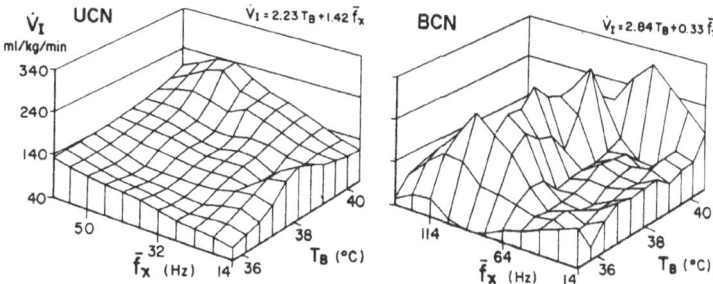

Fig. 4. Plotting of mean inspiratory minute volume ($\dot{V}_I$; y-axis) against mean frequency of chemosensory discarges (f_x; x-axis) and body temperature (T_B; z-axis). Data from 5 cats with unilateral carotid neurotomy (UCN; n=31 observations), and 11 cats with bilateral carotid neurotomy (BCN; n=74 observations). Note that f_x range depicted along x-axis in BCN had been extended to almost twice that shown in UCN. Insets, equations for data fitted to multiple regression model through MINITAB statistical software for VAX/VMS computer.

375

The peripheral chemosensory drive exerted upon the diaphragm and interchondral was tested by the reduction of their respective IEMGs in response to breathing 100% O_2 for 30 s periods. Fig. 5 illustrates the instantaneous correlations of these IEMGs with V_Ts in one cat. At all temperatures studied, the falls in the IEMGs of the interchondral were more pronounced than those of the diaphragm. Furthermore, the IEMGs of the interchondral during these manoeuvres were significantly correlated with the simultaneously recorded V_Ts while those of the diaphragm were not. This means that the interchondral is highly dependent on the reflex drive originated from the arterial chemoreceptor, along the range of T_Bs studied, while the diaphragm is less influenced by these peripheral afferents.

If an increase in T_B would induce hyperventilation by its direct effect upon central structures (including hypothalamic thermoreceptors), we must expect a proportional reduction in the prevailing reflex influence from arterial chemoreceptors. But, on the contrary, we observed that the chemosensory drive -particularly evidenced by the changes in the interchondral IEMG during hyperoxic tests- is well maintained at high T_B, implying that this reflex influence is also enhanced upon raising T_B.

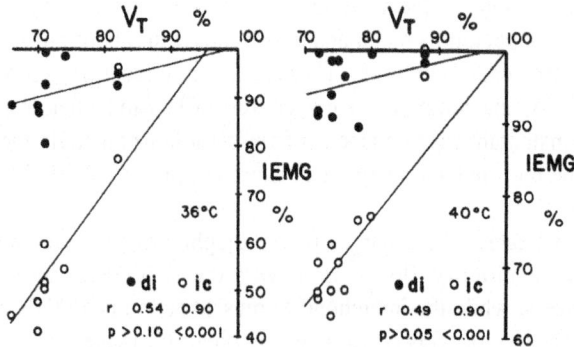

Fig. 5. Correlations of integrated electromyograms (IEMG; y-axes) with tidal volumes (V_T) recorded simultaneously while breathing 100% O_2 for 30 s, at 36 and 40°C, in the same cat. Values expressed in percentages of their respective basal values, recorded while breathing air. IEMGs of diaphragm (di; filled circles) and interchondral (ic; open circles) muscles, and their respective regression lines on V_T. Insets, correlation coefficients (r) and their significances (p).

THE ROLE OF THE CAROTID BODIES IN THERMAL HYPERVENTILATION

The arterial chemoreceptors, particularly the carotid bodies, may contribute to the reflex control of two ventilatory functions. As chemosensors, they participate in the control of alveolar ventilation, which in turn sets the level of chemoreceptor activity. Otherwise, as thermosensors, they take part in the control of pulmonary ventilation, which serves as a mechanism for heat dissipation, thus resetting their level of thermally induced chemosensory activity. However, the common afferent pathway apparently does not provide any clue as to the stimulus nature: chemical or thermal. On the other hand, the respiratory centres also receive inputs from central (medullary) chemoreceptors and central (mainly hypothalamic) thermoreceptors.

The combined input of the carotid afferents with either one or the other of these sources may result in different ventilatory patterns, more appropriate to serve chemostatic or thermostatic purposes.

The compromise between the chemostatic and thermostatic functions of the respiratory system may be achieved by adjusting the ratio of V_T to f_R, which in turn determines the ratio of alveolar to pulmonary ventilation (Albers, 1976). Theoretically, in the presence of hyperthermic tachypnea, alveolar ventilation may be kept constant by reducing V_T in proportion to the increases in f_R. Nevertheless, hyperthermia usually results in alveolar hyperventilation, an indication that the thermostatic control supersedes the chemostatic control.

As to the ventilatory reactions evoked by hyperthermia, the moderate increase in V_T adjusts the blood levels of O_2 and CO_2 to the modified demands of body tissues, while the pronounced increase in f_R enhances heat dissipation. Our observation that only V_T and f_G were clearly dependent on the indemnity of the carotid nerves provides further support to the idea that the arterial chemoreceptors only subserve that part of the enhanced ventilation that maintains the chemostatic balance when the organism is perturbed by hyperthermia. The remaining and important change in f_R evoked by hyperthermia -persisting after BCN- must be mediated by other thermally responsive structures and subserve the thermostatic balance of the animal.

ACKNOWLEDGEMENTS

Our work has been supported by the National Fund for Scientific and Technological Development (grant FONDECYT 499-89), the Catholic University of Chile Research Division (DIUC) and the Gildemeister Foundation. We thank Mr. Ricardo Aravena (Dept. Statistics, Faculty of Mathematics) for suggesting the sequential sum of squares analysis.

REFERENCES

Albers, C., 1976, Respiratory drive in hyperthermia, In: "Acid Base Homeostasis of the Brain Extracellular Fluid and the Respiratory Control System", H.H. Loeschke, ed., Thieme, Stuttgart, p. 112

Dejours, P., 1963, Control of respiration by arterial chemoreceptors, Ann. N. Y. Acad. Sci., 109: 682

Eyzaguirre, C., and Zapata, P., 1984, Perspectives in carotid body research, J. Appl. Physiol., 57 : 931

Fadic, R., Larrain, C., and Zapata, P., 1991, Thermal effects on ventilation in cats. Participation of carotid body chemoreceptors, Respir. Physiol., 86: 51

Gallego, R., Eyzaguirre, C., and Monti-Bloch, L., 1979, Thermal and osmotic responses of arterial receptors. J. Neurophysiol., 42: 665

Gautier, H., Bonora, M., and Gaudy, J.H., 1986, Ventilatory response of the conscious or anesthetized cat to oxygen breathing, Respir. Physiol., 65: 181

Gautier, H., and Gaudy, J.H., 1986, Ventilatory recovery from hypothermia in anesthetized cats, Respir. Physiol., 64: 329

Loyola, H., Fadic, R., Cardenas, H., Larrain. C., and Zapata, P., 1991, Effects of body temperature on chemosensory activity of the cat carotid body in situ, Neurosci. Lett., 132: 251

McQueen, D.S., and Eyzaguirre, C., 1974, Effect of temperature on carotid chemoreceptor and baroreceptor activity, J. Neurophysiol., 37: 1287

Oyarzun, M.J., Iturriaga, R., Donoso, P., Dussaubat, N., Santos, M., Schiappacasse, M.E., Lathrop, M.E., Larrain, C., and Zapata, P., 1991, Factors affecting the distribution of alveolar surfactant during resting ventilation, Am. J. Physiol., 261: L210

Refinetti, R., and Carlisle, H.J., 1989, Thermoregulation during pentobarbital and ketamine anesthesia in rats, J. Physiol., Paris, 83: 300

SECTION V

Developmental and Adaptative
Physiology and Pharmacology

53 CAROTID CHEMOSENSORY RESPONSE TO DOXAPRAM IN THE NEWBORN KITTEN

A. Bairam, F. Marchal, B. Hannhart, J.- P. Crance and S. Lahiri

Laboratoire de Physiologie, Faculté de Médecine de Nancy
Vandoeuvre -Lès- Nancy, France
Unité 272 INSERM, Nancy, France
Department of Physiology, University of Pennsylvania, USA

INTRODUCTION

Doxapram stimulates ventilation by acting on peripheral chemoreceptors and medullary respiratory neurons in a dose-dependent fashion (Hirsh and Wang, 1974 ; Mitchell and Herbert, 1975). In adult cats, doxapram has been found to increase the chemosensory response to hypoxia and hypercapnia (Mitchell and Herbert, 1975 ; Nishino et al., 1982). Several studies have shown that bilateral section of the carotid sinus nerves abolishes or decreases the ventilatory effect of doxapram in adult animals (Hirsh and Wang, 1974 ; Mitchell and Herbert, 1975 ; Folgering et al., 1981 ; Nishino et al., 1982). Despite a frequent use of this drug in the treatment of neonatal apnea (Bairam and Vert, 1986 ; Barrington et al., 1986 ; Tay-Uyboco et al., 1991), its action on carotid chemoreceptor activity is still unclear. We studied the effects of doxapram on carotid chemosensory responses to hypoxia and hypercapnia in newborn kittens. We found that doxapram stimulates the discharge rate of the carotid chemoreceptor afferents at all levels of FiO_2 and $FiCO_2$ tested. In hypoxia, the effect of doxapram is more variable and this variability is not dependent on age.

MATERIALS AND METHODS

Studies were carried out in 9 kittens aged 1 to 11 days, at an average barometric pressure of 740 torr. The kittens were anesthetized with an intraperitoneal injection of sodium thiopental (60 mg/kg), which was supplemented as needed. A femoral vein was cannulated for the administration of drug or saline, and rectal temperature was maintained at approximately 38°C. A tracheostomy was performed and the kittens were artificially ventilated with a respiratory pump and paralysed with pancuronium bromide (1 mg/kg, IV). The inspiratory port of the pump could be easily switched from and to one of the following gas mixtures : 8% O_2 in N_2, air, 100% O_2, 5 and 10% CO_2 in O_2.

Under dissecting microscope, the carotid bifurcation was exposed and the carotid sinus nerve was cut at the level of the petrosal ganglion and dissected into fine filaments under warm mineral oil. The filaments were tested one at a time for neural signal and further divided until one or a few chemoafferents, free of baroreceptor discharge remained. The action potentials were amplified, band pass filtered, fed to a window discriminator and displayed on an oscilloscope and on a chart recorder. The activity was also displayed as impulse/second. The same chemoreceptor preparation was used throughout a given protocol.

To study the effect of intravenous injection of doxapram of 0.4 mg/kg on carotid chemoreceptor response to hypoxia and hypercapnia, the kitten was exposed to the following sequence of inspired gases : 21% O_2, 8% O_2 and 100% O_2 then 5 and 10% CO_2 in O_2. Each gas test was held for about 2 min. Doxapram (0.4 mg/kg) was diluted in saline to a final volume of 0.2 ml and injected when the kitten was breathing 100% O_2. Carotid chemoreceptor activity was recorded before and after doxapram injection.

The chemosensory activity was averaged over consecutive 10 seconds epochs. The data collected at each steady-state level of FiO_2 and $FiCO_2$ before and after drug administration were compared using paired t-test. Differences were considered significant at the $p < 0,05$ level. Data are expressed as a mean $\pm$ SEM.

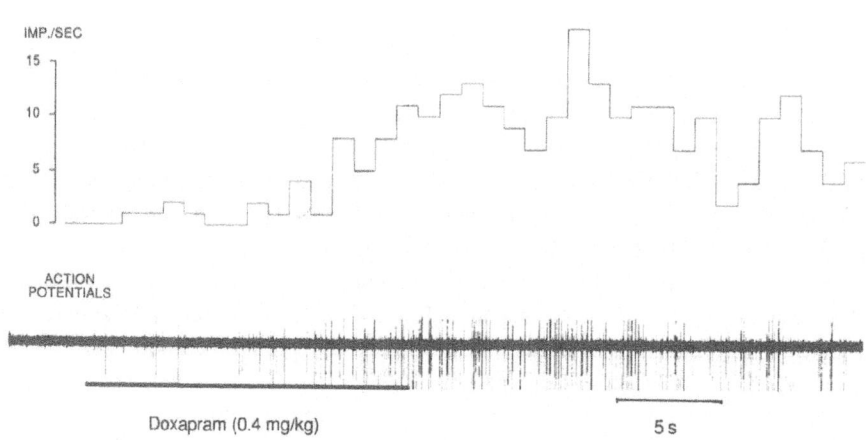

Figure 1. Effect of doxapram on carotid chemosensory response to hyperoxia in a one day old kitten. The injection of doxapram induces an immediate increase in chemoreceptor discharge rate.

RESULTS

An example of doxapram injection in a one day old kitten during hyperoxia is shown in figure 1. Doxapram injection was followed by a prompt increase in carotid chemoreceptor activity. This response was observed in all kittens studied. The mean chemoreceptor activity increased from 0.3 ± 0.2 imp/s to 10.5 ± 2.4 imp/s, 20 seconds after the injection of doxapram, and then decreased gradually to 5.3 ± 0.9 imp/s, one minute after injection. The chemosensory activity was still significantly higher than control 5, 10 and 25 minutes after doxapram (respectively 2.7 ± 0.3, 2.2 ± 0.3 and 1.3 ± 0.2, $p < 0.001$).

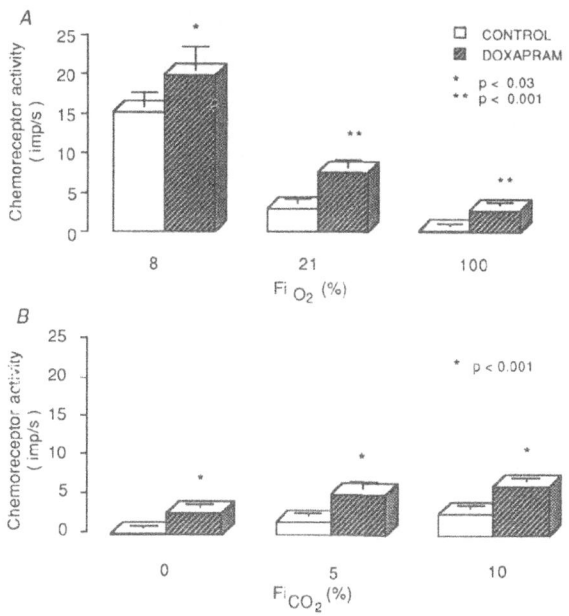

Figure 2. Effects of doxapram on steady-state chemoreceptor activity at three levels of FiO_2 (A) and $FiCO_2$ (B). Doxapram significantly increases the chemosensory response to hypoxia and hypercapnia.

The chemoreceptor activity increased during hypoxia both before and after doxapram. After doxapram, the chemoreceptor activity was significantly increased with respect to control at all levels of FiO_2 tested (fig. 2A). However, during hypoxia, doxapram effect was less consistent than during air or hyperoxia, and no change in chemosensory activity was observed in three kittens aged 1, 10 and 11 days.

Hypercapnia induced an increase in chemoreceptor activity. The mean steady-state chemoreceptor activity in response to hypercapnia, before and after doxapram, is shown in figure 2B. After doxapram, the chemosensory response to hypercapnia increased significantly.

DISCUSSION

In the present study, we demonstrated that the carotid chemosensory response to hypoxia and hypercapnia is stimulated by doxapram, indicating that the mechanisms of response are developed in the newborn kitten.

Decreasing FiO_2 and raising $FiCO_2$ stimulated the chemoreceptor activity. Doxapram had an additive effect to hypoxic and hypercapnic stimuli. However, the effect was more variable during hypoxia. A possible explanation is that in some kittens, hypoxia induced a near maximal increase in the chemosensory activity so that adding another stimulus had no further effect. These results are qualitatively similar to those observed in adult cats where severe hypoxemia - PaO_2 40 torr - limited the effect of doxapram (Mitchell and Herbert, 1974 ; Nishino et al., 1982). In contrast, raising $FiCO_2$ to about 70 torr did not prevent doxapram effect. Also, the carotid chemoreceptor responses to natural stimuli observed in this study are in conformity with our previous report which showed that the chemosensory responses to hypoxia and hypercapnia in the kitten are developed at birth but weak and continue to develop further during the first weeks of postnatal life (Marchal et al., 1992.

The mechanism(s) by which doxapram acts on chemosensory activity is still unclear. Under normoxic conditions, Peers (1991) showed that doxapram inhibits Ca^{2+} activated K^+ conductance of the glomus cell, but the interaction between hypoxia which diminishes K^+ conductance and doxapram is still to be defined. Doxapram is used in the treatment of neonatal apnea. The use of doses as small as 0.25 or 0.50 mg/kg/h was associated with a 50% reduction in the number of apneic spells (Bairam and Vert, 1986; Barrington et al., 1987). Better results were obtained by increasing doses by steps of 0.5 mg/kg/h (Hayakawa et al., 1986 ; Barrington et al., 1987). Doxapram increased minute ventilation and decreased arterial or alveolar PCO_2 (Barrington et al., 1987 ; Tay-Uyboco et al., 1991), but no correlation was found between the ventilatory effects of doxapram and the reduction in apnea frequency. It was suggested that the change in the breathing pattern from irregular or periodic to regular during doxapram therapy, allowed minute ventilation to increase and apnea frequency to decrease (Tay-Uyboco et al., 1991). In newborn kittens, our study clearly shows that doxapram increases the chemosensory responses to O_2 and CO_2. The resulting increase in chemosensory input to the respiratory neurons would thus stabilize the breathing pattern and prevent apnea.

REFERENCES

Bairam, A., and Vert, P., 1986, Low-dose doxapram for apnea of prematurity. Lancet, 5:793.

Barrington, K.J., Finer, N.N., Peters, K.L., and Barton, J.B., 1986, Physiologic effects of doxapram in idiopathic apnea of prematurity. J. Pediatr., 108:125.

Barrington, K.J., Finer, N.N., Torok-Both, G., Jamali, F., and Coutts, R.T., 1987, Dose-response relationship of doxapram in the therapy for refractory idiopathic apnea of prematurity. Pediatrics, 80:22.

Folgering, H., Vis, A., and Ponte, J., 1981, Ventilatory and circulatory effects of doxapram mediated by carotid body chemoreceptors. Bull. Eur. Physiopathol. Resp., 17:237.

Hayakawa, F., Hakamada, S., Kuno, K., Nakashima, T., and Miyachi, Y., 1986, Doxapram in the treatment of idiopathic apnea of prematurity : desirable dosage and serum concentrations. J Pediatr., 109:138.

Hirsh, K., and Wang, S.C., 1974, Selective respiratory stimulating action of doxapram compared to pentylenetetrazol. J. Pharmacol. Exp. Ther., 189:1.

Marchal, F., Bairam, A., Haouzi, P., Crance, J.P., Di Giulio, C., Vert, P., and Lahiri, S., 1992, Carotid chemoreceptor response to natural stimuli in the newborn kitten. Resp. Physiol., 87:183-193.

Mitchell, R.A., and Herbert, D.A., 1975, Potencies of doxapram and hypoxia in stimulating carotid-body chemoreceptors and ventilation in anesthetized cats. Anesthesiology, 42:559.

Nishino, T., Mokashi, A., and Lahiri, S., 1982, Stimulation of carotid chemoreceptors and ventilation by doxapram in the cat. J. Appl. Physiol., 52:1261.

Peers, C., 1991, Effects of doxapram on ionic currents in isolated type I cells of the neonatal rat carotid body. J. Physiol., 438:230.

Tay-Uyboco, J., Kwiatkowski, K., Cates, D.B., Seifert, B., Hasan, S.U., and Rigatto, H., 1991, Clinical and physiological responses to prolonged nasogastric administration of doxapram for apnea of prematurity. Biol. Neonate, 59:190.

54

CAROTID CHEMORECEPTOR RESPONSES TO HYPOXIA AND HYPERCAPNIA IN DEVELOPING KITTENS

J. L. Carroll and R. S. Fitzgerald

The Johns Hopkins Medical Institutions
Baltimore, Maryland

INTRODUCTION

Breathing pattern in the newborn is characterized by irregularity and instability, the ventilatory (VE) defense against hypoxia is not sustained (Bureau et al, 1982, 1984, 1985; Lawson and Long 1983; Martin-Body and Johnston, 1988) and, in some species, the VE response to hypercapnia is not sustained (Schweiler, 1968). Spontaneous oscillation of breathing pattern (Nugent and Finley 1987; Hathorn 1978), post-sigh (or movement) apnea and oscillation (Fleming et al., 1984), periodic breathing, and central, obstructive, and mixed apnea (Waggener et al., 1989) have become hallmarks of newborn breathing control.

All of these features change during development. The defense against hypoxia improves, CO_2 responses increase, and the incidence of periodic breathing (Jansen and Chernick, 1988), spontaneous oscillations in breathing pattern (Fleming et al., 1984; Waggener et al., 1989), apnea, and bradycardia all decrease with age. The mechanisms underlying these developmental changes are not known.

There is evidence that carotid chemoreceptors play a crucial role in the development of breathing control. Hofer has shown that newborn rats deprived of their peripheral chemoreceptors have unstable respiratory control and a high mortality rate (Hofer, 1984, 1986). Piglets deprived of peripheral chemoreceptors become unstable and many die during the neonatal period (Donnelly and Haddad, 1987). Lambs deprived of their carotid chemoreceptors do well until about 3 weeks of life when about half die unexpectedly during the night (Bureau et al., 1985). Fewell et al. have shown that arousal from sleep by rapidly developing hypoxemia is mediated by the peripheral chemoreceptors in immature lambs (Fewell et al., 1989). Thus, a newborn or developing infant with weak carotid chemoreceptor responses would be extremely vulnerable.

We have previously shown in the newborn lamb that the peripheral chemoreceptors are well suited for transient responses (Carroll and Bureau, 1988; Carroll et al., 1991) and transient carotid chemoreceptor responses to CO_2 and O_2 are thought to underlie the tendency of the human infant respiratory control system to oscillate after a transient disturbance such as a sigh (Canet et al., 1989; Fleming et al., 1984). It is our hypothesis that the carotid chemoreceptors are important modulators of the ventilatory response to transient disturbances such as sigh, movement, or apnea in the newborn and that developmental changes in carotid chemoreceptor function are an important determinant in the overall development of respiratory control.

The aim of this study is to characterize carotid chemoreceptor responses to hypoxia

Neurobiology and Cell Physiology of Chemoreception
Edited by P.G. Data *et al.*, Plenum Press, New York, 1993

and hypercapnia in the developing kitten. Results obtained so far indicate that responses to both stimuli are present in 1 week old kittens and both mature with age.

METHODS

Preparation

Kittens were anesthetized with sodium pentobarbital, paralyzed with pancuronium bromide, and mechanically ventilated. Body temperature was maintained at $39 \pm 1°C$ for newborns and $38 \pm 1°C$ for older kittens using an Aqua-K pad. Intravenous fluid was administered to replace insensible water losses, blood glucose was periodically checked by dextrostick, and arterial blood pressure was continuously monitored. Whole nerve activity from the carotid sinus nerve (CSN) was recorded after baroreceptor output was eliminated by mechanical and thermal modification of the carotid sinus. In this manner > 90% of baroreceptor activity could be removed. The ganglioglomerular nerve (GGN) was sectioned.

CSN output was amplified, filtered (Charles Ward Enterprises, BMA831) and input to a moving time averager (Charles Ward Enterprises, 821RSP) and a Gould ES2000 physiologic recorder. The recording system, from the bipolar electrode to the physiologic recorder, was calibrated using a microvolt calibrator (Charles Ward Enterprises Cal, 830).

Subjects and protocol

Kittens were studied at 1 week, 4 weeks, and 8 weeks of age. CSN output was recorded with bipolar electrodes at five levels of CO_2 (3, 5, 7, 9, and 11% end-tidal CO_2) during steady-state hypoxia, normoxia, and hyperoxia. Blood gases were intermittently sampled in order to compare with end-tidal gas measurements.

Steady state CO_2 response curves at the three levels of oxygenation were constructed and carotid chemoreceptor responses to hypoxia were plotted as CSN activity (μV) vs PaO_2.

RESULTS

Hypoxia response

CSN activity increased during hypoxia in kittens of all ages. However, the carotid chemoreceptor neural response was not sustained in all cases. These results are shown in the table below.

Table 1. Carotid Chemoreceptor Response to Hypoxia.

	1 week old	4 weeks old	8 weeks old
# with sustained CSN activity during hypoxia	2	4	5
# unable to sustain CSN activity during hypoxia	2	1	0

In one half of the 1 week old kittens, when exposed to isocapnic hypoxia, CSN neural activity increased initially but then declined back to pre-hypoxia baseline levels of activity within approximately 1 minute. As FIO_2, was decreased in steps, CSN activity increased each time but always returned to pre-hypoxia baseline within several minutes. The decline of CSN activity during constant hypoxia was not related to blood pressure, body temperature, or end-tidal CO_2. In the other half of the one week old kittens, the CSN neural response to hypoxia was sustained, showing no tendency to decline over time (5 minutes). By 4 weeks of age, 4 kittens out of 5 exhibited sustained carotid chemoreceptor neural responses to hypoxia. Eight week old kittens all showed a sustained response to hypoxia.

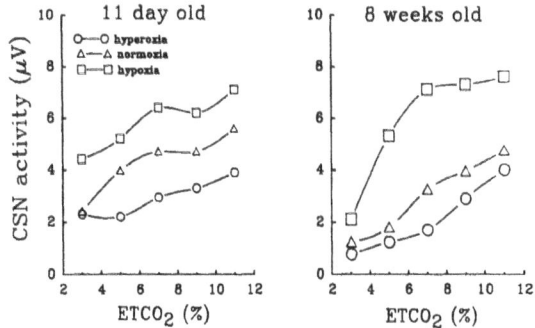

Figure 1.   CSN activity during CO_2 challenge in an 11 day old and an 8 week old kitten.

CO_2 response

All kittens at all ages showed an increase in carotid chemoreceptor neural output when exposed to hypercapnia and a decrease in CSN activity when hyperventilation produced hypocapnia (figure 1). Figure 1 shows an example of an 11 day old kitten and a typical response from an 8 week old. Preliminary results to date indicate that O_2-CO_2 interaction appears to be additive in newborn kittens, becoming multiplicative with age. CO_2 response curves under hypoxia, normoxia, and hyperoxia showed multiplicative O_2-CO_2 interaction in the 8 week old kittens, similar to that described in adult cats (Fitzgerald and Dehghani, 1982).

DISCUSSION

Results to date indicate several potentially important aspects of the development of carotid chemoreceptor function. At least when studied using whole nerve recording, not all newborn kittens showed a sustained neural CSN response to hypoxia. Furthermore, this phenomenon seems to disappear during development. Since the carotid chemoreceptors are necessary for survival of the newborn, the finding that the peripheral chemoreceptor response to hypoxia matures during postnatal development raises many intriguing questions. This finding also increases the likelihood that, in some newborns, the development of the carotid chemoreceptor response to hypoxia may be absent or delayed, possibly leading to adverse consequences or even death.

Since respiratory control system stability may depend to some extent on carotid chemoreceptor responses to CO_2, the finding that the CO_2 response changes with development is also important. It is quite possible that maturational changes in the response gain of the peripheral chemoreceptors are a major determinant of changing control system stability during postnatal development.

These data suggest that even though the mechanisms for the carotid chemoreceptor response to hypoxia are present early on in life, the ability to sustain the response requires time for maturation. Further work is focusing on mechanisms underlying the maturation of hypoxic chemotransduction. Similarly, the CO_2 response and O_2-CO_2 interaction at the carotid chemoreceptors are present early on but also requires time to develop full expression. Understanding the postnatal development of normal carotid chemoreceptor function will allow investigation of how it can go awry.

ACKNOWLEDGMENTS

Supported by NIH Grants KO8HL02543 and HL10342.

REFERENCES

Bureau, M. A., Lamarche, J., Foulon P., and Dalle D., 1985, Postnatal maturation of respiration in intact and carotid body-chemodenervated lambs. J. Appl. Physiol. 59:869.

Bureau, M. A., and Begin, R., 1982, Postnatal maturation of the respiratory response to O_2 in awake newborn lambs. J. Appl. Physiol. 52:428.

Bureau, M.A., Foulon, P., Zinman, R., and Begin, R., 1984, Diphasic ventilatory response to hypoxia in the newborn lamb. J. Appl. Physiol. 56:84.

Canet, E., Carroll, J.L., and Bureau, M.A., 1989, Hypoxia-Induced periodic breathing in newborn lamb. J. Appl. Physiol. 67(3):1226.

Carroll, J.L., and Bureau, M, 1988, Peripheral chemoreceptor CO_2, response during hyperoxia in the 14 day old awake lambs. Respir. Physiol. 73(3):339.

Carroll, J.L., Canet, E., and Bureau, M.A., 1991, Dynamic responses to CO_2 in the awake newborn lamb: Role of the carotid chemoreceptors. J. Appl. Physiol. 71:2198-2205,1991.

Donnelly, D.F., Haddad, G.G., 1987, Severe respiratory changes and neonatal deaths in chemodenervated young piglets. Fed Proc., 46:657.

Fewell, J.E., Kondo, C.S., Dascalu, V., and Filyk, S.C., 1989, Influence of carotid denervation on the arousal and cardiopulmonary response to rapidly developing hypoxemia in lambs. Pediatr. Res. 25(5):473.

Fitzgerald, R.S. and Dehghani, G, 1982, Neural responses of the cat carotid and aortic bodies to hypercapnia and hypoxia <u>J. Appl. Physiol.</u> 52:596.

Fleming, P.J., Goncalves, A.L., Levine, M.R., and Woollard, S., 1984, The development of stability of respiration in human Infants: Changes in ventilatory responses to spontaneous sighs. <u>J. Physiol. (Lond)</u> 347:1.

Fleming, P.J., Levine, M.R., Long, A.M., and Cleave, J.P., 1988, Postnatal development of repiratory oscillations. <u>Ann. N.Y. Acad. Sci</u>. 533:305.

Hathorn, M.KS., 1978, Analysis of periodic changes in ventilation in newborn babies. <u>J. Physiol. (Lond)</u> 285:85.

Hofer, M. A., 1984, Lethal respiratory disturbance in neonatal rats after arterial chemoreceptor denervation. <u>Life Sci.</u> 34:489.

Hofer, M. A., 1986, Role of carotid sinus and aortic nerves in respiratory control of infant rats. <u>Am. J. Physiol</u>. 20:R811.

Jansen, A., and Chernick, V., 1988, Onset of breathing and control of respiration. <u>Semin. Perinatol.</u> 12:104.

Lawson, E.E. and Long, W.A., 1983, Central origin of biphasic breathing pattern during hypoxia in newborns. <u>J. Appl. Physiol.</u> 55:483.

Martin-Body, RL. and Johnston, B.M., 1988, Central origin of the hypoxic depression of breathing in the newborn. <u>Resp. Physiol.</u> 71:25.

Nugent, S.T. and Finley, J.T., 1987, Periodic breathing in infants: A model study. <u>IEEE Trans. Biomed. Eng.</u> BME-34:482.

Schweiler, G.H., 1968, Respiratory regulation during postnatal development in cats and rabbits and some of its morphological substrate. <u>Acta Physiol. Scand. Suppl.</u> 304:1.

Waggener, T.B., Frantz, I.D., Cohlan, B.A., and Stark,A.R., 1989, Mixed and obstructive apneas are related to ventilatory oscillations in premature infants. <u>J. Appl. Physiol</u>. 66:2818.

55

PRESENCE OF CHEMOSENSITIVE SIF CELLS IN THE RAT SYMPATHETIC GANGLIA: A BIOCHEMICAL, IMMUNOCYTOCHE-MICAL AND PHARMACOLOGICAL STUDY

Y. Dalmaz, N. Borghini, J.M. Pequignot and L. Peyrin

URA CNRS 1195, Faculté de Médecine Grange Blanche, 69373 LYON Cedex 08, France

INTRODUCTION

Different functions have been ascribed to the paraganglionic SIF cells located in sympathetic ganglia: they were considered as interneurons, neuroendocrine and/or chemoreceptive cells (Case and Matthews, 1985). In many was, they resemble glomus cells of peripheral chemoreceptors (Kondo, 1977). However, elucidation of their functional role is still difficult because SIF cells are few in number comparatively to the postganglionic neurones, because their incidence varies greatly according to the species and to the ganglia and because they contain numerous putative neurotransmitters.

Previous studies evidenced that, like carotid bodies (Pequignot et al. 1991), superior cervical ganglia respond to hypoxia by decreasing the levels of substance P in rabbit (Hanson et al. 1986) and by increasing the turnover of dopamine (DA) in rat (Dalmaz et al. 1988, 1990) and rabbit (Cheng et al. 1990). Based on these neurochemical data, the existence of some chemosensory elements within ganglia have been suggested. However, morphological studies are not available.

This study was designed to advance in the knowledge of the functional role of SIF cells in ganglia of the rat and referred to the superior cervical and to the coeliac-mesenteric ganglia. Experiments were based on long-term hypoxia exposure ($10\% \ O_2$; 14 days), using neurochemistry of catecholamines, immunocytochemistry of enzymes together with guanethidine treatment in view to eliminate the postganglionic neurones. The experimental model using guanethidine represents a successful tool to study the SIF cells in vivo and gives evidence that SIF cells react to hypoxia by increasing their apparent number and their dopaminergic activity without involvement of neural input.

METHODS

Male Sprague Dawley rats (200-220g) were used.

Guanethidine treatment

Twenty rats were injected with guanethidine monosulfate (Ismelin, CIBA Geigy) daily 5 days a week, at a dose of 50 mg/kg for 3 weeks and at a dose of 25 mg/kg for the following 2 weeks corresponding to the hypoxic exposure. Control rats (n = 20) received 0.9% saline according to the same schedule.

Surgical procedure

Forty rats were subjected to unilateral transection of the carotid sinus nerve between its glossopharyngeal branch and the carotid body. Forty rats were decentralized by transecting the preganglionic nerve 2 mm caudal to the superior cervical ganglion. Surgical experiments were performed one week before hypoxic exposure and the contralateral side was used as control.

Exposure to normobaric hypoxia

Rats were placed for 14 days in a chamber in which the gas composition was maintained at 10% O_2. Expired metabolic water was trapped in a chilled tank. Control rats were exposed to room air (21% O_2).

Catecholamine assay

The last day of hypoxic exposure and 3 h before sacrifice, half of the rats were injected with alpha-methyl-p-tyrosine methylester (200 mg/kg), an inhibitor of the catecholamine synthesis, and the other half was injected with 0.9% saline. After cervical dislocation, the superior cervical and coeliac-mesenteric ganglia were dissected out. The DA and norepinephrine (NE) content was assayed by the HPLC-electrochemical detection procedure (Dalmaz et al. 1988). The turnover of catecholamines was estimated by measuring the decrease in their content after alpha-methyl-para-tyrosine.

Immunocytochemistry of tyrosine hydroxylase (TH)

Ganglia were fixed in 4% paraformaldehyde and frozen. 10 μm sections were collected on slides and incubated overnight in anti-TH (1/4000 dilution) antisera. After the action of PAP, the complex stained with diaminobenzidine was visualized under light microscope (Borghini et al. 1991).

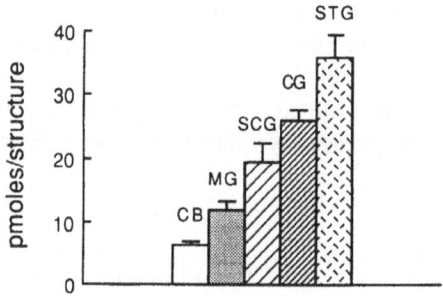

Fig. 1. Content of dopamine in carotid body (CB) (Pequignot et al. 1991), mesenteric ganglion (MG), superior cervical ganglion (SCG), coeliac ganglion (CG), stellate ganglion (STG).

*Table 1. Number of apparent TH immunoreactive SIF cells in the superior cervical and in the coeliac-mesenteric ganglia of the rat exposed to normobaric hypoxia (10% O$_2$) for 14 days. * p < 0.05.*

	superior cervical ganglion	coeliac-mesenteric ganglia
normoxia	578 ± 150 (6)	624 ± 183 (7)
hypoxia	1178 ± 177*(6)	1287 ± 265*(6)

RESULTS

<u>SIF cells in normoxia</u>

The superior cervical, stellate and coeliac-mesenteric ganglia contain substantial amounts of DA (Fig. 1), comparatively higher than those found in carotid bodies (Pequignot et al. 1991). The immunocytochemistry of TH confirms that two distinct cellular populations are immunoreactive and therefore synthetize DA, the first one representing the large and numerous principal noradrenergic neurones, the second one representing the SIF cells, highly immunoreactive, often packed into clusters. These cells average only 3% of the total cells in the superior cervical ganglion (Borghini et al. 1991), but they account until 7% in the coeliac-mesenteric ganglion.

After guanethidine treatment, the NE content is dramatically reduced whilst notable levels of DA are still detectable. Only 5% and 13% of NE are present respectively in the superior cervical and coeliac-mesenteric ganglia against 50% and 30% of DA. Concomitantly, the immunocytochemistry of TH reveals that most of the noradrenergic neurones disappear, whilst SIF cells are still detectable in the coeliac-mesenteric ganglia and moreover increase in the superior cervical ganglion (Borghini et al. 1991). Altogether, these results provide evidence that SIF cells store from 30 to 50% of the total ganglionic DA according to the ganglia and are better visualized after guanethidine .

<u>SIF cells in long-term hypoxia</u>

A striking feature of sympathetic ganglia is to respond to long-term hypoxia by increasing the content and the turnover of DA (2.3 fold in the superior cervical and 1.4

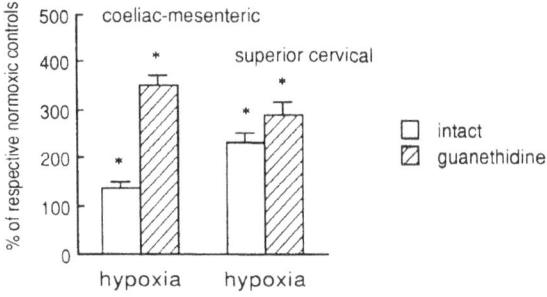

*Fig. 2. Effects of normobaric hypoxia on the DA turnover (pmol/ganglion/3h) in the ganglia of intact rats and of guanethidine-treated rats. * significantly different from respective normoxic values at p <0.05*

fold in the coeliac-mesenteric ganglion) (Fig. 2). The immunocytochemistry of TH reveals that numerous TH immunoreactive SIF cells, packed into clusters, are present in the ganglia after 14 days of hypoxia; their apparent number increases two fold (Table 1, Fig. 3).

After guanethidine treatment, the hypoxia-induced increase in DA in ganglia is maintained or even more increased (Fig. 2), thus demonstrating that the hypoxic response is not located in the neurones but rather in the SIF cells. In fact, concomitantly, the apparent number of TH immunoreactive SIF cells increases also about two fold following 14 days of hypoxia.

Role of neural pathways on the activity of SIF cells

Surgical transection of the carotid sinus nerve does not prevent the hypoxia-induced increase in DA turnover in the superior cervical ganglion. In fact, a two fold rise in the DA turnover is always observed in the superior cervical ganglion of chemodenervated rats (Table 2).

Decentralization of the superior cervical ganglion does not abolish the hypoxia-induced increase in DA turnover. When compared to the respective normoxic control, the magnitude of the DA turnover changes elicited by hypoxia is the same in intact (1.7-fold rise) and in decentralized ganglion (1.9-fold increase). However, because decentralization reduced significantly, by itself, the turnover of DA by 35%, at the end of hypoxia, it set at a lower level in the decentralized ganglion than in the intact ganglion (Table 2).

DISCUSSION

This study gives some further information about SIF cells in sympathetic ganglia of rats. Using neurochemical, immunocytochemical, pharmacological and surgical procedures, in association with long-term hypoxic exposure, three conclusions may be drawn:

1/ The superior cervical and coeliac-mesenteric ganglia are rich in DA, which occurs in two cellular populations, quantitatively of quite different importance. The number of TH immunoreactive SIF cells is very low comparatively to the principal TH immunoreactive neurones (3 to 7% of the total cell population) but SIF cells store a large part of the total ganglionic DA (30 to 50% according to the ganglion). Guanethidine represents a good pharmacological tool to study SIF cells because it does not destroy these cells whilst noradrenergic neurones disappear almost completely. Therefore, SIF cells may be studied using the biochemistry of DA and the immunocytochemistry of TH in guanethidine-treated rats.

2/ The superior cervical and coeliac-mesenteric ganglia, either of intact rats or of guanethidine-treated rats, respond to hypoxia by increasing the turnover of DA and by increasing the number of TH immunoreactive SIF cells. Therefore, hypoxia is able to stimulate SIF cells, which is evidenced by an increase in their biochemical and morphological features.

3/ The superior cervical ganglion is stimulated by hypoxia whithout the involvement of a chemoreceptor brain-reflex nor of preganglionic nervous input. Therefore, the biochemical response of the superior cervical ganglion to hypoxia, which is located in the SIF cells, does not involve preganglionic nervous influence.

Table 2. *Effects of chemodenervation or decentralization on the turnover of DA (pmol/ganglion/3h) in the superior cervical ganglion of rats exposed to hypoxia (10% O2; 14 days). * comparison between respective normoxia and hypoxia at $p < 0.05$; + comparison between intact and decentralized ganglion in normoxia and in hypoxia, at $p < 0.05$.*

ganglion	intact	chemodenervated	decentralized
normoxia	20.9 ± 1.8 (8)	16.8 ± 2.5 (13)	$12.3 \pm 2.4^{+}$ (8)
hypoxia	$35.1 \pm 4.3^{*}$ (8)	$35.4 \pm 4.6^{*}$(9)	$23.5 \pm 5.1^{+*}$(9)

Our results extend data of Madariaga-Domich and Taxi, (1986) and of König and Heym, (1978) in demonstrating that SIF cells of the superior cervical and coeliac-mesenteric ganglia, although few in number, store a great proportion of DA (30 to 50%).

This study points out some functional similarities between SIF cells of ganglia and glomus cells of carotid bodies: long-term hypoxia increases the turnover of DA in ganglia (2 fold rise) and, at a greater extent, in carotid bodies (20 fold increase) (Pequignot et al. 1991); long-term hypoxia induces an hyperplasia of SIF cells (2 fold increase) and morphological changes of glomus cells (Pequignot et al. 1984; Pallot et al. 1990); after hypoxia, numerous SIF cells in ganglia are packed into clusters near capillaries, as previously observed for glomus cells in carotid bodies; the biochemical response to hypoxia in the superior cervical ganglion is not submitted to the presence of nervous preganglionic pathways, like reported for the carotid bodies (Pequignot et al. 1991). The ability of SIF cells to respond to hypoxia in vivo without neural input agrees with results of Cheng et al. (1990) reporting that the superior cervical ganglion of rabbit is able to respond to hypoxia, in vitro, by increasing the synthesis and the release of DA.

In conclusion, the SIF cells of sympathetic ganglia display some chemosensitive properties. As to concern the mechanisms involved in the response of SIF cells to hypoxia, they may be local and/or consecutive to the involvement of humoral factors (adrenaline, glucocorticoids).

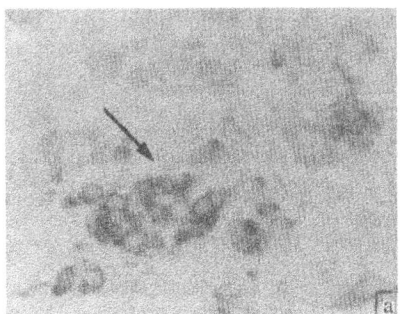 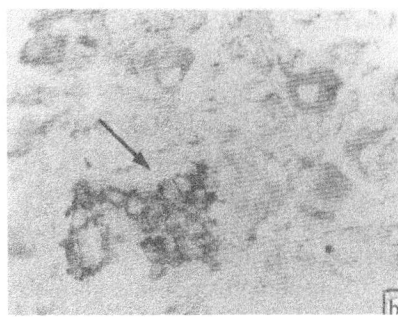

Fig. 3. *Light microscopical pictures of histological sections (10 μm) from ganglia of rats exposed to normobaric hypoxia (10% O₂; 14 days).a/ superior cervical ganglion; b/ coeliac-mesenteric ganglion. Arrows indicate the numerous TH immunoreactive SIF cells, which are packed into clusters. Magnification: x 250.*

ACKNOWLEDGEMENTS

Supported by grants from INSERM (887004), from DRET (88069) and from Région Rhône-Alpes (Neurosciences).

REFERENCES

Borghini, N., Dalmaz, Y., and Peyrin, L., 1991, Effect of guanethidine in small intensely fluorescent cells of the superior cervical ganglion of the rat, J. Auton. Nerv. Syst., 32:13.

Case, C.P., and Matthews, M.R., 1985, A quantitative study of structural features synapses and nearest-neighbour relationships of small, granule-containing cells in the rat superior cervical sympathetic ganglion at various adult stages, Neuroscience, 15:237.

Cheng, G.F., Dinger, B., Hanson, G., and Fidone, S.J., 1990, Effects of hypoxia on catecholamine storage and release in rabbit superior cervical ganglion, in: "Arterial Chemoreception" C. Eyzaguirre, S.J. Fidone, R.S. Fitzgerald, S. Lahiri, and D.M. McDonald, ed., Springer-Verlag, New-York, 55:398.

Dalmaz, Y., Pequignot, J.M., Tavitian, E., Cottet-Emard, J.M., and Peyrin, L., 1988, Long-term hypoxia increases the turnover of dopamine but not norepinephrine in rat sympathetic ganglia, J. Auton. Nerv. Syst., 24:57.

Dalmaz, Y., Borghini, N., Pequignot, J.M., and Peyrin, L., 1990, Involvement of dopaminergic SIF cells in Rat superior cervical ganglion in response to chemoreceptor stimuli, in: "Arterial Chemoreception" C. Eyzaguirre, S.J. Fidone, R.S. Fitzgerald, S. Lahiri, and D.M. McDonald, ed., Springer-Verlag, New-York, 56:404.

Hanson, G., Jones, L., and Fidone, S., 1986, Effects of hypoxia on neuropeptide levels in the rabbit superior cervical ganglion, J. Neurobiol. 17:51.

Kondo, H., 1977, Innervation of SIF cells in the superior cervical and nodose ganglia: An ultrastructural study with serial sections, Biol. Cell. 30:253.

König, R., and Heym, C., 1978, Immunofluorescent localization of dopamine beta hydroxylase in small intensely fluorescent cells in the superior cervical ganglion of the rat, Neurosci Lett. 10:187.

Madariaga-Domich, A., and Taxi, J., 1986, A comparative and quantitative study of small intensely fluorescent cells in the sympathetic ganglia of some small Mammals, Arch. Anat. Micr. Morphol. Exp. 75: 1.

Pallot, D.J., Bee, D., Barer, G.R., and Jacob, S., 1990, Some effects of chronic stimulation on the rat carotid body, in: "Arterial Chemoreception" C. Eyzaguirre, S.J. Fidone, R.S. Fitzgerald, S. Lahiri, and D.M. McDonald, ed., Springer-Verlag, New-York, 40:293.

Pequignot, J.M., Hellström, S., and Johansson, C., 1984, Intact and sympathectomized carotid bodies of long-term hypoxic rats: a morphometric ultrastructural study, J. Neurocytol., 13:481.

Pequignot, J.M., Dalmaz, Y., Claustre, J., Cottet-Emard, J.M., Borghini, N., and Peyrin, L., 1991, Preganglionic sympathetic fibres modulate dopamine turnover in rat carotid body during long-term hypoxia, J. Auton. Nerv. Syst., 32:243.

56 EFFECTS OF CHRONIC HYPOXIA AND COBALT ON MACROPROTEIN PATTERN IN THE RABBIT CAROTID BODY AND SUPERIOR CERVICAL GANGLION : PRELIMINARY OBSERVATIONS

P.G. Data, G. Di Tano, G. Gigante, V. Biondelli, M. Iezzi, C. Di Giulio and L. Morelli

Institute of Physiological Sciences
School of Medicine, "G. D'Annunzio" University
66100-Chieti, Italy

INTRODUCTION

Chronic hypoxic-hypoxia and chronic cobalt administration leads to glomus cells hypertrophy of the carotid body (Mc Gregor et al., 1984; Di Giulio et al., 1991; Pequignot et al., 1984).

Since cellular growth involves an increase rate of protein synthesis, it is of interest to identify the pattern of proteins in the carotid body in animals treated with chronic hypoxia and cobalt.

METHODS

Experiments were performed on 56 rabbits, weighing 1.9-2.8 kg, distributed into seven groups, eight each (Table 1). Control rabbits received blank saline injections (1.0 ml/Kg) daily for 15 days. Group 2 rabbits were exposed to different p_iO_2 levels for different period time: all rabbits were exposed for 48 hrs at p_iO_2 110 torr; one group was exposed for 3 days at p_iO_2 90 torr and other group for 13 days at p_iO_2 70 torr. The rabbits of Group 3 were given intraperitoneal injections of $CoCl_2$ (0.17 umole/Kg) dissolved in saline (0.1 ml) daily for one or 15 days. At the end of treatment, the rabbits were anaesthetized with sodium pentobarbital (30 mg/kg). Carotid body and superior cervical ganglion were dissected out and immediately frozen (- 20° C below 0°). The tissues were homogenized and suspended in 200 µl of buffer diluted at 1:3. Sample buffer (SDS reducing buffer) (Laemmli, 1970), stored at room temperature, contained: 62.5 mM Tris-HCl, pH 6.8, 10% Glycerol, 2% Sodium dodecyl sulfate, 5% Beta-mercaptoethanol. The sample was centrifuged (2 min at 9000 g), the supernatant was removed and denaturated at 56°C.

Polyacrylamide gel electrophoresis, based upon intrinsic charge-to-mass ratio, was applied. The gel pieces were connected through the electrode terminals to the dc power supply regulated at 30 V.

After 2-3 hrs, the power supply was regulated at 90 V to compensate the increase of resistance during elution. Proteins of gel strips were visualized by Bio-rad silver staining or Coomassie brilliant blue R-250. SDS-PAGE standards (low range 14.400 to 97.000 daltons) were used for identification of the proteins molecular weights.

The northern blot, original and/or photocromatic reproduction, was scanned with a densitometer to analyze the primary protein structures.

Table 1. *Experimental Groups, eight rabbits each.*

1.0 ***Control***
1.1 - Control
1.2 - Sham (Saline injections for 15 days)
2.0 ***Hypoxia***
2.3 - p_iO_2 110 Torr for 48 hrs
2.4 - p_iO_2 90 Torr for 5 days
2.5 - p_iO_2 70 Torr for 15 days
3.0 ***Cobalt*** (10 mg/kg, ip)
3.6 - Treated for 24 hrs
3.7 - Treated for 15 days

RESULTS

The original northern blot of carotid body extract is shown in Fig.1. The scanning densitometry of carotid bodies northern blot are summarized in Figures 2 and 3. Band's number (I-IV ; I-V) correspond to scanned macroproteins, and their intensity was calculated in arbitrary units. The rabbits of two control groups were computed together. *Figure 2* shows the macroprotein pattern of the carotid body in rabbits in the chronic state: 48 hrs hypoxia (p_iO_2 110 torr), 24 hrs cobalt (after one ip injection).

These results do not express any significant changes. *Figure 3* shows a significant decrease in two protein bands (III - IV) of carotid bodies in the rabbits treated for 5-15 days. *Figure 4* shows the bands from the superior cervical ganglion. The bands IV and V revealed significant decreases in the chronically treated animals compared to the control.

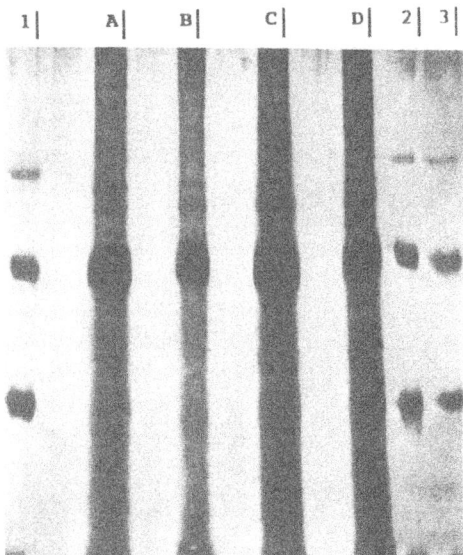

Fig. 1 Original northern blot of carotid bodies.

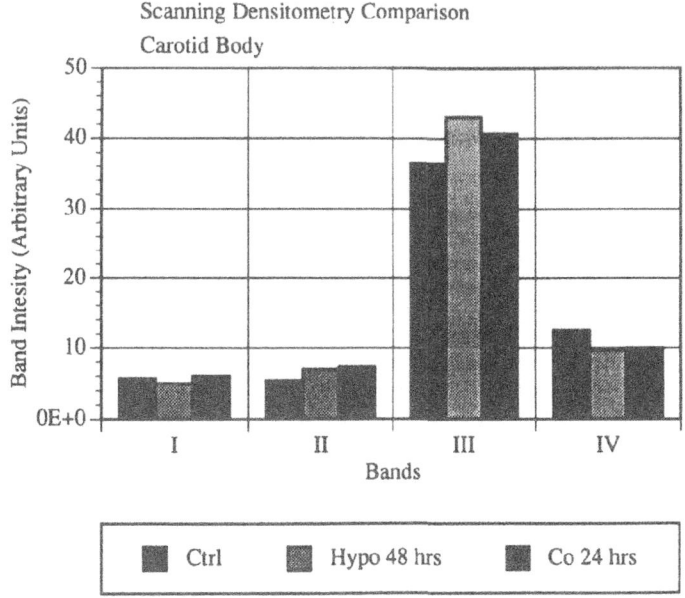

Fig. 2 Scanning densitometry of carotid bodies after 48 hrs hypoxia and 24 hrs cobalt injection.

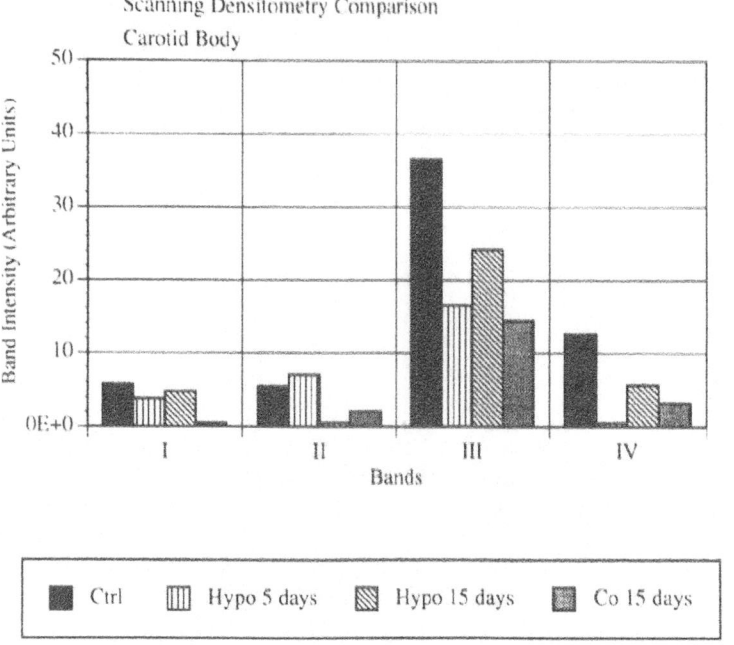

Fig. 3 Scanning densitometry of carotid bodies after hypoxia (5 and 15 days) and cobalt (15 days).

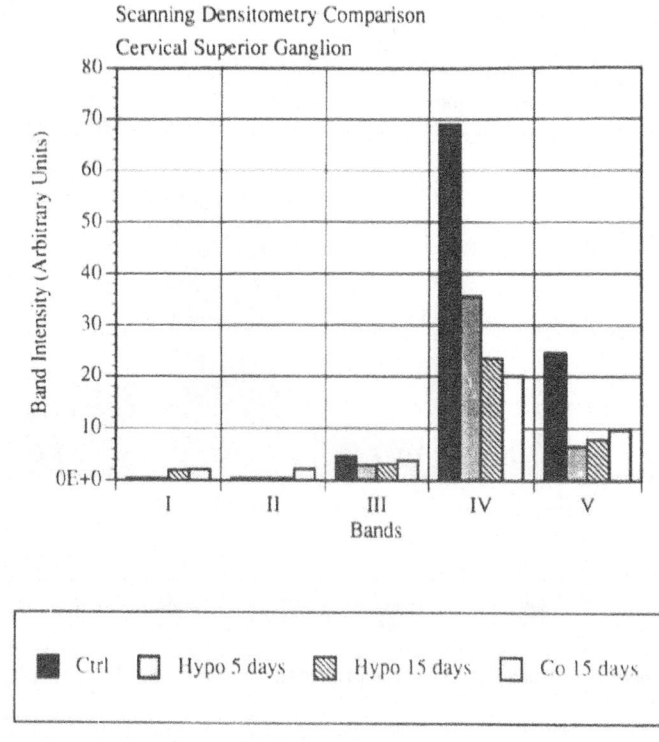

Fig. 4 Scanning densitometry of superior cervical ganglion after hypoxia (5 and 15 days) and cobalt (15 days).

DISCUSSION

The total amount of proteins of carotid body during chronic hypoxia is increased (Dalmaz et al., 1988; Hanbauer et al., 1981). Our study showed that there is a decrease in concentration of some type of macroprotein. The different intensity of the bands is not depending on the Hbe or on the volume, because the macroproteins were extracted with the same volume of solvent. The phenomena might indicate that chronic stimuli induce thesynthesis inhibition of some proteins or their degradation, breakdown and production of smaller peptides, not detectable by our method. The meaning of the macroproteins decrease into superior cervical ganglion and in the context of cellular hypertrophy of the carotid bodies during chronic hypoxia and chronic cobalt administration is not clear.

ACKNOWLEDGMENT

We are grateful to Dr. S. Lahiri for his experimental help and expert consultation and to Dr. I. Bucci for her technical assistance.

REFERENCES

Dalmaz Y., Pequignot J-M., Tacitian E., Cottet-Emart H-M. and Peyrin L., 1988, Long-term hypoxia increases the turnover of dopamine but not norepinephrine in rat sympathetic ganglia, *J. Auton. Nerv. Syst.*, 24:57.

Di Giulio C., Data P.G. and Lahiri S., 1991, Chronic cobalt causes hypertrophy of glomus cells in the rat carotid body, *Am. J. Physiol.*, 261:C102.

Hanbauer I., Karoum F., Hellstrom S. and Lahiri S., 1981, Effects of hypoxia lasting up to one month on the catecholamine content in rat carotid body, *Neuroscience*, 6:81.

Laemmli U.K., 1970, Cleavage of structural proteins during the assembly of the head of bacteriophage T_4, *Nature*, 227:680.

McGregor K.H., Gil J., and Lahiri S., 1984, A morphometric study of the carotid body in chronically hypoxic rats, *J. Appl. Physiol.*, 57:1430.

Pequignot J.M., Hellstrom S. and Johansson C., 1984, Intact and sympathectomized carotid bodies of long-term hypoxic rats: a morphometric ultrastructural study, *J. Neurocy tol.*, 13:481.

57

TIME COURSE OF THE RESPONSE OF CAROTID CHEMORECEPTORS TO SUDDEN RISE OF INSPIRED CO$_2$ IN THE NEWBORN KITTEN

B. Hannhart, A. Bairam and F. Marchal

Unité INSERM 14
Laboratoire de Physiologie, Faculté de Médicine de Nancy
Vandoeuvre-Lès- Nancy, France

INTRODUCTION

The carotid body chemoreceptors are the main sensors for the level of arterial PO$_2$ involved in the ventilatory control. It is well recognized that they also contribute to the ventilatory response to changes in PaCO$_2$. Electrophysiological recordings of carotid chemoreceptor activity in adult animals show that they present a rapid response to sudden step rise of PCO$_2$ in the arterial blood (Dutton et al., 1967; Fitzgerald et al., 1969; Black et al.,1971). It is still appreciable during hyperoxia and is enhanced by hypoxia (Fitzgerald and Parks, 1971;Lahiri et al., 1975, 1982). On the other hand, it has been established that an adaptation of the ventilatory response to CO$_2$ occurs within seconds of a peak response (Dutton et al., 1967). Similarly, the carotid chemoreceptors appear to adapt (Black et al.,1971), at least in hypoxia or under specific stimulation (Lahiri et al. 1982, 1989). However, these results have only been observed in adult animals. In newborns, the weak chemosensory reactivity of carotid body is well documented (Blanco et al., 1984) but no information on the neonatal development either of the dynamic CO$_2$-response of peripheral chemoreceptors or of the O$_2$-CO$_2$ interaction was available.

The purpose of the present study was to examine the dynamic response of the carotid body chemoreceptors in newborn kittens of 1 day to 1 month old when a sudden sustained rise in inspired PCO$_2$ is applied. We also paid attention to the hypoxia-CO$_2$ stimulus interaction by examining the CO$_2$ response of chemoreceptors during hyperoxia and under hypoxic conditions.

MATERIAL AND METHODS

Experiments were performed on 15 kittens of 1 to 31 day old and weighing 80-310 g. They were anesthetized with sodium thiopental (60 mg. kg^{-1}), paralyzed with pancuronium bromide (1 mg. kg^{-1}) and artificially ventilated. A catheter was implanted in the femoral artery to monitor arterial blood pressure (Pa) with a Statham transducer and to sample arterial blood (Radiometer, ABL 300). Core body temperature was maintained at 38 $\pm$ 0.4 °C with a regulated water / heating pad. Inspiratory O$_2$ and expiratory CO$_2$ concentrations were monitored by O$_2$ analyzer (Sensormedics, OM11) and infrared CO$_2$ analyzer (Instrumentation Laboratory, IL200), respectively.

Neurobiology and Cell Physiology of Chemoreception
Edited by P.G. Data *et al.*, Plenum Press, New York, 1993

Carotid body preparation. The proximal trachea and esophagus were reflected back to expose the carotid bifurcation. The sinus nerve was gently stripped of surrounding tissue and cut near the connection with the glossopharyngeal nerve. Carotid body neural output (CBNO) was recorded by platinum bipolar electrodes from the whole desheathed nerve bundle. Carotid sinus was crushed for 15 sec in order to abolish audible cardio-synchronous activity reflecting the baroreceptor discharges. The preparation was covered with warm mineral oil. The amplified signal was processed by amplitude variance analysis as described previously (Dick et al., 1974; Hannhart et al. 1990). The width of the distribution of amplitudes provides an index of whole nerve activity proportional both to the activity of independently firing fibers and to the number of active fibers. Normalization was performed by expressing CBNO as the ratio of the observed amplitude variance on the value measured during pure O_2 breathing (set to 1 unit) .

Protocol: Kinetics of the peripheral chemoreceptor response to CO_2 was analyzed by continuous measurements of CBNO during brisk change in CO_2 concentration in the inspiratory gas mixtures. Since CO_2 response of the peripheral chemoreceptors can be influenced by the O_2 level, the changes were run from either a hyperoxic or hypoxic background. The inspiratory gas could be instantaneously switched between pure O_2 with or without CO_2 ($F_I CO_2$: 0.08), and hypoxia ($F_I O_2$: 0.08) with or without CO_2 ($F_I CO_2$: 0.08). Signals of CBNO, arterial blood pressure (Pa), $F_I O_2$ and $F_E CO_2$ were sampled at a rate of 20 Hz into a micro-computer.

To evaluate the effect of maturation, the data were compared between younger (0-7 days) and older kittens (8-31 days) using unpaired t-test. All values are expresed as mean $\pm$ SEM. A probability of 0.05 was accepted as the minimum level of statistical significance.

RESULTS

In four kittens aged 1, 4, 7 and 8 days, although their CBNO response to hypoxia was normal, no change in CBNO was observed under hypercapnic stimulation, neither in hyperoxia nor in hypoxia. Thus, the time course of the response of CBNO to CO_2 was analyzed in the other 11 kittens.

a) In hyperoxia. During 100% O_2 breathing, arterial blood gases and pH were PaO_2: 228$\pm$ 43 mmHg, $PaCO_2$: 26$\pm$ 3 mmHg and pH: 7.25$\pm$ 0.06, respectively. When 8% CO_2 was mixed with pure oxygen, $PaCO_2$ increased to 67$\pm$ 6 mmHg and pH was 7.00$\pm$0.06.

In hyperoxia, a stepwise addition of CO_2 to the inspired gas was reflected by an exponential increase in $F_E CO_2$ with time and simultaneously, a prompt response of CBNO occurred. In the group of the 6 younger kittens, 5 presented a biphasic response of CBNO with a transient peak of 1.33$\pm$ 0.17 in 16.6$\pm$ 5 sec followed by a partial return in 20$\pm$ 2.5 sec to a steady-state level of 1.24$\pm$ 0.14. A representative response to a step change in $F_I CO_2$ in a 2-day old kitten is illustrated in Fig. 1 (Left panel).

In the group of the 5 older kittens, the biphasic response was observed in only one animal (peak of 1.7). The average steady-state response in this group reached 1.31$\pm$0.06. A representative response to a step change in $F_I CO_2$ in a 31-day old kitten is illustrated in Fig. 2 (Left panel).

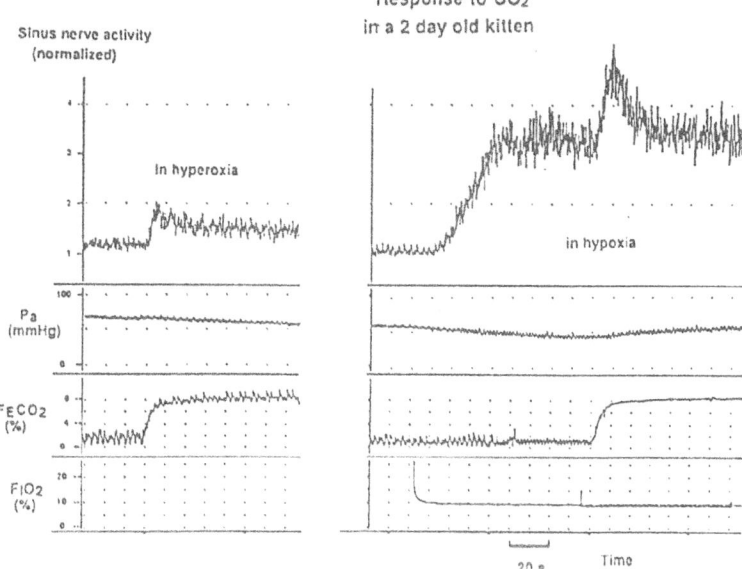

Fig. 1. Representative recording of the response to a step change in F_iCO_2 in a 2-day old kitten. From top to bottom: Carotid body neural output (variance analysis), arterial blood pressure (Pa), F_ECO_2, F_iO_2. Hypercapnic stimulation is reflected by the exponential increase in F_ECO_2. Left panel: in hyperoxia (O_2-signal out of the graph) a slight increase in CBNO is followed by a slight adaptation. Right panel: CBNO was initially increased by hypoxia. With CO_2, a peak response was followed by a consistent adaptation resulting in a very small steady-state response.

b) In hypoxia. During 8% O_2 breathing, arterial blood gases were PaO₂: 38± 3 mg, PaCO₂: 23± 2 mmHg and pH:7.36± 0.06. Under hypercapnic stimulation, PaCO₂ increased to 49± 4 mmHg and pH was 7.10± 0.08.

Hypoxia gave a similar increase in CBNO in both groups to 2.1± 0.2 and 2.0± 0.2, respectively. When CO2 was added to the maintained level of hypoxia, a biphasic response of CBNO was observed in all the kittens of both groups (Fig. 1 and 2, right panels). A prompt response of CBNO occurred with an increase to 2.7± 0.3 and 2.8± 0.3 within 15.1± 2.8 sec and 23.2± 1.9 sec in the younger and older animals, respectively. That peak was followed by a partial return towards a steady level of 2.3± 0.2 and 2.6± 0.3 in the subsequent 19.1± 1.6 sec and 16± 2 sec, respectively.

Mean results in the two separate groups were presented in Fig. 3. Timing and peak response were similar in the two groups but the decline was significantly greater (P<0.05) in the younger kittens and therefore, the steady-state response to CO₂ was significantly higher in the older group (P< 0.05).

DISCUSSION

The present experiments show that in the kitten, the chemosensibility of the carotid body to CO_2 is already developed at birth. At the onset of a sudden change in inspired CO_2,

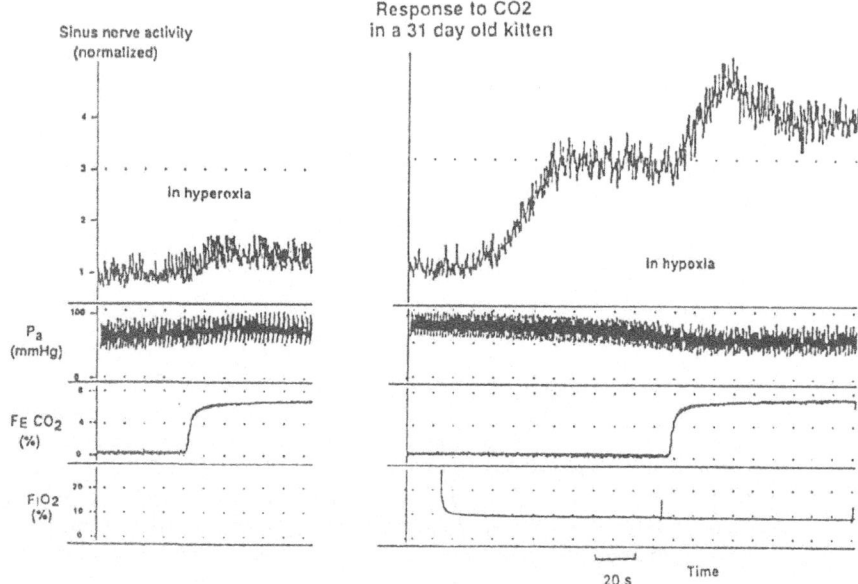

Sinus nerve activity
(normalized)

Response to CO2
in a 31 day old kitten

in hyperoxia

in hypoxia

Pa
(mmHg)

FE CO2
(%)

FiO2
(%)

20 s   Time

Fig. 2. *Representative recording of the response to a step change in F_iCO_2 in a 31-day old kitten (cf. legend of Fig. 1). Left panel: in hyperoxia the response of CBNO did not adapt. Right panel: in hypoxia, the adaptation following the peak response was less complete than for the younger kitten.*

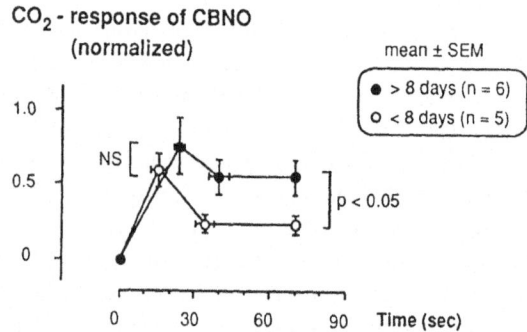

CO$_2$ - response of CBNO
(normalized)

mean ± SEM

● > 8 days (n = 6)
○ < 8 days (n = 5)

1.0

0.5

NS

p < 0.05

0

0   30   60   90   Time (sec)

Fig. 3. *Change in carotid body neural output (CBNO) under hypercapnic stimulation in hypoxia. The animals were partitioned into two groups according to their age (< and > 8 day old)*

the carotid chemoreceptors activity promptly increased, even in hyperoxia. In most animals, the chemosensory discharge reached a maximum within about 20 sec and then returned towards a steady level within the subsequent 20 sec. The peak response appeared independent of age but the youngest animals exhibited the greatest adaptation. The O_2-CO_2 interaction was also efficient, particularly on the transient peak increase.

It has been shown in the newborn, that the ventilatory response to chemical stimulation develops with time postnatally (cf. Hanson, 1986). Peripheral chemoreceptors sensitivity to CO_2 has previously only been described under transient inhalation of CO_2 (Biscoe and Purves, 1967; Hanson, 1986). These works showed that the chemoreceptors responses were weak at birth but their maturation with time has not been described. The present results provide evidence that carotid chemoreceptors activity develop during the first month of life. That could correspond to the postnatal re-setting of chemoreceptors response to hypoxia previously reported in the lamb (Blanco et al. 1984) or in the piglet (Mulligan, 1991).

The decline in the response after the peak was qualitatively similar to that observed for the adult cat (Black et al., 1971; Lahiri et al., 1982, 1989). A biphasic response indicates that the steady-state response was less than the early transient response. Therefore, the large adaptation at birth accounts for the very slight steady-state response to CO_2.

As shown previously in newborn piglet (Mulligan, 1991) the multiplicative interaction between hypoxia and hypercapnia was found to be minimal on the steady-state response to CO_2. However, in agreement with the results reported by Carroll and Bureau (1988), we observed a clear enhancement of the transient peak response by severe hypoxia compared with the response seen in hyperoxia. Thus transient and steady-state responsiveness could depend on different mechanisms.

The blood pressure did not manifest large variations that could explain the time course of chemosensory response to CO_2. Since the carotid sinus nerve was proximally cut, the adaptive response was unlikely to be related to central descending inhibitory efferences but rather reflected an inherent property of the chemoreceptors. But we did not eliminate sympathetic innervation of the carotid body, thus we cannot rule out the possibility that sympathetic modulation could have affected the dynamic response.

The mechanisms of chemoresponsiveness are not known and we can only hypothesize on the possible ones which are responsible of such responses. The biphasic response could involve two processes, an excitatory one followed by an inhibitory one, with different time constants and durations. On the other hand, a process could detect the transient rate of change in PCO_2 independently of another one which would perceive the mean CO_2 stimulus. That could be related to the existence of two types of fibers, the rapid A fibers and the C fibers which respond more slowly (Fidone and Sato, 1969) or of two different discharge patterns of carotid body recently described by Niu et al. (1990) in the goat.

In adult animals, carbonic anhydrase inhibition eliminates the adaptation of the response of chemoreceptors to CO_2 (Black et al., 1971; Gray, 1971; Iturriaga et al., 1991). The question remains whether the relative size of various compartments of distribution of CO_2, H^+ and bicarbonate influences differently the CO_2 responsiveness in newborns and in adults.

In summary, the results of this study indicate that the chemoreceptors present a biphasic response to step change in CO_2 in the first days after birth. Although the steady-state response to CO_2 developed with age, the transient peak response appeared to remain relatively steady during the first month of the life. Both transient and steady state responses were greatly amplified by hypoxia showing the existence of O_2-CO_2 interaction. At birth, in spite of a low CO_2 responsiveness, the peak response developed early and may already play a significant role in the fine tuning of the breathing pattern in the neonate.

REFERENCES

Biscoe, T. J. and Purves, M. J. (1967). Carotid body chemoreceptor activity in the newborn lamb. J. Physiol. 190: 443-454.

Blanco, C. E., Dawes, G. S., Hanson, M. A. and McCooke, H. B. (1984). The response to hypoxia of arterial chemoreceptors in fetal sheep and new-born lambs. J. Physiol. 351: 25-37.

Black, A. M. S., D. I. McCloskey and R. W. Torrance (1971). The response of carotid body chemoreceptors in the cat to sudden changes of hypercapnic and hypoxic stimuli. Respir. Physiol. 13: 36-49.

Carroll, J.L. and M.A. Bureau (1988). Peripheral chemoreceptor CO_2 response during hyperoxia in the 14-day-old awake lamb. Respir. Physiol. 73: 339-350.

Dick, D.E., J.R. Meyer and J.V. Weil (1974). A new approach to quantitation of whole nerve bundle activity. J. Appl. Physiol. 36: 393-397.

Dutton, R. E., A. Hodson, D. G. Davies and V. Chernick (1967). Ventilatory adaptation to a step change in pCO_2 at the carotid bodies. J. Appl. Physiol. 23: 195-202.

Fidone, S.J. and A. Sato (1969). A study of chemoreceptor and baroreceptor A and C fibers in the cat carotid nerve. J. Physiol. (London) 205: 527-548.

Fitzgerald, R.S. L.-M Leitner and M.-J. Liaubet (1969). Carotid chemoreceptor response to intermittent or sustained stimulation in the cat. Respir. Physiol. 6: 395-402.

Fitzgerald, R.S. and D.C. Parks (1971). Effect of hypoxia on carotid chemoreceptor response to carbon dioxide in cats. Respir. Physiol. 12: 218-229.

Gray, B.A. (1971). On the speed of the carotid chemoreceptor response in relation to the kinetics of CO_2 hydration. Respir. Physiol. 11: 235-246.

Hannhart B., C.K. Pickett and L.G. Moore (1990). Effects of estrogen and progesterone on carotid body neural output responsiveness to hypoxia. J. Appl. Physiol. 68: 1909-1916.

Hanson, M.A. (1986). Peripheral chemoreceptor function before and after birth. In "Respiratory Control and Lung Development in the Fetus and Newborn." Ed. B.M. Johnston and P.D. Gluckman. Ithaca, NY, Perinatology Pres, pp. 311-330.

Hanson, M.A., Kumar, P., snd McCooke,H.B. (1987). Postnatal re-setting of carotid chemoreceptor sensitivity in the lamb. J. Physiol. 382: 57P.

Iturriaga, R., S. Lahiri and A. Mokashi (1991). Carbonic anhydrase and chemoreception in the cat carotid body. Am. J. Physiol. 261: C565-C573.

Lahiri, S. and R.G. DeLaney (1975). Stimulus interaction in the responses of carotid body chemoreceptors single afferent fibers. Respir. Physiol. 24: 249-266.

Lahiri, S., E. Mulligan and A. Mokashi (1982). Adaptative response of carotid body chemoreceptors to CO_2. Brain Res. 234: 137-148.

Lahiri, S., A. Mokashi, W. Huang, A.K. Sherpa and C. Di Giulio (1989). Stimulus interaction between CO_2 and almitrine in the cat carotid chemoreceptor. J. Appl. Physiol. 67: 232-238.

Mulligan, E.M. (1991). Discharge properties of carotid bodies: developmental aspects. In: "Developmental Neurobiology of Breathing. Lung Biology in Health and Disease." Ed. G.G. Haddad and J.P. Farber, NY, Marcel Dekker, vol. 53, pp 321-340.

Niu, W.S., M.J.A. Engwall, and G.E. Bisgard (1990). Two discharge patterns of carotid body chemoreceptors in the goat. J. Appl. Physiol. 69: 734-739.

58 THE CAROTID CHEMOSENSORY RESPONSE TO HYPOXIA IN THE DEVELOPING KITTEN

F. Marchal, A. Bairam, P. Haouzi, J.P. Crance and S. Lahiri

Laboratoire de Physiologie, Faculté de Médecine de Nancy
Vandoeuvre -Lès- Nancy, France
Unité 272 INSERM, Nancy, France
Department of Physiology, University of Pennsylvania, USA

In newborns of several mammalian species including infants (Cross and Oppé, 1953, Brady and Ceruti, 1966) and kittens (Bonora et al., 1984), the ventilatory response to hypoxia is weak and frequently biphasic: after a transient increase, ventilation decreases toward or even below prehypoxic level. This response thus appears intermediary between that of the fetus - where hypoxia depresses breathing movements (Boddy et al., 1974) - and that of the adult which is fully developed. The small magnitude of the response may be related to the resetting of arterial chemoreceptor activity to higher PaO_2. In the transition from fetal to neonatal life arterial PO_2 rises, thus temporarily turning off arterial chemoreceptors. In the neonatal lamb, several days are necessary for the carotid chemosensory discharge to adjust to the new PaO_2 level (Blanco et al., 1984a). The non sustained ventilatory response to hypoxia may be related to the central neural depression by hypoxia and/or to the non sustained input of arterial chemoreceptors into the respiratory neurons (Bureau et al., 1985). A biphasic ventilatory response to hypoxia has been shown in piglets (Mulligan, 1991) but not in kittens (Schwieler, 1968, Blanco et al., 1984b). However data in the latter specie are scanty. In this paper, we report the characteristics of dynamic and steady state carotid chemosensory responses to hypoxia in the developping kitten.

MATERIAL AND METHODS

Twenty four kittens aged 0 to 17 days were anesthetized with sodium thiopental (60 mg.kg^{-1}, ip), artificially ventilated with a respiratory pump and paralyzed with pancuronium bromide (1 mg.kg^{-1}). The inspiratory port of the pump could be connected to: air, 100% O_2, or 8% O_2 in N_2. The average barometric pressure was 740 Torr. Femoral vessels were catheterized and rectal temperature was maintained between 37 and 38°C with a heating pad. The carotid sinus nerve (CSN) was prepared to select chemosensory fibers as described previously for adult cats (Lahiri and DeLaney, 1975). Briefly, the nerve was separated into several filaments which were tested for the neural signals. The filaments were then further subdivided and tested until non-rhytmic discharges of a single or a few fibers remained and were responsive to inspired PO_2 changes: transition from air to hypoxia increased, and subsequently to hyperoxia eliminated or strikingly diminished the activities. The signals were amplified, band pass filtered, fed to a window discriminator and displayed on an oscilloscope. The signal output from the window discriminator was counted and averaged every second. The signals were also displayed on a chart recorder. The quantitative chemosensory responses to the changes of inspired PO_2 were tested each for about 2-3 min, in the following sequence : PiO_2 of 145 Torr, 55 Torr and 690 Torr. When possible, near the end of the exposure to each gas mixture arterial blood was directly sampled in a glass capillary tube (80 μl) for the measurement of PaO_2.

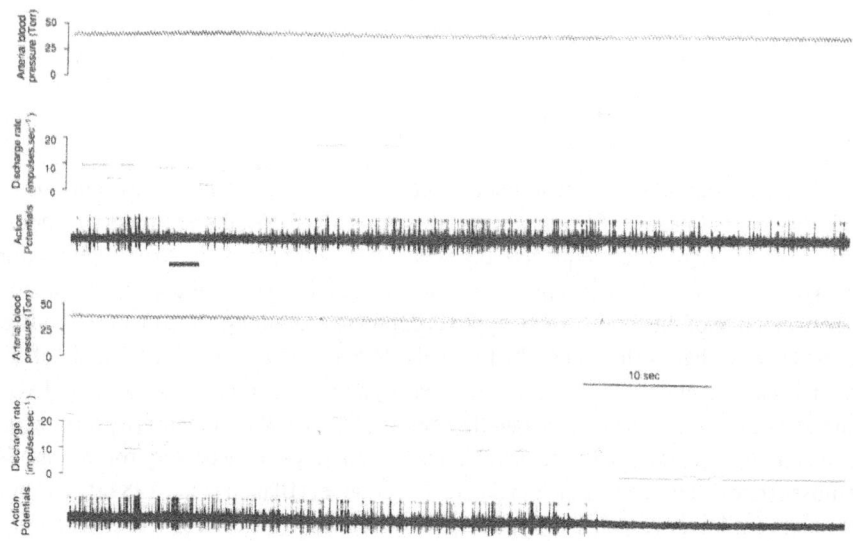

Figure 1. Representative recording of the response to hypoxia in a 2 day old kitten. Upper and lower panels are continuous. From top to bottom: arterial blood pressure, discharge frequency and action potentials from two carotid chemoreceptor afferents. Upper panel: PiO_2 was changed from 155 to 55 Torr at horizontal bar. The chemosensory activity reached a maximum some 30 seconds after the change of PiO_2, and declined gradually to a stable value. Lower panel : the chemosensory activity remained stable until the inspired PO_2 was changed to 690 Torr (horizontal bar), which diminished the discharge rate rapidly.

The dynamic chemosensory response to hypoxia was studied using the signal output of the counter and averaging it over consecutive epochs of 10 seconds, until a steady state was reached. When single fibers could be clearly identified from a preparation, their respective discharge frequency was calculated manually over periods of 10 seconds, in each condition of inspired gas mixture, at steady state. In order to evaluate the effect of maturation, the data from single fiber activity were compared between young (group 1: 0-9 days) and older kittens (group 2: 10-17 days). Data are expressed as mean ± SEM, unless otherwise indicated.

RESULTS

The dynamic response to hypoxia was studied in 24 kittens. A typical biphasic response of two chemosensory afferents to a PiO_2 of 55 Torr in a two-day old kitten is illustrated in fig. 1. Note that arterial blood pressure remains fairly constant throughout the exposure. A similar pattern of response was observed altogether in 11 kittens, 8/15 young kittens and 3/9 older ones. In these kittens the mean afferent activity on air (4.2 ± 0.7 impulse.sec^{-1}) increased to a peak value of 16.0 ± 1.9 impulse.sec^{-1} and declined to a stable value of 11.4 ± 1.2 impulse.sec^{-1}. Thirteen other kittens showed a sustained increase in chemosensory activity from 3.2 ± 0.6 impulse.sec^{-1} in room air to 14.2 ± 1.5 impulse. sec^{-1} in hypoxia.

The steady state response to hypoxia as a function of PaO_2 in 11 kittens of group 1 (21 fibers) and 4 kittens of group 2 (14 fibers) is presented in fig. 2. Both groups show characteristic curvilinear responses, but the average curve for group 1 is shifted to the left of group 2. The difference in $PaCO_2$ (36.3 ± 4.0 Torr in group 1 vs 28.2 ± 5.0 Torr in group 2) was not statistically significant.

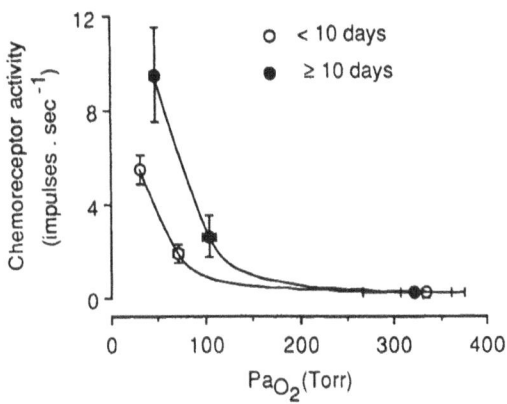

Figure 2. The steady state chemosensory discharge is plotted against PaO_2 in group 1 (○) and 2 (●). The response curve of the younger kittens is placed to the left of the older ones.

DISCUSSION

The carotid chemoreceptors were found to be responsive to hypoxia, even in very newborn animals. However, their responses to hypoxia were quantitatively different from those reported in adult cats (Lahiri and DeLaney, 1975). Indeed, the dynamic response was biphasic in 45% of the kittens studied, and the steady state response was weak, and increased significantly over the first two weeks of postnatal life.

The incidence of biphasic responses to hypoxia appeared to be higher in the younger animals, but the difference between group 1 and 2 was not statistificant. Bonora et al

417

(1984) have shown that the ventilatory response to hypoxia may be biphasic throughout the first month of life in behaving kittens. The fact that the biphasic chemosensory response is not systematically observed may explain why it has not been reported in previous studies where limited numbers of animals were used (Schwieler, 1968, Blanco et al., 1984b). Also, the variability in the dynamic chemosensory response to hypoxia is in keeping with the large intersubject variability reported for the neonatal ventilatory response to hypoxia (Bureau et al., 1985). The mechanisms of the biphasic response to hypoxia is unclear. Using an elegant technique of microelectrode recording from the petrosal ganglion and an intact CSN preparation, Mulligan (1991) recently described a non sustained response to hypoxia of some carotid chemosensory afferents in the piglet. The fact that the biphasic chemoreceptor response is also observed when the carotid sinus nerve is cut - as in the present study - suggests that it is related to the properties of the neonatal glomus cell rather than to a descending inhibition.

Hypoxemia has been shown to suppress breathing movements of the fetal sheep in utero (Boddy et al., 1974), and arterial chemoreceptors play little role in their maintainance during gestation (Moore et al., 1989). Considering the low PaO_2 of the fetus, the basal carotid chemosensory activity is small but responsive to maternally induced hypoxia and umbilical cord clamping (Blanco, 1984a). Therefore, the carotid chemosensory activity progressively adjusts to the new PaO_2 during the first few weeks of postnatal life in lambs (Blanco et al., 1984a), piglets (Mulligan, 1991) and kittens (fig.2).

The findings of the present study indicate that the biphasic ventilatory response to hypoxia may be explained - in part - by the non sustained response of the carotid chemosensory afferents. Simultaneous recording of carotid chemoreceptor and phrenic nerve activities are needed to establish the temporal relationship between biphasic chemosensory and ventilatory responses to hypoxia.

ACKNOWLEDGEMENTS

The authors thank N. Bertin, B. Chalon, G. Colin and J. Beyrend for their skiful technical assistance and M. Schaller for typing manuscript. This work was supported by grants from : le District de l' Agglomération Nanceienne, la Fondation pour la Recherche Médicale and l' Institut National de la Santé et de la Recherche Médicale.

REFERENCES

Blanco, C.E., Dawes G.S., Hanson M.A. and McCooke H.B. (1984a), The response to hypoxia of arterial chemoreceptors in fetal sheep and newborn lambs. J. Physiol., 351:25-37.

Blanco, C.E., Hanson M.A., Johnson P. and Rigatto H. (1984b), Breathing pattern of kittens during hypoxia. J. Appl. Physiol., 56:12-17.

Boddy, K., Dawes G.S., Fisher R., Pinter S. and Robinson J.S., Foetal respiratory movements, electrocortical and cardiovascular responses to hypoxaemia and hypercapnia in sheep. J. Physiol. (London), 1974, 243, 599-618.

Bonora, M., Marlot D., Gautier H. and Duron B., 1984, Effects of hypoxia on ventilation during postnatal development in conscious kittens. J. Appl. Physiol., 56:1464-1471.

Brady June, P. and Ceruti E., 1966, Chemoreceptor reflexes in the newborn infant:

Effects of varying degrees of hypoxia on heart rate and ventilation in warm environment. J. Physiol., 184, 631-645.

Bureau, M.A., Lamarche J. Foulon P. and Dalle D., 1985, The ventilatory response to hypoxia in the newborn lamb after carotid body denervation. Respir. Physiol., 60:109-119.

Cross, K.W. and Oppé, T.E., 1953, The effect of inhalation of high and low concentrations of oxygen on the respiration of the premature infant. J. Physiol., 117, 38-55.

Lahiri, S. and DeLaney R.G., 1975, Stimulus interaction in the responses of carotid body chemoreceptor single afferent fibers. Respir. Physiol., 24:249-266.

Lahiri, S., Smatresk, N.J. and Mulligan E., 1983, Responses of peripheral chemoreceptors to natural stimuli. In: Physiology of the peripheral arterial chemoreceptors, edited by Acker H. and O'Regan R.G. Amsterdam, Elsevier, pp. 221-256.

Moore, P.J., Parkes, M.J., Nijhuis, J.G. and Hanson, M.A., 1989, The incidence of breathing movements of fetal sheep in normoxia and hypoxia after peripheral chemodenervation and brain-stem transection. J. Develop. Physiol., 11, 147-151.

Mulligan, E.M., 1991, Discharge properties of carotid bodies: developmental aspects. In: Developmental Neurobiology of Breathing. Lung Biology in Health and Disease, edited by Haddad G.G. and Farber J.P., vol. 53, New-York, Dekker M., pp 321-340.

Schwieler, G.H., 1968, Respiratory regulation during postnatal development in cats and rabbits and some of its morphological substrate. Acta Physiol. Scand., 72 (suppl. 304):1-123.

59 ACTIONS OF DOXAPRAM ON K⁺ CURRENTS IN ISOLATED TYPE I CELLS OF THE NEONATAL RAT CAROTID BODY

C. Peers

Department of Pharmacology, Worsley Medical and Dental Building
Leeds University, Leeds LS2 9JT, UK

INTRODUCTION

Stimuli which excite afferent chemosensory fibres of the carotid sinus nerve have also been shown to cause release of transmitters (particularly dopamine) from type I cells[1]. The close correlations observed, in both time course and magnitude, between transmitter release and afferent nerve activity have indicated that transmitter release is a fundamental step in the chemotransductive process [2]. Such release is Ca^{2+}-dependent [3], and over the past few years several groups have used isolated type I cells (or clusters of such cells) to examine cellular processes which may underly transmitter release. These studies have been advanced by the introduction of microfluorimetric and patch-clamp techniques to investigate respectively the control of intracellular ion levels and the activity of ion channels in the type I cell plasma membrane. Ion channels (particularly K^+ channels) have been the focus of much recent attention, as their activity has been shown to be regulated by a variety of physiological and pharmacological chemostimuli[4-9]. These effects may be important in regulating membrane potential and hence Ca^{2+} entry through voltage-gated (L-type) Ca^{2+} channels.

Here, I report the effects on type I cell K^+ channels of doxapram (1-ethyl-4-(2-morpholinoethyl)-3,3-diphenyl-2-pyrrolidinone hydrochloride hydrate), a commonly used respiratory stimulant which increases ventilation due chiefly to an action on the carotid body [10]. Its effects are compared with those of hypoxia and acidity, the two major physiological stimuli of the carotid body.

METHODS

Procedures used for isolating type I carotid body cells and obtaining whole-cell patch-clamp recordings have previously been described in detail elsewhere [8,9]. In brief, carotid bodies were removed from halothane-anaesthetised rat pups (aged 8 to 11 days), chopped into pieces and incubated in nominally Ca^{2+}- and Mg^{2+}-free phosphate-buffered saline (PBS) containing collagenase (0.03% to 0.05%) and trypsin (0.020% to 0.025%) for 20 to 30 min at 37°C. The digested tissue was then triturated, centrifuged and resuspended in Ham's F-12 culture medium supplemented with 10% fetal calf serum. The cells were plated onto polylysine-coated coverslips and kept in a humidified incubator (5% CO_2) overnight.

Neurobiology and Cell Physiology of Chemoreception
Edited by P.G. Data *et al.*, Plenum Press, New York, 1993

Fragments of coverslip were transferred to a continually-perfused recording chamber for patch-clamp studies. The perfusate contained (in mM): NaCl 135, KCl 5, MgSO$_4$ 1.2, CaCl$_2$ 2.5, HEPES 5 (21-24°C, pH 7.40). Whole-cell patch-clamp recordings[11] were made from type I cells using patch electrodes (2-7 MΩ resistance) filled with (in mM): KCl 107, CaCl$_2$ 1, MgSO$_4$ 2, NaCl 10, K-EGTA 11, HEPES 11, ATP 2 (pH 7.20). Cells were voltage-clamped at -70mV and 50msec depolarizing steps to various test potentials were applied at 0.2Hz. Outward K$^+$ currents evoked in response to these steps were measured for amplitude (using VCAN software, John Dempster, Strathclyde University) as previously described[8].

RESULTS

Figure 1 illustrates the effects of doxapram on K$^+$ currents recorded in an isolated type I cell. In this experiment, the cell membrane was repeatedly depolarized (0.2Hz) from -70mV to +20mV for 50msec, and each point plotted is an evoked K$^+$ current amplitude. Doxapram was seen to cause dose-dependent inhibitions of these K$^+$ currents, and its effects were rapid in onset and readily reversible. K$^+$ currents recorded from type I cells under these conditions (in the absence of doxapram) are known to consist of at least two different components; a Ca^{2+}-dependent and a Ca^{2+}-independent component. The Ca^{2+}-dependent component arises because Ca^{2+} influx through its own (L-type) Ca^{2+} channels activates high-conductance ('maxi-K' or 'BK') Ca^{2+}-dependent K$^+$ channels[8]. These channels can be inhibited directly by charybdotoxin (and not apamin[8]), or indirectly by blocking Ca^{2+} entry with Ca^{2+} channel blockers such as cadmium ions or organic Ca^{2+} channel antagonists (e.g. nifedipine[7]), or by recording in extracellular solutions containing low Ca^{2+} concentration (0.1mM) and raised Mg^{2+} concentration (6mM)[12].

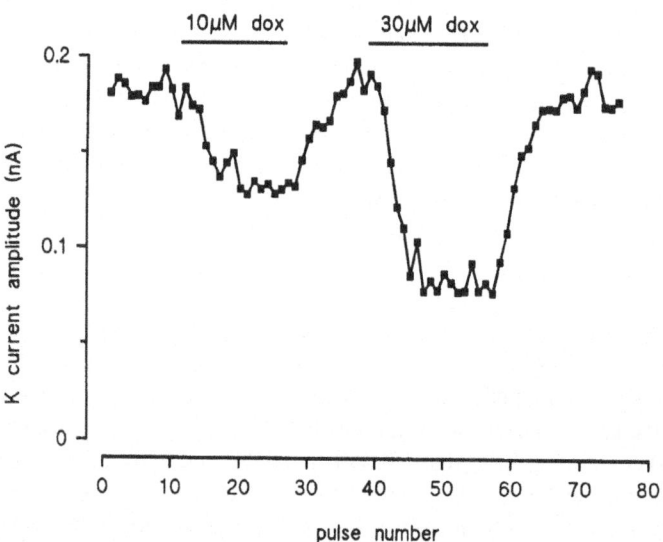

Figure 1. Time series study in which a type I cell was repeatedly depolarized to +20mV (0.2Hz). Each plotted point is a measured K$^+$ current amplitude. Horizontal bars show periods of application of doxapram (dox) to the perfusate.

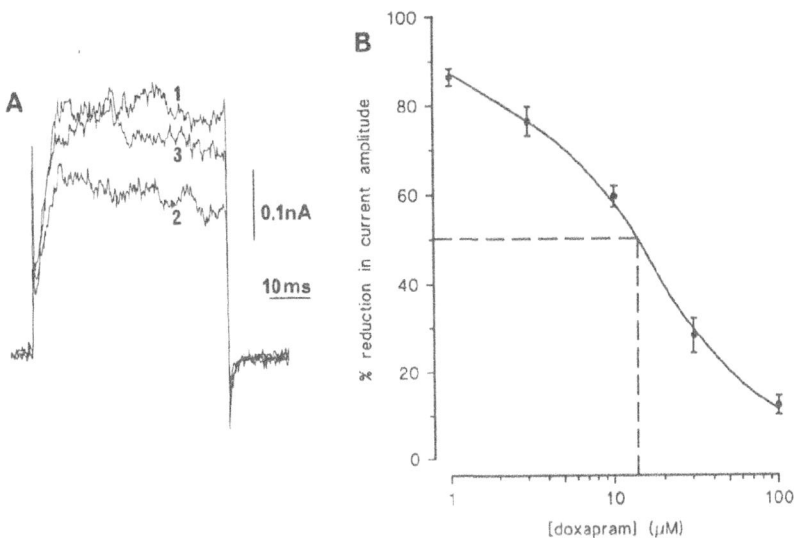

Figure 2. (A) Example K⁺ currents recorded in a type I cell before (1), during (2) and after (3) bath application of 10μM doxapram. Test potential +30mV, normal bath solution (see Methods). (B) Dose-response curve for doxapram-induced inhibition of K⁺ currents in type I cells. Each plotted point is the mean with vertical standard error bars, taken from between 6 and 10 cells. IC_{50} approximately 13μM. Doxapram caused significant reductions ($p<0.02$ to $p<0.0001$) at all concentrations tested.

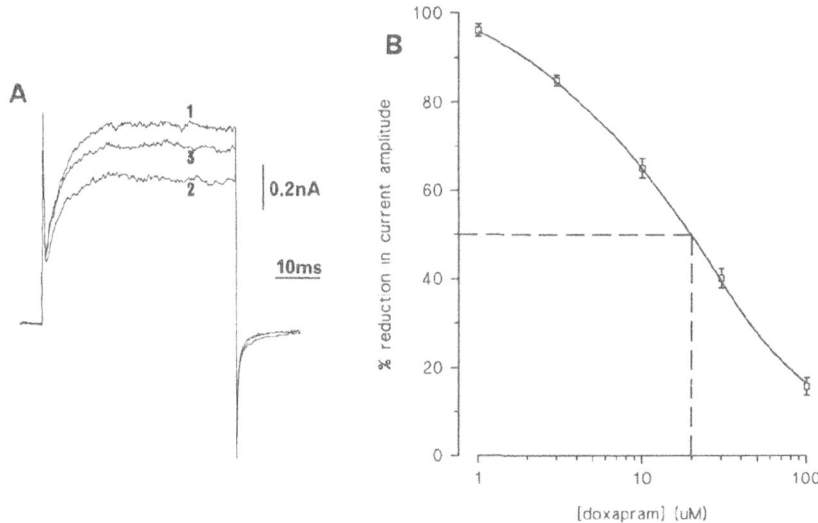

Figure 3. (A) Example K⁺ currents recorded in a type I cell using high Mg^{2+}, low Ca^{2+} solutions before (1), during (2) and after (3) bath application of 10μM doxapram. Test potential +30mV. (B) Dose-response curve for the effects of doxapram on K⁺ currents recorded in high Mg^{2+}, low Ca^{2+} solutions. Each plotted point is the mean with vertical standard error bars, obtained from between 9 and 14 cells. IC_{50} approximately 20μM. Doxapram caused significant reductions ($p < 0.0005$ to $p < 0.0001$) at all but the lowest dose tested (1μM).

Example traces of K^+ currents recorded from a type I cell using normal extracellular solutions before, during and after bath application of 10μM doxapram are shown in figure 2A. Figure 2B shows a dose-response curve for doxapram under these conditions, and indicates that the IC_{50} for doxapram was approximately 13μM. Thus doxapram, like the physiological stimuli to the carotid body (hypoxia and acidity [4-9]) causes reductions in K^+ current amplitudes in type I cells. However, these physiological stimuli cause selective inhibition of the Ca^{2+}-dependent component of K^+ currents in these cells [7-9], which only accounts, on average, for approximately 30% to 50% of the total whole-cell K^+ current. At high doses (figure 2B) doxapram caused much greater inhibitions of K^+ current amplitudes, indicating that the drug also inhibits the Ca^{2+}-independent K^+ current in type I cells.

The effects of doxapram on the Ca^{2+}-independent K^+ current in type I cells, recorded in isolation using high $[Mg^{2+}]$, low $[Ca^{2+}]$ solutions (see earlier) are shown in figure 3. Example traces obtained from a type I cell under these conditions, and the effect of doxapram are shown in figure 3A. Figure 3B is a plot of the dose-response curve for doxapram under these conditions, which yields an IC_{50} value of approximately 20μM. Thus doxapram inhibited both Ca^{2+}-dependent and Ca^{2+} independent K^+ currents in type I carotid body cells. However, it should be noted that at the lowest dose tested (1μM) doxapram was without significant effect under these conditions. Indeed, at all but the highest dose tested (100μM), the inhibitory actions of doxapram were significantly less ($p < 0.04$ to $p < 0.002$) on the Ca^{2+}-independent K^+ current recorded in isolation than on the total K^+ current recorded under control conditions (figure 2). This indicates that doxapram is more potent in inhibiting the Ca^{2+}-dependent K^+ current in type I cells, and an IC_{50} value for the effects of doxapram on the Ca^{2+} dependent K^+ current is estimated (taking into account the effects shown in figures 2B and 3B, together with the knowledge that the Ca^{2+}-dependent component of the K^+ current accounts for approximately 30 to 50% of the total K^+ current [8]) to be approximately 5μM.

DISCUSSION

Doxapram has been shown to cause increases in discharge of afferent chemosensitive fibres and an elevation of ventilation in the anaesthetised cat [10]. The doxapram-induced increases in ventilation were markedly reduced by cutting the carotid sinus nerve, suggesting that doxapram has a direct stimulatory effect on the carotid body [10]. However, the mechanism by which excitation occurred is still completely unknown. Results presented here demonstrate that doxapram inhibits two types of K^+ channel in neonatal rat type I carotid body cells, and that the Ca^{2+}-dependent component is more sensitive to the drug than the Ca^{2+}-independent component. These findings raise the question of how doxapram-induced K^+ channel inhibition in type I cells can be linked to excitation of the intact organ. This question is best approached by comparing the effects of doxapram with those of the physiological stimuli (hypoxia and acidity) of the carotid body. Table 1 summarises the effects of all these stimuli on ionic currents recorded in type I cells, and shows that all agents inhibit Ca^{2+}-dependent K^+ currents, but none affect the inward Ca^{2+} current. In these two respects, the actions of doxapram resemble those of physiological stimuli, but the drug has the additional effect of inhibiting the Ca^{2+}-independent K^+ current.

Hypoxia and acidity have previously been shown to depolarize type I cells [5,13], and in the case of hypoxia at least, the depolarization appears to be a direct consequence of

424

K$^+$ channel inhibition [5]. This would be expected to lead to opening of the voltage-gated L-type Ca^{2+} channels, hence Ca^{2+} influx and Ca^{2+}-dependent transmitter release. Indeed, hypoxia-induced transmitter release has been shown to be inhibited by nitrendipine, an L-type Ca^{2+} channel antagonist [14]. It is tempting to speculate that doxapram may also excite the carotid body by inducing type I cell depolarization and hence causing Ca^{2+}-dependent transmitter release, given the findings presented here and summarized in table 1. However, such speculation must be made with caution, since it is not even known whether doxapram evokes release of transmitters from type I cells.

Table 1. Effects of carotid body stimuli on ionic currents found in neonatal rat type I cells. IK$_{Ca}$, Ca^{2+}-dependent K$^+$ current; IK$_V$, Ca^{2+}-independent K$^+$ current; I$_{Ca}$, Ca^{2+} current. Downward arrow indicates inhibition.

Effects of chemostimuli on ionic currents in neonatal rat type I cells

	K$_{Ca}$	K$_V$	Ca	ref.
hypoxia	↓	no effect	no effect	7
extracellular acidosis	↓	no effect	no effect	8
intracellular acidosis	↓	no effect	no effect	9
doxapram	↓	↓	no effect	15

Furthermore, there are alternative models for hypoxia and acidity-induced transmitter release from type I cells (based on microfluorimetric and radiolabelled dopamine release studies [16,17]) in the adult rabbit carotid body, which do not involve Ca^{2+} influx through its own channels. Therefore effects of doxapram on intracellular Ca^{2+} stores, on membrane transporter systems and indeed on transmitter release itself are required before doxapram's inhibition of K$^+$ channels reported here can be incorporated into a mechanism for excitation of the intact carotid body.

ACKNOWLEDGEMENTS

This work was supported by the Wellcome Trust.

REFERENCES

1. S.J. Fidone and C. Gonzalez, Initiation and control of chemoreceptor activity in the carotid body, in: "Handbook of Physiology, III: The Respiratory System, vol. 2" A.P. Fishman, N.S. Cherniack, J.G. Widdicome and S.R. Geiger, ed.s, American Physiological Society, Bethesda (1986).

2. *R. Rigual, E. Gonzalez, S. Fidone and C. Gonzalez*, Effects of low pH on synthesis and release of catecholamines in the cat carotid body in vitro, <u>Brain Res.</u>, 309: 178-181 (1984).

3. *M.C. Fishman, W.L. Greene and D. Platika*, Oxygen chemoreception by carotid body cells in culture, <u>Proc. Natl. Acad. Sci.</u> USA, 82: 1448-1450 (1985).

4. *J. Lopez-Barneo, J.R. Lopez-Lopez, J. Urena and C. Gonzalez*, Chemotransduction in the carotid body: K^+ current modulated by PO_2 in type I chemoreceptor cells, <u>Science</u>, 241: 580-582 (1988).

5. *M.A. Delpiano and J. Hescheler*, Evidence for a pO_2-sensitive K^+ channel in the type-I cell of the rabbit carotid body, <u>FEBS Lett.</u>, 249: 195-198 (1989).

6. *A. Stea and C.A. Nurse*, Whole-cell and perforated-patch recordings from O_2-sensitive rat carotid body cells grown in short- and long-term culture, <u>Pflugers Archiv.</u>, 418: 93-101 (1991).

7. *C. Peers*, Hypoxic suppression of K^+ currents in type I carotid body cells: selective effect on the Ca^{2+}-activated K^+ current, <u>Neurosci. Lett.</u>, 119: 253-256 (1990).

8. *C. Peers*, Selective effect of lowered extracellular pH on Ca^{2+}-dependent K^+ currents in type I cells isolated from the neonatal rat carotid body, <u>J. Physiol.</u>, 422: 381-395 (1990).

9. *C. Peers and F.K. Green*, Intracellular acidosis inhibits Ca^{2+}-dependent K^+ currents in isolated type I cells of the neonatal rat carotid body. <u>J. Physiol.</u>, 437: 589-602 (1991).

10. *T. Nishini, A. Mokashi and S. Lahiri*, Stimulation of carotid chemoreceptors and ventilation by doxapram in the cat, <u>J. Appl. Physiol.</u>, 52: 1261-1265 (1982).

11 *O.P. Hamill, A. Marty, E. Neher, B. Sakmann and F.J. Sigworth*, Improved patch-clamp techniques for high-resolution current recording from cells and cell-free membrane patches, <u>Pflugers Archiv.</u>, 391: 85-100 (1981).

12. *C. Peers*, Effects of D600 on hypoxic suppression of K^+ currents in isolated type I carotid body cells of the neonatal rat, <u>FEBS Lett.</u>, 271:37-40 (1990).

13. *C. Eyzaguirre*, An overview of mechanisms associated with the onset of sensory discharges in the carotid nerve, <u>in</u>: "Arterial Chemoreceptors", C. Belmonte, D.J. Pallot, H. Acker and S. Fidone, ed.s, Leicester University Press, Leicester (1981).

14. *A. Obeso, S. Fidone and C. Gonzalez*, Pathways for calcium entry into type I cells: significance for the secretory response, <u>in</u>: "Chemoreceptors in Respiratory Control", J.A. Ribeiro and D.J. Pallot, ed.s, Croom Helm, London (1987).

15. *C. Peers*, Effects of doxapram on ionic currents recorded in type I cells of the neonatal rat carotid body, <u>Brain Res.</u>, 568: 116-122 (1991).

16. *T.J. Biscoe and M.R. Duchen*, Responses of type I cells dissociated from rabbit carotid body to hypoxia, <u>J. Physiol.</u>, 428: 39-59 (1990).

17. *A. Rocher, A. Obeso, C. Gonzalez and B. Herreros*, Ionic mechanisms for the transduction of acidic stimuli in rabbit carotid body glomus cells, <u>J. Physiol.</u>, 433: 533-548 (1991).

60 STIMULATORY EFFECT OF LONG-TERM HYPOXIA ON THE POSTERIOR PART OF A2 NORADRENERGIC CELL GROUP IN NUCLEUS TRACTUS SOLITARIUS OF RAT

J.M. Pequignot, V. Soulier, J.M. Cottet-Emard, Y. Dalmaz, N. Borghini and L. Peyrin

URA 1195 CNRS, Faculté de Médecine Grange-Blanche
F-69373 Lyon Cedex 08, France

INTRODUCTION

The nucleus tractus solitarius has been demonstrated to receive the primary chemoreceptor afferents of the carotid sinus nerve. The chemosensory fibres terminate in a relatively restricted area which extends caudal to the obex and includes the commissural nucleus and the medial subnucleus (Donoghue et al., 1984; Housley et al., 1987, Housley and Sinclair, 1988). This region seems to correspond to the anatomic location of an aggregate of noradrenergic neurones that make up the A2 cell group. In fact, the A2 group extends both caudal and rostral to the obex and, as referred above, only the posterior part could receive the peripheral chemosensory inputs.

Located in the nucleus tractus solitarius are neurones that project to other noradrenergic regions of the brainstem, the cell groups A1, A5, A6, and A7. These cell groups may be of particular importance in the adaptative responses to hypoxia. Indeed, A1 is involved in control of vasopressin release (Sawchenko et al., 1982), while A5 and A7 regulate the peripheral sympathetic outflow (Byrum and Guyenet, 1987; Neil and Loewy, 1982). The locus coeruleus (A6) is considered as a major component of adaptative responses to changes in internal or external environment (Foote et al. 1983).

The aim of this study was to determine the functional effects of long-term hypoxia on the noradrenergic cell groups A1, A2 (subdivided into anterior and posterior parts), A5 and A6, and to define if the observed changes may be mediated by the carotid chemoreceptors. For this purpose, an attempt was made to investigate the possible involvement of norepinephrine (NE) in the chemoreceptive pathway by estimating the turnover of NE, which provides a good index of functional activity of noradrenergic neurones (Sharman, 1981). The experiments were carried out in rats either intact or subjected to bilateral transection of the carotid sinus nerve.

METHODS

Animals

Intact and chemodenervated male Sprague-Dawley rats (200-220g) were kept for 2 weeks in a plexiglass chamber supplied with 10% O_2 in nitrogen. The gas composition inside the chamber was maintained at $10 \pm 0.5\%$ O_2. The CO_2 expired by rats was eliminated by circulating the gas mixture from the chamber through soda lime. Metabolic water contained in expired gases was continuously trapped in a chilled tank.

Chemodenervation was performed one week before the exposure by bilateral transection of the carotid sinus nerve at its point of branching from the glossopharyngeal nerve and at the cranial pole of the carotid body.

Intact and chemodenervated control rats were maintained in normoxic air in the same room as the hypoxic rats and sacrificed 2 weeks after beginning the experiment.

Estimation of norepinephrine turnover

The NE turnover was assessed in catecholamine cell groups by blocking the catecholamine biosynthesis with α-methyl-p-tyrosine. Each experimental group was subdivided into two subgroups: one subgroup received α-methyl-p-tyrosine (250 mg/kg ip, 2.5 h before sacrifice), while the other received 0.9% saline. First, basal level of NE in the different cell groups was measured in saline-treated rats. Then, the amount of NE remaining 2.5 h after a-methyl-p-tyrosine was subtracted and the difference afforded the NE turnover.

Tissue preparation

The rats were sacrificed by cervical dislocation and then decapited. The brain was rapidly dissected out and frozen on dry ice. The brainstem was cut into serial frontal slices of 480 μm thickness and the noradrenergic cell groups, A1, A2, A5 and A6, were punched out (Palkovits and Brownstein, 1987). The A2 cell group was subdivided into two half parts, corresponding to areas located respectively caudal and rostral to the obex. Punches were placed in 100-200 μl of 0.4 M perchloric acid containing 2.7 mM disodic ethylenediaminetetraacetic acid. Norepinephrine was assayed using high performance liquid chromatography coupled with electrochemical detection (Pequignot et al., 1986).

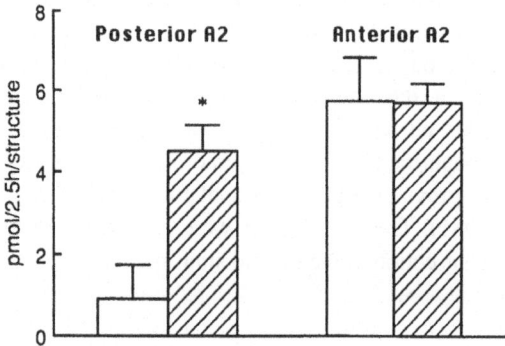

Fig. 1. Effect of normobaric hypoxia lasting for 14 days on the turnover of norepinephrine in the A2 noradrenergic cell group. A2 was subdivided caudorostrally into two equal parts referred as "posterior A2" and "anterior A2" (mean ± SEM). * P < 0.05, hypoxia (hatched bars) vs. normoxia (clear bars).

RESULTS

Effects of long-term hypoxia on the posterior and anterior parts of A2

Hypoxia lasting for 14 days elicited a 5-fold increase in turnover of NE in the posterior part of A2. In contrast, the NE turnover in the anterior part of A2 was not altered by hypoxia (Fig. 1).

Effects of long-term hypoxia on the A1, A5, and A6 cell groups

Long-term hypoxia had different effects according to the cell group. In the A1 cell group, hypoxia elicited a non-significant increase in NE turnover ($30 \pm 11\%$, $P < 0.10$). In both A5 and A6 cell groups, hypoxia induced a decrease in NE turnover, $-46 \pm 13\%$ and $-35 \pm 12\%$ respectively (Fig 2).

Effects of chemodenervation

The NE turnover estimated in bilaterally chemodenervated rats exposed to hypoxia did not differ from the normoxic levels, in all noradrenergic cell groups studied.

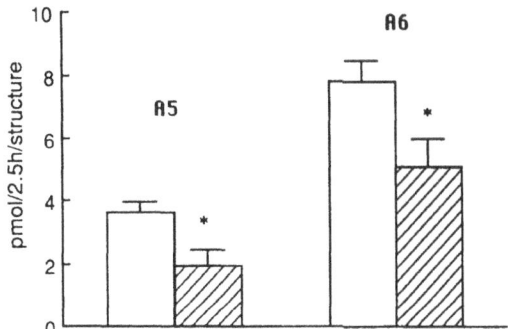

Fig. 2. *Effects of normobaric hypoxia lasting for 14 days on the turnover of norepinephrine in the A5 and A6 noradrenergic cell groups (mean $\pm$ SEM). * P < 0.05, hypoxia (hatched bars) vs. normoxia (clear bars).*

DISCUSSION

The present data show that central NE is a neurotransmitter involved in the chemosensory adaptation to hypoxia. A long-term hypoxic exposure influences the noradrenergic cell groups of rat brainstem according to different patterns, thus suggesting a specific organization of the responses to hypoxia.

First, the results indicate that the anterior and posterior parts of the A2 cell group are differentially affected by hypoxia. Indeed, only the part of A2 caudal to the obex was stimulated by hypoxia, while the part of A2 rostral to the obex remained unaffected. This would suggest a functional suborganization of A2. Transection of the carotid sinus nerves, which in the rat vehiculate the bulk of peripheral chemosensory inputs to the central nervous system (Sapru and Krieger, 1977), eliminated the noradrenergic response to hypoxia. This result implicates peripheral chemoreceptors in the hypoxic activation of noradrenergic neurones located in the posterior part of A2 cell group. In fact, the present data give a neurochemical support in awake rats to the previous finding that the region of nucleus tractus solitarius just caudal to the obex is the site of projection for peripheral chemosensitive nerve fibres (Donoghue et al., 1984; Housley and Sinclair, 1988; Onai et al., 1987). Taken together this finding and our own results would implicate the posterior part of A2 in central integration of peripheral chemosensory inputs. On the other hand, the part of nucleus tractus solitarius just rostral to the obex is the primary site of projection for peripheral barosensory fibres (Donoghue et al., 1984; Housley et al., 1987). Thus, it is tempting to speculate that the anterior part of A2 participates to central integration of peripheral barosensory inputs.

The present study reveals a hypoxia-mediated decrease in noradrenergic activity in A5 and A6 cell groups, a result which contrasts with that obtained in A2. A direct input to sympathetic preganglionic neurones originates from the A5 group (Loewy et al., 1979) and changes in blood pressure evoked by stimulation of the A5 area support the view that A5 cells might subserve a sympathomodulatory function (Byrum and Guyenet, 1987; Neil and Loewy, 1982). Thus, the inhibitory effect of hypoxia on A5 might indicate that this cell group is involved in the control of sympathetic outflow in this condition.

Locus coeruleus, the A6 cell group, is considered as an alarm system which arouses the brain by augmenting the central nervous activity in response to rapid changes in internal or external environment (Foote et al., 1983). Indeed, the firing rate of A6 neurones is altered in response to acute hypoxia (Elam et al., 1981). The present study showed a persistent effect of long-term hypoxia, thus suggesting a lack of steady-state adaptation after 2 weeks of exposure. Interestingly, a similar conclusion was drawn from analysis of the pattern of hypoxia-induced neurochemical changes in carotid body and peripheral sympathetic target organs (Dalmaz et al., 1989).

Like in A2, the chemodenervation was able to abolish the hypoxia-induced changes in noradrenergic activity in both A5 and A6, thus showing that the carotid body chemoreceptors are essential for these responses.

An unexpected result was the lack of a clearcut effect of long-term hypoxia on noradrenergic activity of the A1 cell group. This group is a major component of the central regulation of vasopressin secretion through a direct projection of A1 neurones to the vasopressin-secreting cells located in the hypothalamic paraventricular and supraoptic nuclei (Sawchenko et al., 1982). Stimulation of carotid body chemoreceptors is able to increase the release of vasopressin from the supraoptic nucleus (Harris, 1979). However, it was recently shown that specific stimulation of the arterial carotid chemoreceptors did not alter the activity of neurones located in A1 (Jamieson and Harris, 1989). Thus, it might be that the A1 cell group is not a major structure for regulation of vasopressin during hypoxia.

In conclusion, central NE appears as a neurotransmitter involved in adaptation to hypoxia. A long-term exposure stimulates the posterior part of A2, an area which is the primary site of projection of chemosensory fibres, but does not affect the anterior part, an area which is supposed to be the primary site of projection of barosensory fibres. In contrast, long-term hypoxia exerts an inhibitory effect on the noradrenergic functions of

A5 and A6 groups and apparently fails to alter the noradrenergic activity of A1. The observed hypoxia-induced changes appear dependent on peripheral chemosensory inputs.

ACKNOWLEDGEMENTS

This work was supported by grants from Direction des Recherches, Etudes et Techniques (Contrat nr. 88-069), Région Rhône-Alpes (Contrat Neurosciences 1991), INSERM (Contrat nr. 88.40 07) and CNRS.

REFERENCES

Byrum, C. E., and Guyenet, P. G., 1987, Afferent and efferent connections of the A5 noradrenergic cell group in the rat, J. Comp. Neurol., 261: 529.

Dalmaz, Y., Pequignot , J. M., Cottet-Emard, J. M., Tavitian E., and Peyrin, L., 1989, Changes of dopamine, norepinephrine and epinephrine turnover in sympathetic tissues after long-term hypoxic stress, in: "Stress: Neurochemical and Humoral Mechanisms", G. R Van Loon, R. Kvetnansky, R. M. McCarty, J. Axelrod, eds, Gordon and Breach Science, New-York .

Donoghue, S., Felder, R. B., Jordan, D., and Spyer, K.M., 1984, The central projections of carotid baroreceptors and chemoreceptors in the cat: A neurophysiological study, J. Physiol., 347: 397.

Elam, M., Yao, T., Thoren, P., and Svensson, T.H., 1981 Hypercapnia and hypoxia: chemoreceptor-mediated control of locus coeruleus neurons and splanchnic, sympathetic nerves, Brain Res., 290: 281.

Foote, S. L., Bloom, F. E., and Aston-Jones, G., 1983, Nucleus locus coeruleus: new evidence of anatomical and physiological specificity, Physiol. Rev., 63: 844.

Harris, M. C., 1979, Effects of chemoreceptor and baroreceptor stimulation on the discharge of hypothalamic supraoptic neurones in rats, J. Endocrinol., 82: 115.

Housley, G. D., Martin-Body, R. L., Dawson, N. J., and Sinclair, J.D., 1987, Brainstem projections of the glossopharyngeal nerve and its carotid sinus branch in the rat, Neuroscience, 22: 237.

Housley, G. D., and Sinclair, J.D., 1988, Localization by kainic lesions of neurones transmitting the carotid chemoreceptor stimulus for respiration in rat, J. Physiol., 406: 99.

Jamieson, S. M., and Harris, M.C., 1989, Stimulation of carotid body chemoreceptors does not influence the discharge of A1 neurons projecting to the forebrain, Neuroscience, 32: 227.

Loewy, A. D., McKellar, S., and Saper, C.B., 1979, Direct projections from the A5 catecholamine cell group to the intermediolateral cell column, Brain Res., 174: 309

Neil, J. J., and Loewy, A.D., 1982, Decreases in blood pressure in response to L-glutamate microinjections into the A5 catecholamine cell group, Brain Res., 241: 271.

Onai, T., Saji, M., and Miura, M., 1987, Functional subdivisions of the nucleus tractus solitarii of the rat as determined by circulatory and respiratory responses to electrical stimulation of the nucleus, J. Auton. Nerv. Sys., 21: 195.

Palkovits, M., and Brownstein, M.J., 1987, "Maps and Guide to Microdissection of the Rat Brain", Elsevier, Amsterdam.

Pequignot J. M., Cottet-Emard, J. M., Dalmaz, Y., de Haut de Sigy, M., and Peyrin, L., 1986, Biochemical evidence for norepinephrine stores outside the sympathetic nerves in rat carotid body, Brain Res., 367: 238.

Sapru, H. N., and Krieger, A.J., 1977, Carotid and aortic chemoreceptor function in the rat, J. Appl. Physiol., 42: 344.

Sawchenko, P. E., and Swanson, L.W., 1982, The organization of noradrenergic pathways from the brainstem to the paraventricular and supraoptic nuclei in the rat, Brain Res. Rev., 4: 275.

Sharman, D. F., 1981, The turnover of catecholamines, in: "Central Neurotransmitter Turnover", C. J. Pycock, P. V. Taberner, eds, Croom Helm, London, p. 431.

61 TAURINE INTERACTION WITH THE CAT CAROTID BODY FUNCTION IN VITRO

M. Pokorski*, J. Albrecht, I. Kolsicka and
R. Strosznajder*

*Department of Neurophysiology and Department of Neuropathology,
Polish Academy of Sciences Medical Research Center, Warsaw,
Poland

INTRODUCTION

It is generally accepted that transmission between the carotid body (CB) chemoreceptor cell and the sinus nerve afferent fiber is chemically mediated (Gomez-Nino et al., 1990). The identity of the afferent neurotransmitter has yet to be established, but dopamine (DA), which is abundant in the dense-cored granules of the chemoreceptor cell and is usually released by both natural (e.g. hypoxia) (Gomez-Nino et al., 1990; Shaw et al., 1989) and pharmacological (e.g. cyanide) (Obeso et al., 1989) stimuli, has been implicated as playing a major role. The range of other presumptive transmitters identified in the CB suggests that the chemosensory process involves neuroactive substances other than, or complementary to, DA.

Taurine (TAU), a sulfur-containing β-amino acid, is distributed in numerous tissues. TAU is released by hypoxia and offers protection against hypoxic damage in the brain (Lehmann et al., 1988), the retina (Tseng et al., 1990), and the heart (Hochachka, 1986). The possibility thus arises that TAU may modulate neurotransmissory functions of CB, a chemosensory organ for which hypoxia is the most potent natural stimulus. An important cue for subserving such a role would be the release of TAU from the CB under stimulation. In the present study we tested this possibility in an vitro preparation of CB using a cyanide (CN⁻) pulse as stimulant. We present evidence that TAU is a new neuromodulator candidate for the chemosensory processing in the carotid body.

METHODS

Twenty five pairs of CB were used for the study. Each experiment was performed on a pair of homologous CB dissected from a normoxic cat anesthetized with α-chloralose and urethane (30 and 800 mg/kg i.p., respectively). One CB always served as control to the other subjected to a given maneuver.

CB were incubated for 20 min at 37 °C in 1.5 ml of a standard Krebs-Henseleit buffer of pH 7.4, containing 10 mM glucose and 1.2 mM $CaCl_2$ and supplemented with 1 μCi of [³H]TAU or [³H]DA (Amersham, U.K.). After incubation, CB were placed on Whatman GF/C glass fiber filters, cut to fit a 25 mm Swinnex filter unit. Continuous superfusion was then carried out with the Krebs-Heneleit solution at a rate of 0.5 ml/min, which yielded a constant basal efflux of radioactivity after 20 min.

Neurobiology and Cell Physiology of Chemoreception
Edited by P.G. Data *et al.*, Plenum Press, New York, 1993

Table 1. Effects of taurine (TAU), cyanide (CN⁻), and CN⁻ plus TAU on [³H]dopamine release from the carotid body.

Pulse	Dopamine Release (% over basal efflux)	n
TAU	4 ± 2	12
CN⁻	26 ± 5 *	12
CN⁻ + TAU	59 ± 6 **	12

*Results are means ± SE. * P < 0.05 vs. TAU;*
** P < 0.05 vs. CN⁻ and TAU.*

TAURINE RELEASE

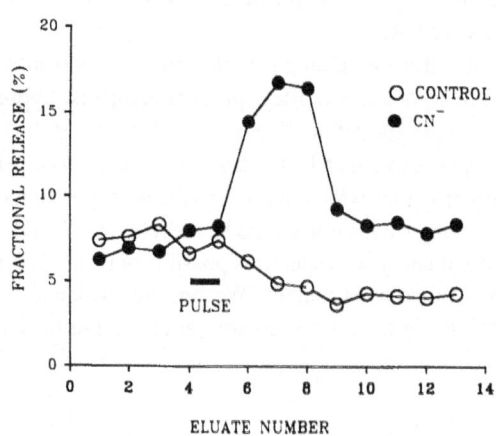

Fig. 1. An experiment on a pair of carotid bodies preloaded with [³H]TAU. A CN⁻ pulse (5x10⁻⁴ mM) increased [³H]TAU release (solid circles) over its basal efflux (open circles).

The release test consisted of applying a 2 min pulse of 5×10^{-4} CN⁻, 40 mM TAU, or both to the superfusate of one CB, the other serving as control, while both CB were preloaded with the same neuromodulator. The test was performed in duplicate and was followed by standard washing. Eluates were gathered every 2 min. The eluted radioactivity and that left over on the filters after the end of superfusion were counted.

Release was expressed as a fraction of radioactivity lost in sequential eluates. Results are expressed as means ± SE and analyzed either by a paired t-test or by a one-way analysis of variance and Scheffe's test.

RESULTS

Taurine Release from the Carotid Body

An example of the experiment in which a CN⁻ pulse was applied to a CB preloaded with [³H]TAU is shown in Fig. 1. The CN⁻ pulse caused about 2-fold increase in [³H]TAU release over its basal efflux. In contrast, the unstimulated CB showed a spontaneous [³H]TAU efflux decreasing with time. On average, the increase in TAU release due to the CN⁻ stimulus over the basal release amounted to 72 ±9% in 13 trials.

Dopamine Release from the Carotid Body

The release of [³H]DA preloaded into CB was studied with the following combination of stimulant pulses: cyanide - 5×10^{-4} mM, cold taurine - 40 mM, cyanide plus cold taurine in the foregoing concentrations. The effects on DA release of these pulses are set out in Table 1. These data show that the cyanide and taurine challenges produced a diverse response. Whereas

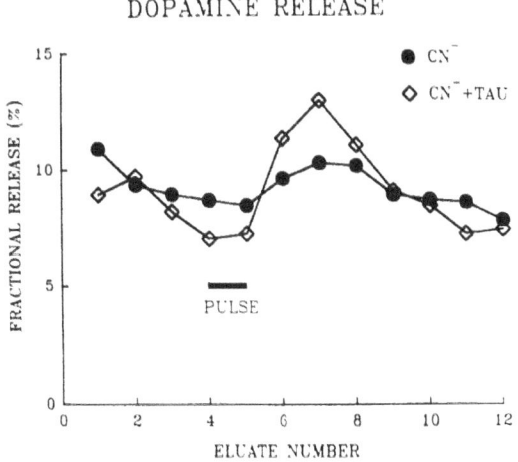

Fig. 2. Comparison of [³H]DA release from two carotid bodies treated with CN⁻ alone (5×10^{-4} mM; solid circles) and CN⁻ plus TAU (40 mM) pulses (open diamonds). TAU added to CN⁻ augmented the CN⁻ -induced [³H]DA release.

the CN⁻-treated carotid bodies exhibited a significant increase in [³H]DA release, the TAU-treated ones did not. TAU, however, caused a striking augmentation of the stimulatory effect of CN⁻ on [³H]DA release. This augmentation is exemplified by an experiment shown in Fig. 2, in which [³H]DA release from a CN⁻ + TAU- treated CB was more than double that of a CN⁻ alone-treated CB.

DISCUSSION

In this study we investigated the potential neuromodulatory role of taurine in the chemosensory processing in the carotid body. The basic premise for playing such a role relies on the hypothesis that a neuromodulator is released from a tissue in response to a stimulus. The results fulfill this criterion by showing that the preloaded [³H]TAU was released from the carotid body stimulated by CN⁻. Furthermore, we corroborated previous studies showing that DA extrusion from the carotid body is involved in the chemosensory response (Obeso et al., 1989) and presented evidence that the CN⁻ induced DA release was markedly enhanced by taurine. An argument therefore can be made that taurine is an authentic neuromodulator of the chemosensory transmission.

How taurine would interact with the chemosensory response is unclear. Most studies dealing with DA metabolism in the carotid chemoreceptor cell show the release of DA from the intracellular dense-cored vesicles by CB stimuli, the process being calcium-dependent (Shaw et al., 1989). Taurine interferes with calcium binding by excitable membranes (Pasantes-Morales and Gamboa, 1980; Huxtable, 1989), which may modify the extrusion of the vesicular DA.

The neuromodulatory action of taurine is usually attributed to its transmembrane carrier-mediated transport system (Okamoto and Namima, 1978). A similar active efflux system releases dopamine from its nonvesicular cytoplasmic pool in the neural tissue (Hurd and Ungerstedt, 1989; Debler et al., 1991). The main feature of such a system is that movement of a substrate into a compartment stimulates the efflux of a second substrate from that compartment. It is therefore tempting to suggest that in our preparation the CN stimulus activated the carrier-mediated efflux of nonvesicular dopamine from the chemoreceptor cell, the efflux being augmented by taurine entering the cell, a second substrate for the system, added to the superfusate. Taurine alone would remain without effect in the unstimulated state.

This study cannot discern between the vesicular and nonvesicular mechanisms of DA release from CB. We believe however the results hint at a novel, in regard to the carotid body, mechanism of a carrier-mediated efflux process being involved in DA release. In conclusion, taurine acts as a neuromodulator of the chemosensory process in the carotid body. The exact determinants of this action remain to be elucidated.

REFERENCES

Debler, E. A., Sershen, H., Hashim, A., Lajtha A., Reith M. E. A., 191, Carrier-mediated efflux of [³H]dopamine and [³H]1-Methyl-4-Phenylpyridine: effect of ascorbic acid, Synapse 7:99.

Gomez-Nino, A., Dinger, B., Gonzalez, C., and Fidone, S. J., 1990, Differential stimulus coupling to dopamine and norepinephrine stores in rabbit carotid body type I cells, Brain Res. 525:160.

Hochachka, P. W., 1986, Defensive strategies against hypoxia and hypothermia, <u>Science</u>. 231: 234.

Hurd, Y. L., and Ungerstedt, U., 1989, Influence of a carrier transport process on in vivo release and metabolism of dopamine: dependence on extracellular Na^+, <u>Life Sci.</u>. 45:283.

Huxtable, R. J., 1989, Taurine in the central nervous sytem and the mammalian actions of taurine, <u>Prog. Neurobiol.</u>. 32:471.

Lehmann, A., Hagberg, H., Andine, P., and Ellren, K., 1988, Taurine and neuronal resistance to hypoxia, <u>FEBS Lett.</u>. 233:437.

Obeso, A., Almaraz, L., and Gonzalez, C., 1989, Effects of cyanide and uncouplers on chemoreceptor activity and ATP content of the cat carotid body, <u>Brain Res</u> 481:250.

Okamoto, X., and Namima, M., 1978, Uptake, release, and homo- and hetero-exchange diffusions of inhibitory amino acids in guinea pig cerebellar slices, <u>J. Neurochem.</u>. 31:1393.

Pasantes-Morales, H., and Gamboa, A., 1980, Taurine effects on calcium transport in nervous tissue, <u>in</u>: "Natural Sulfur Compounds. Novel Biochemical and Structural Aspects," D. Cavallini, G. E. Gaull, and V. Zappia, ed., Plenum Press, New York and London.

Shaw, K., Montague, W., and Pallot, D. J., 1989, Biochemical studies on the release of catecholamines from the rat carotid body in vitro, <u>Biochim. Biophys. Acta.</u> 1013:42.

Tseng, M. T., Liu, K. N., and Radtke, N. R., 1990, Facilitated ERG recovery in taurine-treated bovine eye, an ex vivo study, <u>Brain Res.</u>. 509:153.

CONTRIBUTORS

Abramovici Armand - Department of Pathology, Sackler Medical School, University of Tel Aviv, Tel Aviv, Israel

Acker Helmut - Max-Planck-Institut für Systemphysiologie, Dortmund 1, F.R.G.

Albrecht J. - Department of Neuropathology, Polish Academy of Sciences Medical Research Center, Warsaw, Poland

Alexander A. - Department of Biology, McMaster University, Hamilton, Ontario, Canada

Almaraz Laura - Departamento de Fisiología y Bioquímica, Facultad de Medicina, Universidad de Valladolid, Valladolid, Spain

Bairam Aida - Laboratoire de Physiologie, Faculté de Médecine, Unité INSERM 272, Université de Nancy I, Nancy, France

Bartels Else Marie - University Laboratory of Physiology, Oxford University, Oxford, U.K.

Bee Denise - Department of Experimental Medicine , University of Sheffield Medical School, Sheffield, U.K.

Behm Reinhard - Department of Physiology, University of Rostock, Rostock, F.R.G.

Biondelli Vincenzo - Institute of Physiological Sciences, School of Medicine, "G. D'Annunzio" University, Chieti, Italy

Bisgard Gerald E. - Department of Comparative Biosciences, School of Veterinary Medicine, University of Wisconsin, Madison, Wisconsin, U.S.A.

Bölling Brigitte - Max-Plank-Institut für Systemphysiologie, Dortmund 1, F.R.G.

Borghini N. - URA CNRS 1195, Faculté de Médecine Grange-Blanche, Lyon, France

Bradford Aidan J. - Department of Physiology, Royal College of Surgeons in Ireland, Dublin 2, Ireland

Buckler Keith James - University Laboratory of Physiology, Oxford, U.K.

Carroll John Lee - Johns Hopkins Hospital, Pediatric Pulmonology, Baltimore, Maryland, U.S.A.

Cerretelli Paolo - Department of Physiology, University of Geneva, Geneva, Switzerland

Chen J. - Department of Physiology, University of Utah School of Medicine, Salt Lake City, Utah, U.S.A.

Cherniack Neil S. - Department of Medicine, Case Wastern Reserve University, Cleveland, Ohio, U.S.A.

Clarke J.A. - Department of Anatomy, Queen Mary and Westfield College, London, U.K.

Costa M. - I.T.B.A., C.N.R., Milano, Italy

Cottet-Emard J.M. - URA CNRS 1195, Faculté de Médecine, Grange- Blanche, Lyon, France

Cragg Patricia Ann - Department of Physiology, University of Otago Medical School, Dunedin, New Zealand

Crance Jean-Pierre - Laboratoire de Physiologie, Faculté de Médecine, Unité INSERM 272, Université de Nancy I, Nancy, France

Cutz E. - Hospital For Sick Children, Toronto, Ontario, Canada

Dagerlind A. - Department of Histology and Neurobiology, Karolinska Institute, Stockholm, Sweden

Dalmaz Yvette - Departement de Physiologie, Faculté de Médecine Grange-Blanche, Lyon, France

Daly Michael de Burgh - Department of Physiology, Royal Free Hospital School of Medicine, London, U.K.

Dashwood Michael - Department of Physiology, Royal Free Hospital School of Medicine, London, U.K.

Data Pier Giorgio - Institute of Physiological Sciences, School of Medicine, "G.D'Annunzio" University, Chieti, Italy

Delpiano Marco Antonio - Max-Planck-Institut für Systemphysiologie, Dortmund, F.R.G.

de Muinck Keizer W. H. - Department of Pharmacology and German Institute for High Blood Pressure Research, Ruprecht-Karls University, Heidelberg, F.R.G.

Di Giulio Camillo - Institute of Physiological Sciences, School of Medicine, "G.D'Annunzio" University, Chieti, Italy

Di Tano Guglielmo - Institute of Physiological Sciences, School of Medicine, "G D'Annunzio" University, Chieti, Italy

Dinger Bruce G. - Department of Physiology, University of Utah School of Medicine, Salt Lake City, Utah, U.S.A.

Dufau Evelyne - Max-Plank-Institut für Systemphysiologie, Dortmund 1, F.R.G.

Ead Harold William - Department of Physiology, Royal Free Hospital, London, U.K.

Edelman Norman H. - Department of Medicine, UMDNJ-Robert Wood Johnson Medical School, New Brunswick, New Jersey, U.S.A.

Evrard Y. - Institut de Recherchers Internationales Servier, Paris, France

Eyzaguirre Carlos - Department of Physiology, University of Utah School of Medicine, Salt Lake City, Utah, U.S.A.

Fadic R. - Laboratory of Neurobiology, Catholic University of Chile, Santiago 1, Chile

Feely S. - Department of Physiology and Histology, University College, Dublin 2, Ireland

Ferretti G. - Department of Physiology, University of Geneva, Geneva, Switzerland

Fidone Salvatore J. - Department of Physiology, University of Utah School of Medicine, Salt Lake City, Utah, U.S.A.

Finley James C.W. - Department of Medicine, Case Western Reserve University School of Medicine, Cleveland, Ohio, U.S.A.

Fitzgerald Robert S. - Department of Environmental Health Sciences, Johns Hopkins Medical Institutions, Baltimore, Maryland, U.S.A.

Forster II Robert E. - Department of Physiology, University of Pennsylvania School of Medicine, Philadelphia, Pennsylvania, U.S.A.

Franz U. - Department of Physiology, University of Rostock, Rostock, F.R.G.

García-Sancho J. - Departamento de Bioquímica y Fisiología, Facultad de Medicina, Universidad de Valladolid, Valladolid, Spain

Gigante Gabriele - Institute of Physiological Sciences, School of Medicine, "G. D'Annunzio" University, Chieti, Italy

Giles S. - Department of Physiology and Histology, University College, Dublin 2, Ireland

Gómez-Niño A. - Departamento de Bioquímica y Fisiología, Facultad de Medicina, Universidad de Valladolid, Valladolid, Spain

Gonzalez C. - Departamento de Bioquímica y Fisiología, Facultad de Medicina, Universidad de Valladolid, Valladolid, Spain

Görlach Agnes - Max-Plank-Institut für Systemphysiologie, Dortmund 1, F.R.G.

Grassi Bruno - I.T.B.A., C.N.R., Milano, Italy

Habeck Jörg-Olaf - Department of Pathology, Kuchwald Hospital, Chemnitz, F.R.G.

Hannhart Bernard - Laboratory of Respiration Physiopathology, INSERM U-14, Vandoeuvre-lé-Nancy, France

Hanson G. - Department of Physiology, University of Utah School of Medicine, Salt Lake City, Utah, U.S.A.

Haouzi P. - Laboratoire de Physiologie, Faculté de Médecine, Unité INSERM 272, Université de Nancy I, Nancy, France

He L. - Department of Physiology, University of Utah School of Medicine, Salt Lake City, Utah, U.S.A.

Hertzberg T. - Department of Pediatrics, Karolinska Institute, Stockholm, Sweden

Hökfelt T. - Department of Histology and Neurobiology, Karolinska Institute, Stockholm, Sweden

Holgert Hans - Department of Pediatrics, Karolinska Institute, Stockholm, Sweden

Holtermann Georg - Max-Plank-Instutut für Systemphysiologie, Dortmund 1, F.R.G.

Honda Yoshiyuki - Department of Physiology, School of Medicine, Chiba University, Chiba Inohana, Japan

Huang W-X - Department of Physiology, School of Medicine, University of Pennsylvania, Philadelphia, Pennsylvania, U.S.A.

Iezzi Manuela - Institute of Physiological Sciences, School of Medicine, "G. D'Annunzio" University, Chieti, Italy

Iturriaga Rodrigo - Laboratory of Neurobiology, Catholic University of Chile, Santiago, Chile

Katz David M. - Department of Neurosciences, Case Western Reserve University School of Medicine, Cleveland, Ohio, U.S.A.

Kennedy M. - Department of Physiology and Histology, University College, Dublin 2, Ireland

Kolsicka I. - Department of Neuropathology, Polish Academy of Sciences Medical Research Center, Warsaw, Poland

Kondo Hisatake - Department of Anatomy, School of Medicine, Kanazawa University, Kanazawa, Japan

Kou Yu-Ru - Department of Medicine, Case Western Reserve University, Cleveland, Ohio, U.S.A.

Koyano Hayao - Department of Physiology, Akita University School of Medicine, Akita, Japan

Kummer Wolfgang - Institute of Anatomy and Cell Biology, University of Heidelberg, Heidelberg, F.R.G.

Lagadic-Gossmann D. - University Laboratory of Physiology, Oxford, England

Lagercrantz Hugo - Department of Pediatrics, Karolinska Institute, Stockholm, Sweden

Lahiri Sukamhai - Department of Physiology, University of Pennsylvania School of Medicine, Philadelphia, Pennsylvania, U.S.A.

Larrain C. - Laboratory of Neurobiology, Catholic University of Chile, Santiago 1, Chile

Leitner Louis-Michel - Laboratoire de Physiologie, Faculté de Médecine, Toulose, France

López-López Jose Ramon - Departamento de Bioquímica y Fisiología, Facultad de Medicina, Universidad de Valladolid, Valladolid, Spain

Loyola H. - Laboratory of Neurobiology, Catholic University of Chile, Santiago 1, Chile

Marchal Francois - Laboratoire de Physiologie, Faculté de Médecine, Unité INSERM 272, Université de Nancy I, Nancy, France

Marconi C. - I.T.B.A., C.N.R., Milano, Italy

McKeogh Donogh - Department of Physiology and Histology, University College, Dublin 2, Ireland

McLaren A. J. - University Laboratory of Physiology, Oxford University, Oxford, U.K.

McQueen Daniel - Department of Pharmacology, University of Edinburgh Medical School, Edinburgh, Scotland, U.K.

Melton Joseph E. - Department of Medicine, University of New Jersey, R W J Medical School, New Brunswick, New Jersey, U.S.A.

Mewes H. - Department of Physiology, University of Rostock, Rostock, F.R.G.

Meyer M. - Max-Plank-Institute für Experimental Medicine, Göttingen, F.R.G.

Mitchell G. - Department of Comparative Biosciences, University of Wisconsin, Madison, Wisconsin, U.S.A.

Mokashi A. - Department of Physiology, School of Medicine, University of Pennsylvania, Philadelphia, Pennsylvania, U.S.A.

Morelli Luca - Institute of Physiological Sciences, School of Medicine, "G.D'Annunzio" University, Chieti, Italy

Muddle J.R. - Neurological Sciences, Royal Free Hospital School of Medicine, London, U.K.

Neubauer Judith - Department of Medicine, UMDNJ-Robert Wood Johnson Medical School, New Brunswick, New Jersey, U.S.A.

Newman Charlotte - Department of Pathology, University of Toronto, Toronto, Ontario, Canada

Niu W. - Department of Comparative Biosciences, University of Wisconsin, Madison, Wisconsin, U.S.A.

Nolan Pilip - Department of Physiology and Histology , University College, Dublin 2, Ireland

Nurse Colin - Department of Biology, McMaster University, Hamilton, Ontario, Canada

Nye Piers C.G. - University Laboratory of Physiology, Oxford University, Oxford, U.K.

O'Regan Ronan - Department of Physiology and Histology, University College, Dublin 2, Ireland

Obeso Ann - Departamento de Fisiología y Bioquímica, Facultad de Medicina, Universidad de Valladolid, Valladolid, Spain

Pallot David - Department of Anatomy, University of Leicester Medical School, Leicester, U.K.

Paterson David J. - University Laboratory of Physiology, Oxford University, Oxford, U.K.

Peers Christopher - Department of Pharmacology, Worsley Medical and Dental Building, University of Leeds, Leeds, U.K.

Pequignot Jean-Marc - Laboratoire de Physiologie, Faculté de Médecine Grange-Blanche, Lyon, France

Pérez-García M.T. - Departamento de Bioquímica y Fisiología, Facultad de Medicina, Universidad de Valladolid, Valladolid, Spain

Peyrin L. - URA CNRS 1195, Faculté de Médecine Grange-Blanche, Lyon, France

Pizarro J. - Department of Comparative Biosciences, University of Wisconsin, Madison, Wisconsin, U.S.A.

Pokorski Mieczyslaw - Department of Physiology, Polish Academy of Sciences Medical Research Center, Warsaw, Poland

Polak Joseph - Department of Neurosciences, Case Western University School of Medicine, Cleveland, Ohio, U.S.A.

Prabhakar Nanduri R. - Department of Medicine, Case Western Reserve University, Cleveland, Ohio, U.S.A.

Ramirez B. - Laboratory of Neurobiology, Catholic University of Chile, Santiago 1, Chile

Rettig R. - Department of Pharmacology and German Institute for High Blood Pressure Research, Ruprecht-Kals University, Heidelberg, F.R.G.

Rocher A. - Departamento de Bioquímica y Fisiología, Facultad de Medicina, Universidad de Valladolid, Valladolid, Spain

Rumsey William L. - Department of Radiopharmaceuticals, Bristol-Myers Squibb, Pharmaceutical Research Institute, New Brunswick, New Jersey, U.S.A.

Sato Minoru - Department of Physiology, Akita University School of Medicine, Akita, Japan

Schamel Abdenbi - Laboratoire de Cytologie, Université de Bordeaux II, Talence, France

Schena F. - Department of Physiology, University of Geneva, Geneva, Switzerland

Shirahata Machiko - Department of Environmental Health Sciences, Johns Hopkins Medical Institutions, Baltimore, Maryland, U.S.A.

Soulier V. - URA 1195 CNRS, Faculté de Médecine Grange-Blanche, Lyon , France

Spergel Daniel - Department of Physiology, University of Pennsylvania School of Medicine, Philadelphia, Pennsylvania, U.S.A.

Spyer K.M. - Department of Physiology, Royal Free Hospital School of Medicine, London, U.K.

Stea Anthony - Department of Biology, McMaster University, Hamilton, Ontario, Canada

Stensaas L.J. - Department of Physiology, University of Utah School of Medicine, Salt Lake City, Utah, USA

Strosznajder R. - Department of Neurophysiology, Polish Academy of Sciences Medical Research Center, Warsaw, Poland

Sykes R.M. - Department of Physiology, Royal Free Hospital School of Medicine, London, U.K.

Tanaka Michiko - Department of Physiology, School of Medicine, Chiba University, Chiba, Japan

Torrance Robert William - University Laboratory of Physiology, St. John's College, Oxford, U.K.

Unger Th. - Department of Pharmacology and German Institute for High Blood Pressure Research, Ruprecht-Karls University, Heidelberg, F.R.G.

Vaughan-Jones Richard David - University Laboratory of Physiology, Oxford, U.K.

Verna Alain - Laboratoire de Cytologie, Université de Bordeaux II, Talence, France

Vollmer C. - Department of Biology, McMaster University, Halmilton, Ontario, Canada

Vouillarmet A. - URA CNRS 1195, Faculté de Médecine Grange-Blanche, Lyon, France

Wang D. - Research Institute, University of Toronto, Toronto, Ontario, Canada

Wang W.-J. - Department of Physiology, University of Utah School of Medicine, Salt Lake City, Utah, U.S.A.

Wang Z.-Z. - Department of Physiology, University of Utah School of Medicine, Salt Lake City, Utah, U.S.A.

Warner M. - Department of Comparative Biosciences, University of Wisconsin, Madison, Wisconsin, U.S.A.

Wilson David F. - Department of Biochemistry and Biophysics, University of Pennsylvania School of Medicine, Philadelphia, Pennsylvania, U.S.A.

Xi L. - Department of Physiology, University of Geneva, Geneva, Switzerland

Yamamoto Miyuki - Department of Anatomy, School of Medicine, Tohoku University, Sendai, Japan

Yoshizaki Katsuaki - Department of Physiology, Akita University School of Medicine, Akita, Japan

Zapata Patricio - Laboratory of Neurobiology, Catholic University of Chile, Santiago 1, Chile

The manufacturer's authorised representative in the EU is Springer
Nature Customer Service Centre GmbH, Europaplatz 3, 69115 Heidelberg,
Germany. If you have any concerns regarding our products, please
contact ProductSafety@springernature.com

Printed and bound by CPI Group (UK) Ltd, Croydon, CR0 4YY
23/04/2026
02095624-0015